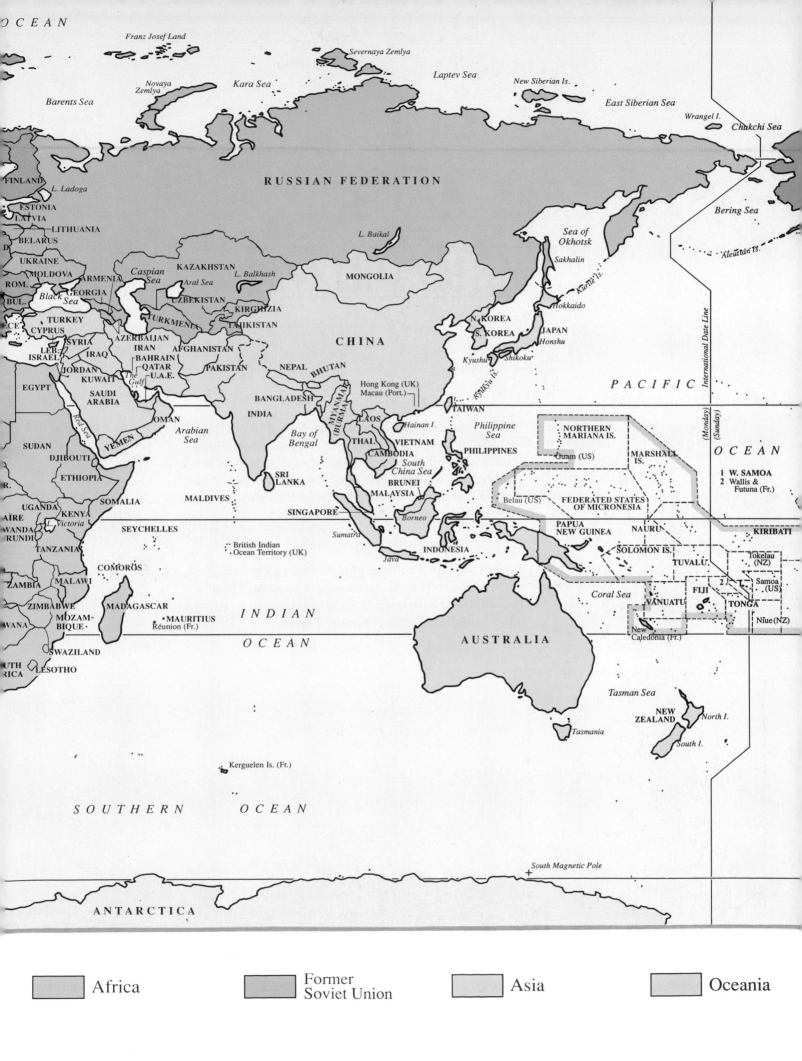

WWF

ATLAS of the ENVIRONMENT

Geoffrey Lean

○

Don Hinrichsen

Helicon

Helicon Publishing Ltd
42 Hythe Bridge Street
Oxford OX1 2EP

British Library Cataloguing in Publication Data. A catalogue record
for this book is available from the British Library.

Hardback: ISBN 0 09 177434 9
Paperback: ISBN 0 09 177433 0

A Banson production
3 Turville Street
London E2 7HR

General editor: Geoffrey Lean
Contributing editors: Don Hinrichsen, Adam Markham,
                          Michael Green, WCMC; Martin Jenkins,
                          WCMC; Mary Lean; Paul S. Wachtel,
                          WWF; Sandra Woods, WWF.
Project editors: Jane Lyons, Reet Nelis
Research editor: Ben Barkow
Cartographic editor: David Burles
Maps drawn by: Richard Natkiel Associates
                          David Burles, Richard Cowburn,
                          Craig Hildrew
Designed by: Tangent Graphic Design
Printed and bound by: Butler and Tanner Ltd, Frome

Front cover pictures: (top row, left to right) UNEP, Somsak
Sripolklung, Chusack Uthaipanumas, Martin Wright/Still Pictures,
UNEP, UNEP; (middle row) Mark Edwards/Still Pictures, UNEP,
UKAEA, UKAEA; (bottom row) UNEP, Mark Edwards/Still Pictures,
Mark Edwards/Still Pictures, UNEP.

Contents

Introduction

In June 1992, 118 heads of government – the greatest gathering of power ever to come together in one place – met in Rio de Janeiro, Brazil, for the Earth Summit. They adopted two treaties – to combat global warming and to conserve the world's biological diversity – each of which were signed by 153 countries by the end of the meeting. They agreed a declaration of environmental principles. And they drew up an 800-page plan of action for the 1990s and the 21st century, called Agenda 21, to move towards sustainable development, aimed at meeting the needs of the world's present population without so eroding its natural resources that future generations are unable to satisfy theirs. The Summit succeeded beyond immediate expectations, despite meeting in the midst of severe worldwide recession, but it was no more than a tentative first step in what is becoming a race against time.

The 1990s will decide whether the world can act fast enough to stop the crisis escalating out of control. The warming of the world's climate brought about by the greenhouse effect, the destruction of the planet's protective ozone layer, the felling of the tropical rainforests, the worldwide loss of fertile topsoil, the accelerating loss of wild species – and the growth of destitution among the poor of the earth – present a new and remarkable challenge. If the trends of the 1970s and 1980s are allowed to continue, they will become as destructive as nuclear war. They can only be checked through international cooperation on a scale never before seen; it may be that the Earth Summit has begun to pave the way for this, though immense changes still have to be made. One way or another, the course will be set in this decade.

This atlas sets out to present the basic facts about the natural and human environments, and what is happening to them, putting flesh on the bones of Agenda 21. It contains an unprecedented amount and range of information, and presents it in a new way. Earlier atlases have portrayed nature as it affects humanity; the maps in this volume show how people are affecting nature. The book describes the state of the world's forests, wetlands, deserts, croplands, mountains, rivers and seas, and details the threats to them. It examines pollution and wildlife, energy and food production, health and education, development and disasters. It presents solutions as well as problems, But, above all, it avoids polemic, seeking instead to provide information so that readers can make up their own minds.

The chapters are written to be accessible to those with no previous background in the subject, while providing enough new information to be useful to the expert. The editors have brought together information from a wide variety of sources, which are listed in the bibliography. They have drawn on the most authoritative available data, concentrating particularly on those collected by the United Nations and its agencies, the World Bank, the World Resources Institute, the Worldwatch Institute, the Organization for Economic Cooperation and Development, the International Institute for Environment and Development, IUCN–The World Conservation Union and WWF. The information is the latest that could be found without compromising accuracy.

If we are to solve the crisis threatening our environment, it is important that we all understand it. This book presents the basics from which to build this understanding.

The editors would like to express their gratitude to the following people who have assisted substantially in the compilation of this book:

Debra Adams, Greenpeace; A. Blanco-Bazán, International Maritime Organization; Claire Billington, World Conservation Monitoring Centre; Mario Camileon, US Office of Foreign Disaster Assistance; Nanda na Champassak, Office of the UN High Commissioner for Refugees; Sue Cooper, Greenpeace; Eric Cowell; Robert Cubey, World Conservation Monitoring Centre; Steve Davis, IUCN, Kew; D.J. Drewry, British Antarctic Survey; Ray Gamble, International Whaling Commission; Catherine Grey, International Tanker Owners Pollution Federation; R.K. Headland, Scott Polar Research Institute; Larry Heligman, United Nations Population Division; Tim Inskipp, World Conservation Monitoring Centre; Charu Jasani, Uranium Institute; Tim Jones, RAMSAR Bureau UK; Charles Kepelman, UN Office of the Disaster Relief Co-ordinator; Margaret Klinowska, Cambridge University; Arlin J. Krueger, NASA; Lars Ludvigsen, UN Centre for Human Settlement (HABITAT); Richard Luxmore, Wildlife Trade Monitoring Unit, WCMC; Adam Markham, WWF International; Alex Marshall, United Nations Population Fund (UNFPA); Hans F. Meyer, International Atomic Energy Agency; P. Micovic, World Health Organization; François Nectoux; Claire Nihoul, Oslo Commission; Bob Nixon, Forest Planning Canada; Fred Pearce; Steve Price, WWF-Canada; Bob Reid, International Board for Plant Genetic Resources; Peter Saunders; Courtney Scott, NASA; Karen Simpson, UNESCO; Paul Stapleton, International Board for Plant Genetic Resources; Jorgen Thomsen, TRAFFIC; Richard Walker; Martin Wattam, UN Information Office, UK.

Political divisions shown throughout are those generally recognized as at May 1, 1992.

All figures are in US dollars.

Major Biomes, Climatic Regions and Land Use

All life exists in a thin layer wrapped around the globe, caught between the molten heat of the earth's interior and the cold immensities of space. The biosphere, the only part of the entire universe known to support life, stretches from the depths of the oceans to the highest point, some 15 kilometers up in the atmosphere, at which living things have been found. Proportionately, it is no thicker than the shine on a billiard ball.

The uniqueness, fragility and beauty of the biosphere were first described and photographed by astronauts in the late 1960s. Few of humanity's works can be seen from space: instead the earth is revealed as a mosaic of the delicate and intricate patterns of nature.

From this perspective, the earth's marbled surface, dominated by blue sea and white cloud and ice, suggests that this is the water planet. Just over 70 per cent of it is covered by water. All life came from the oceans, and they still make the planet habitable, regulating its climate and sustaining it as an oasis in the limitless black desert of the universe.

Land: the other 30 per cent

From space, one can also see how little of the land surface can support human life. The sandy brown deserts cover much of the globe; in all about a third of the land surface is arid, or semi-arid. Another 11 per cent is permanently under ice; another 10 per cent is tundra. Over most of the rest, the soil is too thin, too poor, or too wet to be of much use. Only about 11 per cent of the total ice-free land area, some 1.5 billion hectares, presents no serious obstacles to cultivation, and almost all of that is already in use. Up to another 13 per cent could possibly be used, but only with great effort and expense.

But the world is losing ground: every year some 26 billion metric tons of topsoil are eroded from the land, a result of humanity's misuse of it. A single centimeter of top soil can take up to 1,000 years to form; in many parts of the world, it is being destroyed in less than five.

Biomes

Complex interactions of climate, geology, soil types, water resources and latitude together determine what types of plant and animal life thrive in different places. The world can be divided into 14 major ecological regions, called "biomes": these are distributed across five major climatic regions and eight zoogeographical realms. Together they support about a trillion metric tons of vegetation. Forests account for three quarters of this; the tropical rainforests alone contain a third of the world's plant matter, although they cover only about 7 per cent of its surface. Deserts and tundra, by contrast,

each support about 1 per cent of the world's vegetation though they cover vast tracts of the land.

Biomes describe what the world would be like if people had not altered the natural environment. Hence a map of biomes shows Europe covered in temperate broadleaf forest, though people long ago cleared almost all the tree cover, converting the land to agriculture, pastureland and urban sprawl. The map shows what types of vegetation the region would naturally support, not what actually grows there.

The changing face of the earth

Humanity is making enormous changes to the face of the earth. The deserts are expanding. One third of the world's land surface is threatened with desertification: every year 6 million hectares of productive lands are lost beyond hope of recovery, and another 21 million are so impoverished that it is no longer worth growing crops or grazing animals upon them. The forests are falling. Ten thousand years ago about half the world was covered with forests, but more than three quarters of them have now been destroyed or degraded. Tropical rainforests, the oldest and richest terrestrial habitats, are being felled at a rate of between 16 and 20 million hectares a year. Wetlands, coral reefs, mangrove swamps and upland watersheds are all being rapidly degraded and destroyed.

As the habitats disappear, species are becoming extinct at 25,000 times the natural rate. By the turn of the century, a million species will have vanished forever; vast holes will be torn in the web of life. Most will die unknown, without ever having been classified or named. Almost all will disappear without anyone knowing their potential value. Wild species provide vital genetic resources for safeguarding the world's crops, supply valuable raw materials for industry, and are behind 40 per cent of the prescriptions dispensed in the United States each year. Yet only a fraction of 1 per cent of them have been examined for the benefits they could provide.

Pollution provides further stress. Burning fossil fuels – coal, oil and gas – pumps vast quantities of pollutants into the atmosphere. Some form acids which crumble the facades of buildings and have killed tens of thousands of lakes in Europe and North America. Some play a part in *Waldsterben*, forest death, which affects 45 million hectares of temperate forests worldwide. Some affect human health: 30,000 people are thought to die each year in the US alone from pollution emitted from vehicle exhausts. One – carbon dioxide – is the main source of the greenhouse effect which threatens, within the lifetime of today's children, to make the world warmer than at any time in the last 100,000 years. This global warming is likely to be the greatest single

Major biomes

Biomes are ecological regions, defined in terms of their plant and animal life, and usually identified with the prevailing vegetation types. Unesco has designated 14 major biomes, distributed across eight zoogeographical realms: the Nearctic, the Neotropical, the Palearctic, the Afrotropical, the Indomalayan, the Australian, the Oceanian and the Antarctic.

Climatic regions

Polar
(Ice cap & tundra)

Cooler humid
(Subarctic & continental)

Warmer humid
(Marine west coast, humid subtropical & Mediterranean)

Dry
(Steppe & desert)

Tropical humid
(Savanna & rain forest)

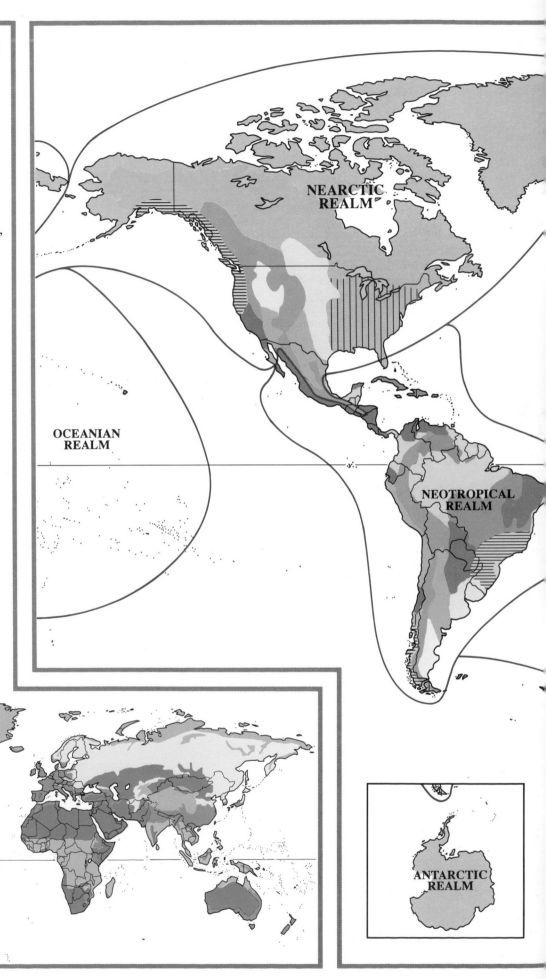

NEARCTIC REALM

OCEANIAN REALM

NEOTROPICAL REALM

ANTARCTIC REALM

Times, 1989

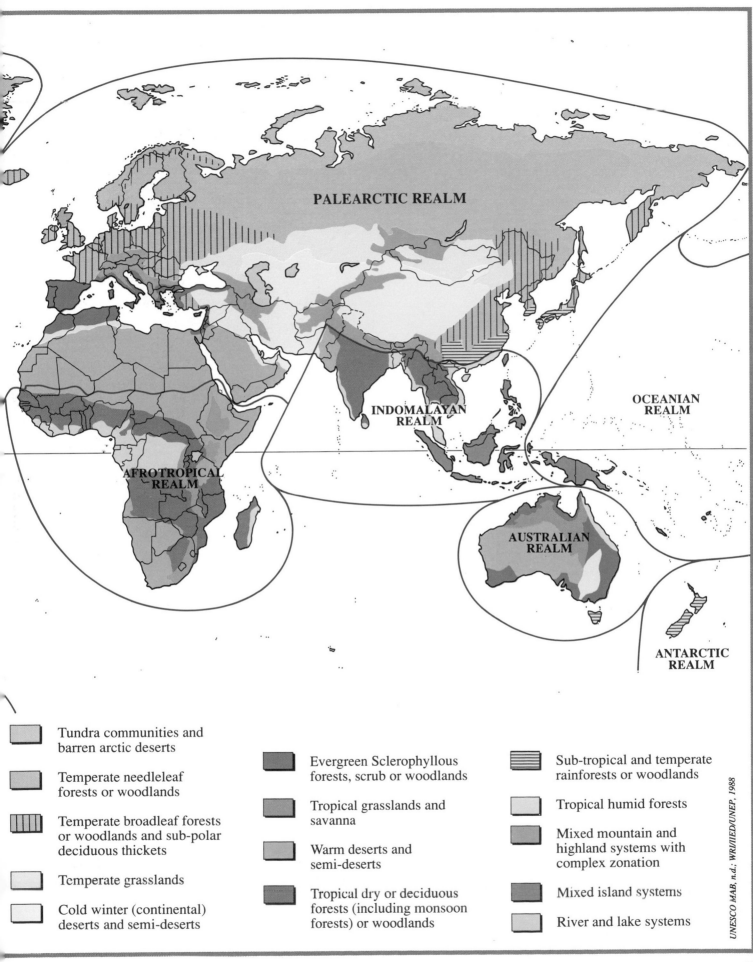

PALEARCTIC REALM

INDOMALAYAN REALM

OCEANIAN REALM

AFROTROPICAL REALM

AUSTRALIAN REALM

ANTARCTIC REALM

Tundra communities and barren arctic deserts

Temperate needleleaf forests or woodlands

Temperate broadleaf forests or woodlands and sub-polar deciduous thickets

Temperate grasslands

Cold winter (continental) deserts and semi-deserts

Evergreen Sclerophyllous forests, scrub or woodlands

Tropical grasslands and savanna

Warm deserts and semi-deserts

Tropical dry or deciduous forests (including monsoon forests) or woodlands

Sub-tropical and temperate rainforests or woodlands

Tropical humid forests

Mixed mountain and highland systems with complex zonation

Mixed island systems

River and lake systems

UNESCO MAB, n.d.; WRI/IIED/UNEP, 1988

Land use

The map shows cultivated land and livestock production. The major crop which is grown in each area (including cash crops) is marked. Cattle and sheep rearing is also indicated, although pasture lands themselves are not. The largely small-scale cultivation of potatoes and rearing of pigs are not shown.

Further information on uncultivated areas may be found elsewhere in the book. Forest areas are marked on the maps on pages 66-67 (tropical) and 82-83 (temperate). Deserts and areas at risk of desertification are shown on pages 54-55 and tundra and arctic deserts on the major biomes map on pages 12-13.

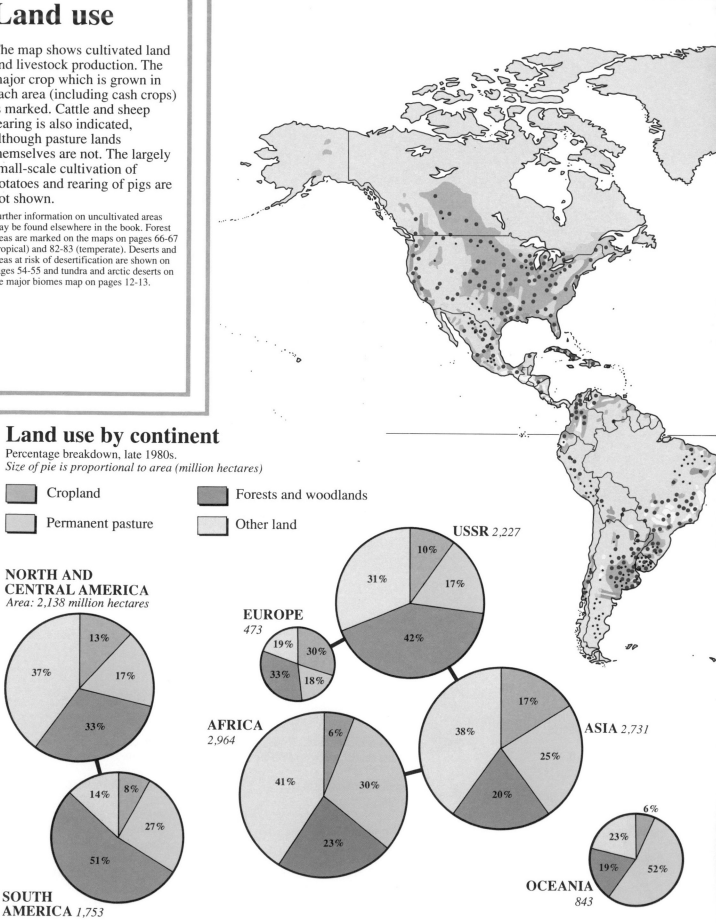

Land use by continent

Percentage breakdown, late 1980s.
Size of pie is proportional to area (million hectares)

- Cropland
- Forests and woodlands
- Permanent pasture
- Other land

NORTH AND CENTRAL AMERICA
Area: 2,138 million hectares

13%
17%
33%
37%

SOUTH AMERICA *1,753*

8%
27%
51%
14%

EUROPE *473*

19% 30%
33% 18%

USSR *2,227*

10%
17%
42%
31%

AFRICA *2,964*

6%
30%
23%
41%

ASIA *2,731*

17%
25%
20%
38%

OCEANIA *843*

6%
52%
19%
23%

WRI/IIED/UNEP, 1992

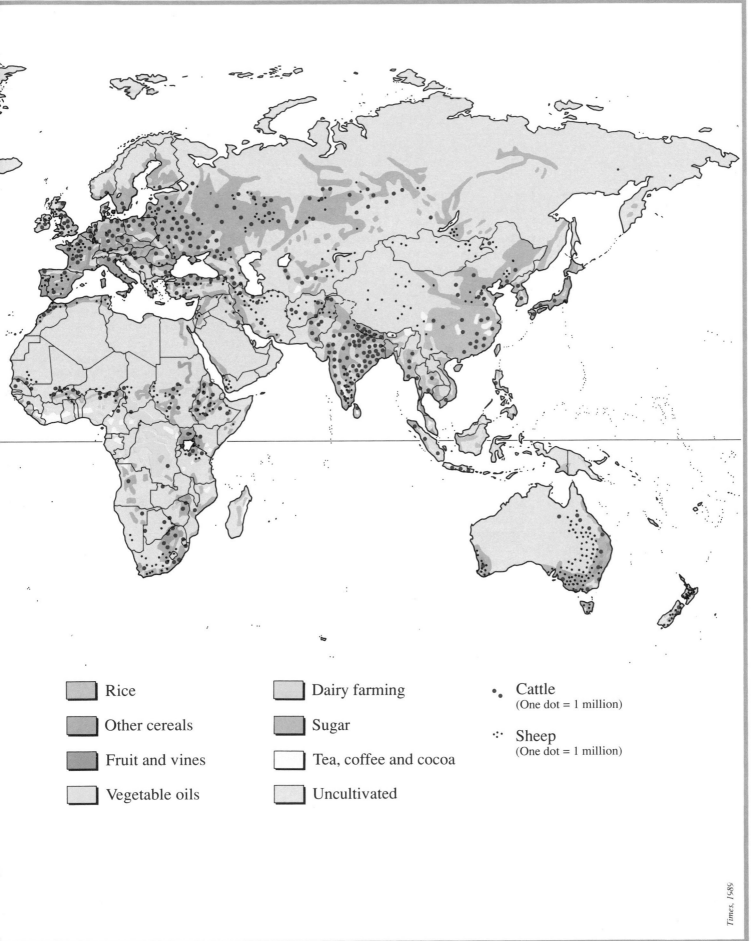

	Rice		Dairy farming	•	Cattle
					(One dot = 1 million)
	Other cereals		Sugar		
	Fruit and vines		Tea, coffee and cocoa	∴	Sheep
					(One dot = 1 million)
	Vegetable oils		Uncultivated		

Times, 1989

influence on the future face of the earth; as it progresses, sea levels will rise, and biomes will shift towards the poles, causing immense disruption.

The people factor

Already 1.2 billion people – nearly a quarter of humanity – live in absolute poverty, unable to meet their basic needs for food, shelter and clothing. More than 400 million of them get only four fifths of the food they need just to maintain their bodies, and so are condemned to stunted growth and constant danger of disease and death. Nearly 30 per cent of the people in the Third World do not have safe water to drink; waterborne diseases kill 25,000 people a day. Some 14 million children under five die needlessly each year from hunger and preventable illness.

Population growth helps to fuel both environmental destruction and human destitution. There are now 5.4 billion people on earth, and another 95 million – more than the population of Mexico – are added every year. The world's population is expected at least to double by the year 2100; if present trends continue, it could reach 14 billion or more, about treble today's numbers. More than 90 per cent of the growth is taking place in developing countries, those least able to cope with the consequences. The populations of many developing countries are doubling in less than 25 years.

Yet the style of economic activity, in both industrialized and developing countries, is even more destructive than the growth in numbers. Almost everywhere the rich get richer, and the poor get poorer: both trends intensify the environmental crisis. In the Third World, poverty is the most destructive agent. Governments are over-exploiting their natural resources in a vain attempt to pay off crippling debt burdens. And the poorest people, denied access to the good land held unequally by the rich, are forced to fell the forests and to degrade marginal soils as they attempt to feed their families. Population growth is itself fuelled by poverty. Poor people usually want plenty of children to help them earn a living and provide security in old age. When living standards increase, and fewer children die, then people are more prepared to have smaller families.

Meanwhile the rich contribute the main burden of pollution. Industrialized nations, with a fifth of the world's population, contribute two thirds of the emissions of the gases that cause global warming. They emit more than 85 per cent of chlorofluorocarbons (CFCs), which both destroy the world's protective ozone layer and help cause the greenhouse effect, and they are responsible for most of the damage being done by acid rain.

Sustainable development

Rising poverty in poor countries and increasing pollution in rich ones are both unsustainable: they are destroying the earth's life support systems and ensuring ultimate economic collapse. Economic growth, if it is to continue, has to be sustainable, meeting present needs without compromising either the fate of the planet or the future of coming generations. Slowly the concept of sustainable development is gaining ground, both in the textbooks and at the grassroots.

All over the world there are heartening initiatives, some small-scale, some larger, to tackle the environmental crisis: conserving soil, saving forests, reducing pollution. Public concern in Western countries has, for the first time, forced environmental issues to the top of the political agenda. Similar concern has been at the root of many of the democratic movements in Eastern Europe and the former Soviet Union. Grassroots activism is making itself felt in a growing number of Third World countries. And there have been encouraging international agreements, most notably against the pollution that damages the ozone layer. But time is getting short.

The 1990s are the crucial decade: the last chance to prevent global warming running out of control, to repair the ozone layer, to safeguard what remains of the tropical rainforests, to halt the march of the deserts and to address the destitution of the world's poorest people.

Human Numbers

The world's population is about 5.4 billion and it is expanding rapidly. Every day we share the earth and its resources with over 250,000 more people than the day before; every year, there are another 95 million mouths to feed. It is the equivalent of adding a Philadelphia to the world population every week; a Los Angeles every two weeks; a Mexico every year; and a US and Canada every three years.

Though fertility rates are dropping, the sheer momentum of population growth ensures that at least another 3 billion people will be added to the planet between now and the year 2025; it could be as high as 4 billion. At present growth rates, 1 billion people are added to the human ark every 11 years. The six-billionth inhabitant will be born sometime during 1998.

If current trends are not reversed, or at least slowed down, we could be facing a global population of around 10 billion by the year 2050. But the problem is not population growth *per se*. It is that over 90 per cent of births now take place in the countries least able to cope with the resource and environmental consequences of burgeoning populations. Between now and the turn of the century, the number of people in the Third World will grow by over 900 million, or 24.6 per cent. Meanwhile the population of industrialized countries will grow by only 56 million, or 5.2 per cent.

The demographic transition

The population explosion began in the West, around the middle of the 17th century. Until then the numbers of people in the world had grown, but slowly, from about 150 million at the time of Christ to somewhere around 500 million. Births and deaths had more or less cancelled each other out. But then the rate of increase quickened dramatically: by 1850 there were some 1,200 million people on earth and the growth rate continued to rise. The reason was not that people had more babies but that they lived longer, as food supplies increased, public health improved and, eventually, proper sanitation spread. Birth rates stayed much the same as before, but death rates fell, causing population to grow. As Peter Adamson, a consultant to the United Nations Children's Fund (UNICEF), once said, "It's not that people suddenly started breeding like rabbits: it is just that they stopped dying like flies."

Eventually, as people in industrialized countries became more prosperous, birth rates fell until once again they virtually matched the number of deaths. Populations in Europe and North America have all but stabilized: in almost every country they are growing at less than 1 per cent. Were it not for immigration, most of Europe's population would hardly grow at all during the 1990s. Three countries – Austria, Belgium and Italy – have already achieved zero population growth; while in Germany and Hungary the population is actually declining. Industrialized countries have completed the "demographic transition".

Third World nations have yet to do so, and it is this that is fuelling the world's present prodigious growth of numbers. Since the Second World War, death rates have fallen dramatically in developing countries, partly as a result of the fight against killer diseases like smallpox and malaria. Indeed, they fell much faster than they had done during Europe's population explosion. But the fall in death rates has not been preceded by an equivalent agricultural revolution, nor accompanied by similar economic development – and there are no colonies to take the overflow. Birth rates have declined somewhat, but they remain high, and may not fall enough to complete the transition.

High birth rates in the Third World

The populations of most developing countries are growing at well over 2 per cent a year, many top 3 per cent – which means that their numbers will double in less than 23 years. During the first half of the 1990s, Africa's population as a whole will increase by 3 per cent a year – the highest regional growth rate the world has ever seen. Zambia and Côte d'Ivoire are both growing annually by 3.8 per cent, a rate that will double their numbers in 18 years. A Nigerian child born in 1990 and living to be 70 would see his country's population grow eight times over to about 900 million.

The average woman in the Middle East and Africa bears between six and eight children, while her equivalent in industrialized countries bears one or two. There are good reasons for big families. Children are seen as economic assets, they do useful work for the family from the age of six or seven and by the time they reach 10 or 12 they are often producing more for the family than they consume. They provide security in old age. And while infant mortality remains high, parents need to have a lot of babies to make sure that enough survive.

But though having lots of children may often make sense for individual families, it is ruinous to the societies in which they live. Rapidly growing populations hamper economic development, and Third World countries simply cannot provide for their burgeoning numbers.

The results are only too evident: mounting unemployment and underemployment; spreading slums and squatter settlements; lack of access to education, health care, drinking water and sanitation, and family planning services. Poverty deepens and more people are pushed to the edge of survival.

There is another disturbing trend. According to the United Nations Population Fund (UNFPA), in 1985

Population growth rates

Average annual % growth rate, 1990 – 1995
World average = 1.7%

- **3.5% and over**
  (Population doubles every 19 years or less)
- **3.0 – 3.49%**
  (Population doubles every 20 – 23 years)
- **2.0 – 2.99%**
  (Population doubles every 24 – 34 years)
- **1.0 – 1.99%**
  (Population doubles every 35 – 70 years)
- **0 – 0.99%**
  (Population doubles every 71 years or more)
- **Population decline**
- **Insufficient data**

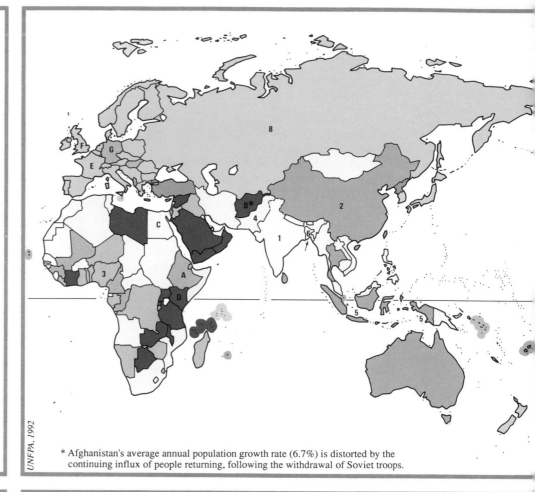

UNFPA, 1992

* Afghanistan's average annual population growth rate (6.7%) is distorted by the continuing influx of people returning, following the withdrawal of Soviet troops.

Infant mortality

Infant deaths per 1,000 live births, 1990 – 1995
World average = 63

- **160 and over**
- **120 – 159**
- **80 – 119**
- **40 – 79**
- **0 – 39**
- **Insufficient data**

Infant mortality is an indicator of the level of development. Much of the developing world has high rates of infant deaths. Those countries with the fastest growing populations tend to have high rates of infant mortality.

UNFPA, 1992

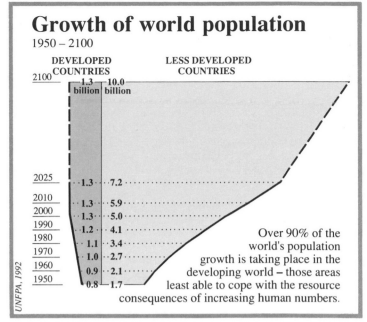

Growth of world population
1950 – 2100

	DEVELOPED COUNTRIES	LESS DEVELOPED COUNTRIES
2100	1.3 billion	10.0 billion
2025	1.3	7.2
2010	1.3	5.9
2000	1.3	5.0
1990	1.2	4.1
1980	1.1	3.4
1970	1.0	2.7
1960	0.9	2.1
1950	0.8	1.7

UNFPA, 1992

Over 90% of the world's population growth is taking place in the developing world – those areas least able to cope with the resource consequences of increasing human numbers.

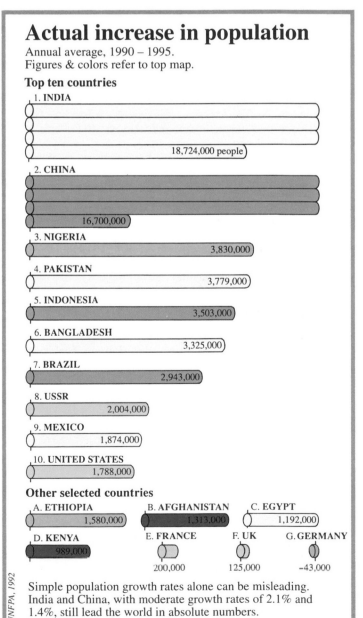

Actual increase in population

Annual average, 1990 – 1995.
Figures & colors refer to top map.

Top ten countries

1. INDIA — 18,724,000 people
2. CHINA — 16,700,000
3. NIGERIA — 3,830,000
4. PAKISTAN — 3,779,000
5. INDONESIA — 3,503,000
6. BANGLADESH — 3,325,000
7. BRAZIL — 2,943,000
8. USSR — 2,004,000
9. MEXICO — 1,874,000
10. UNITED STATES — 1,788,000

Other selected countries

A. ETHIOPIA — 1,580,000
B. AFGHANISTAN — 1,313,000
C. EGYPT — 1,192,000
D. KENYA — 989,000
E. FRANCE — 200,000
F. UK — 125,000
G. GERMANY — -43,000

UNFPA, 1992

Simple population growth rates alone can be misleading. India and China, with moderate growth rates of 2.1% and 1.4%, still lead the world in absolute numbers.

around 37 per cent of the total population in the developing world were children below the age of 15. In Africa, children make up 45 per cent of the population. This number of dependants puts added strain on the productive members of society.

High fertility: high mortality

The terrible contrast between the rich and poor worlds is dramatized by comparisons of life expectancy at birth and infant mortality rates. Even after the improvements of the last 40 years, Africans can expect to live, on average, only 54 years. Gambians are lucky if they reach the age of 43 and half of all Angolans die before reaching 47. Asian life expectancy is 65 and Latin American 68. Europeans, on the other hand, live on average to 75, Americans to 76 and Canadians to 77.

Infant mortality is, perhaps, an even more revealing indicator. In Africa, the average number of infant deaths per 1,000 live births is 94, reaching a peak at 159 in Mali and 143 in Sierra Leone. The infant mortality rate for Latin America is 52 and for Asia 64. In Western Europe, North America, Singapore, Australia, New Zealand and Japan, infant mortality is 10 or less per 1,000 live births.

Every year some 14 million children die in developing countries before they reach their fifth birthday.

Every year, too, half a million mothers die in pregnancy or childbirth – all but 1 per cent of them in the Third World. An African woman is 500 times more likely to die from giving birth than her counterpart in one of the richer developed countries. Another 100,000 to 200,000 women die each year as a result of illegal abortions, and again the great majority are from developing countries.

Almost all these women and children die unnecessarily. Breastfeeding, immunization programs, simple mixtures of salts to prevent dehydration from diarrhea, basic health care, education on health and family planning, and – above all – clean water and enough food, would prevent most of the infant and child deaths. Maternal health care, and spacing and reducing the numbers of births would help reduce maternal mortality. Some of these are simple and cheap; all ought to be available as a basic human right.

The World Fertility Survey, completed in 1984, showed that many mothers in developing countries did not want any more children, but were not able to get access to proper information or contraceptive aids. Birth rates would fall heavily if all the women who said they wanted no more children actually succeeded in stopping their childbearing: the number of births would be cut by about a quarter in Africa and about a third in Asia and Latin America. There is clearly a great unmet need for family planning, but it, alone, is not enough.

If people want to have children, however, even the best contraceptive is useless. Rapid population growth is linked to poverty, and both must be tackled together. Family planning programs that ignore social conditions have rarely succeeded. Nations as diverse as Colombia, China, Sri Lanka, Chile and Cuba – and the Indian state of Kerala – which not only provided their citizens with family planning and health care centers, but also addressed poverty, have achieved massive declines, cutting fertility by a third to half between 1960 and 1985.

The crucial role of women

The status of women is particularly crucial for bringing down population growth rates. One of the most important factors in reducing fertility is the level of women's education. Literate women with a secondary school education are much more likely to take advantage of family planning and maternal and child health care facilities than illiterate women with little or no education. Poorly educated women in Brazil, for instance, have an average of six children each; those with secondary education only two. In Thailand, where women have exceptional opportunities for a Third World country, a vigorous family planning program has helped cut fertility from six children per woman to three in only two decades. Between 1960 and 1985, Costa Rica achieved a 53 per cent decline: 70 per cent of its women use contraceptives despite little effort to spread family planning. The reason seems to be that it has a good record in promoting health and education and in tackling poverty.

One way or another, population growth will slow down because many developing countries simply cannot sustain their escalating numbers. It will either happen through family planning and development, or by famine, disease and war brought about by collapsing economies.

The Urban Revolution

The earth is witnessing an urban revolution, as people worldwide crowd into towns and cities. In 1800 only some 50 million people lived in urban areas: by 1985 the number of urban dwellers had risen to 2 billion. In 1800 only 5 per cent of the world's population were urban dwellers, now the proportion has risen to more than 45 per cent, and by the year 2010 more people will live in towns and cities than in the countryside. Humanity will, for the first time, have become a predominantly urban species.

Though the world is getting more crowded by the day, absolute numbers of population are less important than where people concentrate, and whether these areas can cope with them.

Simple calculations of population density can be misleading. Africa as a whole, for example, has an average of only 22 people per square kilometer. But vast regions are virtually uninhabitable and people are concentrated in the more fertile areas, along coasts and in towns and cities. Small countries with big populations – like Burundi with a population density of 213 per square kilometer, Mauritius with 585 and Rwanda with 293 – give a better picture of the population density.

Asia has 114 people per square kilometer, Europe 105. The most crowded of their countries is tiny Malta, with 1,100 inhabitants per square kilometer, followed by Bangladesh with 888 inhabitants per square kilometer, Bahrain with 759, the Netherlands with 441, and Japan with 328 per square kilometer. However, densities tell nothing about the quality of the infrastructure – roads, housing, job creation, for example – or the availability of crucial services.

The move to the towns

In themselves these figures give us no idea of the relative condition of the people of these countries. The main question is not how many people there are in a given area, but how well their needs can be met. Density figures have to be set beside measurements of wealth and employment, the quality of housing and the availability of education, medical care, clean water, sanitation and other vital services.

The urban revolution is taking place mainly in the Third World, where it is hardest to accommodate it. It took London 130 years to go from 1 million to 8 million inhabitants; by contrast, Mexico City zoomed from 1 million to 20 million in less than 50. Between 1950 and 1985, the number of city dwellers has grown more than twice as fast in the Third World as in industrialized countries. During this period, the urban population of the developed world increased from 477 million to 838 million, less than double; but it quadrupled in developing countries, from 286 million to 1.14 billion.

The difference is widening. Urban growth rates are slowing down in industrialized countries, and in some the number of city dwellers is actually declining; as a whole, the rate of increase is now down to 0.8 per cent a year. In the Third World, by contrast, towns and cities are increasing by an average of 3.6 per cent a year, four and a half times as fast. Every year they must absorb 80 million more people. Africa's urban population is racing along at 5 per cent a year on average, doubling city numbers every 14 years. By the turn of the century, three in every four Latin Americans will live in urban areas, as will two in every five Asians and one in every three Africans. Developing countries will have to increase their urban facilities by two thirds by then, if they are even to maintain their present inadequate level of services and housing.

The urban challenge

In 1940, only one out of every 100 of the world's people lived in a really big city, one with a population of over a million. By 1980, this proportion had already risen to one in 10. Two of the world's biggest cities, Mexico and São Paulo are already bursting at the seams, and their populations are doubling in less than 20 years. By the turn of the century, Mexico City will have almost 26 million people and São Paulo some 24 million, while 19 of the world's largest 25 cities will be in developing countries. Transport, communication, health and sanitation systems are likely to break down as these megacities grow; Alexandria's sewage system, built for a million people, now strains to carry the wastes of 4 million. Political and social systems are likely to be destabilized.

About a third of the people of the Third World's cities now live in desperately overcrowded slums and squatter settlements. Many are unemployed, uneducated, undernourished and chronically sick. In Bombay and Rio de Janeiro, over 3 million people are squeezed into slums and shantytowns; 60 per cent of the entire population of Bogota and Kinshasa – and 79 per cent of the people of Addis Ababa – live in slums. In India, they are called *bustees*, in Brazil *favelas*, in Chile *callampas* (mushrooms), in Tunisia *gourbevilles*, in Argentina – most graphically – *villas miserias*. They receive tens of millions of new people every year, flocking in from the countryside in what is the greatest mass migration in history. Pushed out of the countryside by rural poverty, drawn to the cities in the hope of a better life, they find no houses waiting for them, no water supplies, no sewerage, no schools – and no welcome, for they are resented by wealthier citizens and ignored, at best, by the authorities. They throw up makeshift hovels built of whatever they can find – sticks, fronds, cardboard, tar-paper, straw, petrol tins

World cities, 1985

▮ Cities over 5 million
● (Bar proportional to population)

● Cities over 2 million

The 75 largest cities
(Numbered on map)   Millions

1	Tokyo/Yokohama (Japan)	18.82
2	Mexico City (Mexico)	17.30
3	São Paulo (Brazil)	15.88
4	New York (United States)	15.64
5	Shanghai (China)	11.96
6	Calcutta (India)	10.95
7	Buenos Aires (Argentina)	10.88
8	Rio de Janeiro (Brazil)	10.37
9	London (UK)	10.36
10	Seoul (Republic of Korea)	10.28
11	Greater Bombay (India)	10.07
12	Los Angeles (United States)	10.05
13	Osaka/Kobe (Japan)	9.45
14	Beijing (China)	9.25
15	Moscow (Russian Federation)	8.97
16	Paris (France)	8.68
17	Jakarta (Indonesia)	7.94
18	Tianjin (China)	7.89
19	Cairo/Giza (Egypt)	7.69
20	Tehran (Iran)	7.52
21	Delhi (India)	7.40
22	Milan (Italy)	7.22
23	Manila/Quezon (Philippines)	7.03
24	Chicago (United States)	6.84
25	Karachi (Pakistan)	6.70
26	Bangkok (Thailand)	6.07
27	Lima/Callao (Peru)	5.68
28	Madras (India)	5.19
29	Hong Kong	5.13
30	St Petersburg (Russian Fed.)	5.11
31	Dhaka (Bangladesh)	4.89
32	Madrid (Spain)	4.71
33	Bogotá (Colombia)	4.49
34	Baghdad (Iraq)	4.42
35	Philadelphia (United States)	4.18
36	Santiago (Chile)	4.16
37	Naples (Italy)	4.11
38	Pusan (Republic of Korea)	4.11
39	Shenyang (China)	4.08
40	Bangalore (India)	3.97
41	Detroit (United States)	3.83
42	Sydney (Australia)	3.78
43	Caracas (Venezuela)	3.74
44	Lahore (Pakistan)	3.70
45	Rome (Italy)	3.69
46	Lagos (Nigeria)	3.65
47	Wuhan (China)	3.38
48	Guangzhou (China)	3.30
49	San Francisco (United States)	3.30
50	Katowice (Poland)	3.27
51	Belo Horizonte (Brazil)	3.25
52	Barcelona (Spain)	3.20
53	Toronto (Canada)	3.16
54	Melbourne (Australia)	3.15
55	Ahmadabad (India)	3.14
56	Hyderabad (India)	3.12
57	Istanbul (Turkey)	2.94
58	Alexandria (Egypt)	2.93
59	Washington DC (United States)	2.91
60	Ankara (Turkey)	2.90
61	Birmingham (UK)	2.87
62	Montreal (Canada)	2.84
63	Houston (United States)	2.83
64	Guadalajara (Mexico)	2.77
65	Pôrto Alegre (Brazil)	2.74
66	Recife (Brazil)	2.74
67	Rangoon (Myanmar)	2.73
68	Boston (United States)	2.71
69	Chongqing (China)	2.70
70	Casablanca (Morocco)	2.69
71	Kinshasa (Zaïre)	2.69
72	Dallas (United States)	2.68
73	Athens (Greece)	2.68
74	Chengdu (China)	2.67
75	Algiers (Algeria)	2.66

UN Population Division, 1987

22

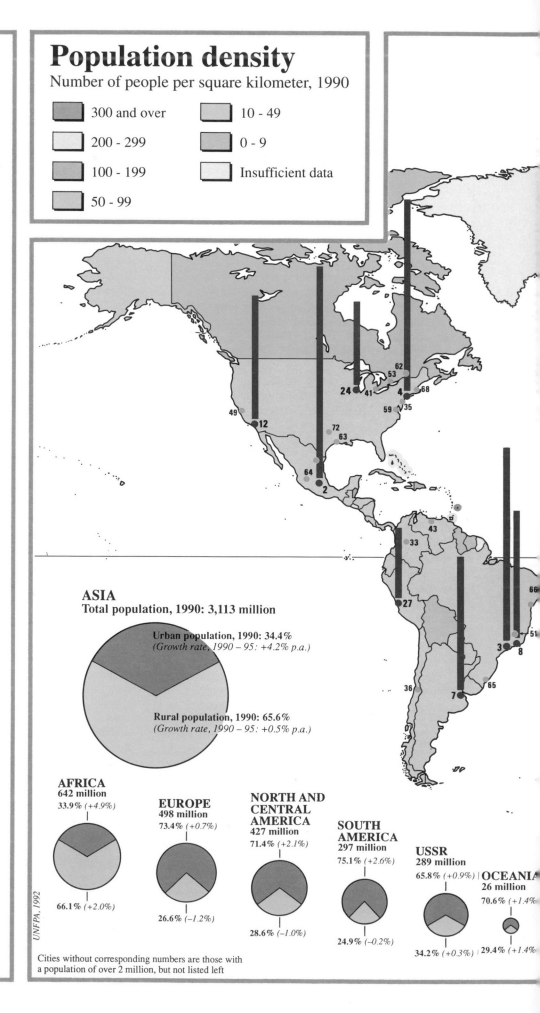

Population density
Number of people per square kilometer, 1990

▮ 300 and over
▮ 200 - 299
▮ 100 - 199
▮ 50 - 99
▮ 10 - 49
▮ 0 - 9
▮ Insufficient data

ASIA
Total population, 1990: 3,113 million

Urban population, 1990: **34.4%**
(Growth rate, 1990 – 95: +4.2% p.a.)

Rural population, 1990: **65.6%**
(Growth rate, 1990 – 95: +0.5% p.a.)

AFRICA
642 million
33.9% (+4.9%)
66.1% (+2.0%)

EUROPE
498 million
73.4% (+0.7%)
26.6% (−1.2%)

NORTH AND CENTRAL AMERICA
427 million
71.4% (+2.1%)
28.6% (−1.0%)

SOUTH AMERICA
297 million
75.1% (+2.6%)
24.9% (−0.2%)

USSR
289 million
65.8% (+0.9%)
34.2% (+0.3%)

OCEANIA
26 million
70.6% (+1.4%)
29.4% (+1.4%)

UNFPA, 1992

Cities without corresponding numbers are those with
a population of over 2 million, but not listed left

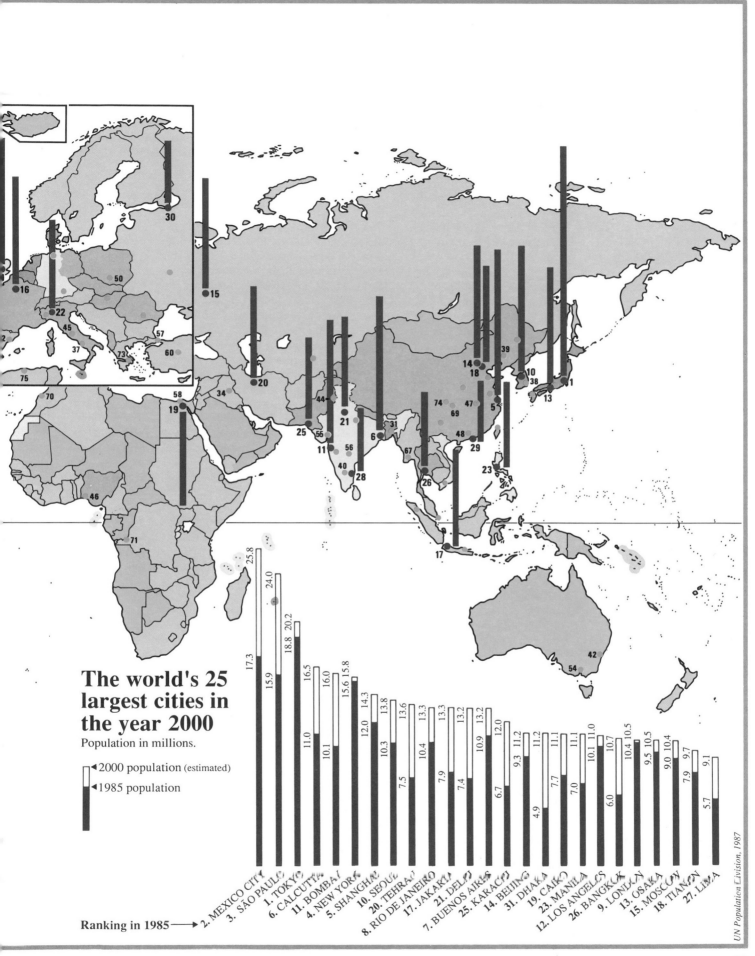

The world's 25 largest cities in the year 2000

Population in millions.

◄ 2000 population (estimated)
◄ 1985 population

Ranking in 1985 ➤

2. MEXICO CITY 17.3 25.8
3. SÃO PAULO 15.9 24.0
1. TOKYO 18.8 20.2
6. CALCUTTA 11.0 16.5
11. BOMBAY 10.1 16.0
4. NEW YORK 15.6 15.8
5. SHANGHAI 12.0 14.3
10. SEOUL 10.3 13.8
20. TEHRAN 7.5 13.6
8. RIO DE JANEIRO 10.4 13.3
17. JAKARTA 7.9 13.3
21. DELHI 7.4 13.2
25. KARACHI 6.7 13.2
14. BEIJING 9.3 12.0
31. DHAKA 4.9 11.2
19. CAIRO 7.7 11.2
23. MANILA 7.0 11.1
12. LOS ANGELES 10.1 11.1
26. BANGKOK 6.0 11.0
9. LONDON 10.4 10.7
13. OSAKA 9.5 10.5
15. MOSCOW 9.0 10.5
18. TIANJIN 7.9 10.4
27. LIMA 5.7 9.7
9.1

UN Population Division, 1987

23

and, if they are lucky, corrugated iron. They have to take the land no-one else wants, that is too wet, too dry, too steep or too polluted for normal habitation.

Yet all over the world, the inhabitants of these apparently hopeless slums show extraordinary enterprise in improving their lives. While many settlements remain stuck in apathy, many are gradually improved through the vigor and cooperation of their people, who turn flimsy shacks into solid buildings, build schools, lay out streets, and put in electricity and water supplies. Barrio women in Bogota now grow vegetables by hydroponics, without soil and using very little water, bringing them a steady income. Some of the earliest *favelas* in Rio are now respectable suburbs. Governments can help by giving the squatters the right to the land that they usually occupied illegally, giving them the incentive to improve their homes and neighborhoods.

Changing outmoded building codes and urban planning systems could have an immediate impact on the acute housing crisis afflicting so many Third World cities. In Nairobi's slums, for instance, simple homes could be put up for around $340 each, but the city's by-laws, dating from the colonial period, prevent it. Legally, it is impossible to build a house for less than 10 times that amount, since every dwelling must be accessible by car. As a result, 100,000 slum houses and squatter shacks cannot be improved by legal means, while just one tenth of Nairobi's residents occupy two thirds of its land.

The most important way, however, to ameliorate the effects of the Third World's exploding cities is to slow down migration. This involves correcting the bias most governments show towards cities and towns and against the countryside.

Whether because of favoritism, elitism, or political pressure, national development expenditures tend to be concentrated in urban areas, particularly capital cities. This makes them more attractive to rural residents seeking to improve their social and economic conditions. Food prices are kept artificially low to please urban dwellers, reducing rural incomes further. With few sources of hard currency, many governments in developing countries concentrate their limited development efforts in cities and towns, rather than the countryside, where many of the most destitute live. Food production falls as the countryside slides ever deeper into depression.

The demanding city

Since the process of urbanization concentrates people, the demand for basic necessities like food, energy, drinking water and shelter is also increased, which can exact a heavy toll on the surrounding countryside. One of the major contributors to deforestation in Africa has not been rural fuelwood use, but the wholesale conversion of wood to charcoal for sale to people living in towns and cities.

Urban areas use up arable land; a particularly serious problem in Egypt where only 4 per cent of the country is cultivable – limited to thin strips of land along the river Nile and the Mediterranean coast. Between 1980 and the end of this century, urban areas in the Third World will more than double in size – from about 80,000 square kilometers to more than 170,000 square kilometers. High-quality agricultural land is shrinking in many regions; taken out of production because of overuse and mismanagement. Creeping urbanization could aggravate this situation, further constricting economic development.

The only hope of ameliorating the flight to the cities is to reverse national priorities in many countries, and concentrate more resources in the countryside. This would be the most effective way of tackling poverty because most poor people still live in rural areas. It would boost food production and build national economies more securely. But ultimately the choice of priorities comes down to a question of power. The people of the countryside are powerless beside those of the towns; the destitute of the countryside will starve in their scattered millions, while the poor concentrated in urban slums pose a constant threat of disorder. In all but a few developing countries the bias towards the cities will continue, and so will the migrations that are swelling their numbers beyond control.

Food Production and Consumption

For four decades after the Second World War, food production steadily outpaced demand. The Green Revolution helped boost grain production in the Third World, while stunning technological advances improved yields in developed countries. But a billion people are still hungry and there are signs that the long agricultural boom may be ending.

Enough to feed the world

Grain production – which provides about half the world's calories – increased from around 700 million metric tons in 1950 to over 1.8 billion metric tons in 1986. It grew at around 3 per cent a year, outstripping population growth. Similarly, meat, milk and fish production rose by 2 per cent annually between 1965 and 1986, while the harvest of vegetables, pulses and fruit grew by 2.5 per cent a year.

The World Commission on Environment and Development, reporting in 1987, attributed the increase mainly to the development of high-yielding new seed varieties, a ninefold increase in the use of chemical fertilizers, a 32-fold rise in pesticide applications and a doubling of the world's irrigated cropland, from 135 million hectares in the 1960s to 271 million hectares in 1985.

Distributed evenly, the 1986 harvest could support 6 billion people – the projected population of the earth for the year 1998. But there are huge inequalities in both production and consumption.

All developed regions have greatly increased their per capita food production since the 1960s. So has Asia. Western Europe, where population growth has stabilized, now produces about 30 per cent more food for each of its people than in the mid-1960s. Africa has also increased its food production in absolute terms, but not enough to keep up with population growth; it now produces 27 per cent less food for each African than in 1967.

The hungry billion

On average, people in the richest countries eat 30-40 per cent more calories than they need. The people of the poorest nations on average get 10 per cent less than this basic minimum. And this conceals wide differences: Kenyans on average get 92 per cent of what they need, but the poorest 40 per cent of the rural people suffer serious malnutrition, attempting to subsist on less than three quarters of their requirements.

Over 1 billion people – about one in every five on earth – do not get enough food to lead fully productive lives. At least 400 million of them get less than 80 per cent of their basic needs, and are condemned to stunted growth and constant danger of serious illness. Two thirds live in Asia, another fifth in Africa. Two thirds are probably under 15 years old. And their numbers are growing.

Every year about 11 million children under the age of five die from hunger or hunger-related diseases. Those that survive may never reach their full potential. One third of Peru's children are so underfed that their growth is stunted. And if a child does not get enough to eat in its first years of life, its brain will not develop properly. One study followed up malnourished Indian children under five for the next 17 years of their lives – and found that their capacity for work was 30 per cent less than that of children from the same class and the same villages who had had enough to eat.

Poverty and hunger

People are hungry in a world that produces more than enough because they are too poor to buy or grow the food they need. So merely increasing food production does not in itself tackle hunger; consumption also has to rise.

India is a production success story – and a consumption disaster. Its wheat harvest more than doubled under the impact of the Green Revolution between 1965 and 1972; the most spectacular increase in history. It provided food aid to the newly-emergent Bangladesh and for a while became the world's second biggest donor after the United States. By the mid-1980s, it had a grain surplus of 24 billion metric tons. Nevertheless, it still had about half of all the hungry people on earth. Consumption of grain per head failed to increase over the period of the production "miracle" and nearly half its people are too poor to buy enough to eat.

Crops and cash

If their own countrypeople cannot afford to buy food, landowners will naturally grow more cash crops – such as cotton, coffee, tea, sugar or tobacco – for export. Governments, saddled with huge debt burdens, will encourage them to earn the foreign exchange. There is vigorous debate as to whether the growth of cash crops has reduced food production; but there is little doubt that they tend to take up the best land, pushing subsistence farmers onto areas with poorer soil and rainfall; yields are lower on this marginal land, so the farmers have to exploit more of it, increasing the spread of the deserts. Cash crops also receive most of the Third World's credit, fertilizers and pesticides, and agricultural advice.

As food production has fallen, particularly in Africa, more and more has had to be imported. In 1984, 140 million Africans – more than a quarter of the continent's population – were fed with grain from overseas; though neither they nor their countries could afford to buy

Food production and consumption

Calorie intake per day, 1988

■	3,200 and over
■	2,800 – 3,199
☐	2,400 – 2,799
☐	2,000 – 2,399
■	Under 2,000
☐	Insufficient data

According to the United Nations, the amount of food consumed every day by the **average** global citizen should amount to no less than 2,400 calories. Below this level, chronic malnutrition may result. But like all averages this one conceals important variations in global diet. Take Eskimos, for example. Because they live for most of the year in cold, dark climates, they need to consume more than the average caloric intake in order to survive; perhaps as much as 3,000 calories a day. On the other hand, the average resident of the dry tropics needs somewhat less than this, perhaps as little as 2,000 calories a day, in order to lead a healthy life.

There may be some dispute as to whether or not anyone consuming between 2,000 and 2,399 calories a day is actually malnourished, depending on where they live. But people consuming fewer than 2,000 calories a day on a **regular basis** are considered chronically malnourished.

World Bank, 1991

WRI/IIED/UNEP, 1992

Areas suffering soil degradation

■	Areas of serious degradation
■	Areas of some degradation
☐	Stable soils
☐	Areas without vegetation

The map on the left shows areas where soils are so degraded by agricultural and industrial activities that they cannot regain their productivity without major reclamation work beyond the means of farmers, and areas which are less affected but have suffered soil degradation.

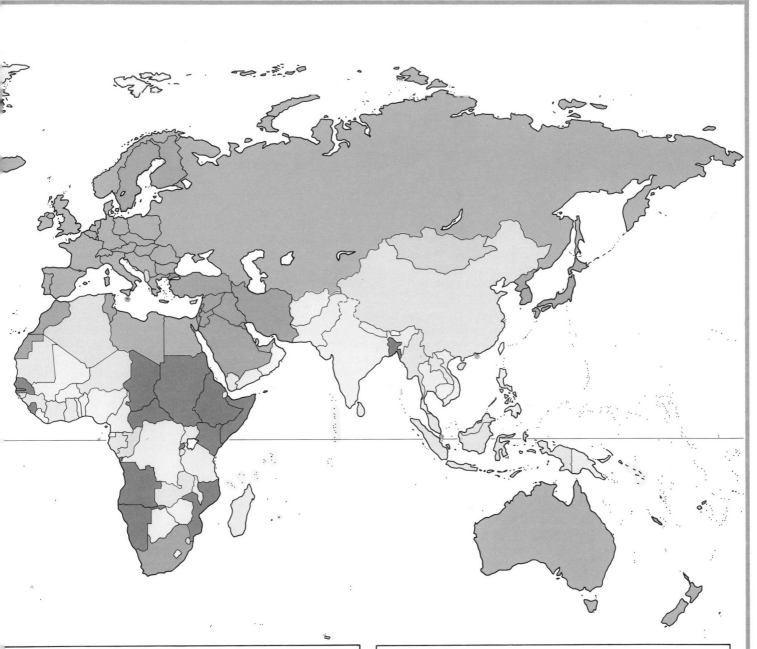

World per capita food production

Index, 1965=100

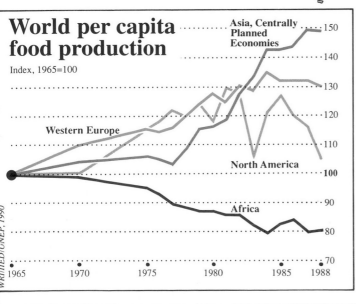

Asia, Centrally Planned Economies

Western Europe

North America

Africa

150
140
130
120
110
100
90
80
70

1965   1970   1975   1980   1985   1988

WRI/IIED/UNEP, 1990

Percentage of population engaged in agriculture

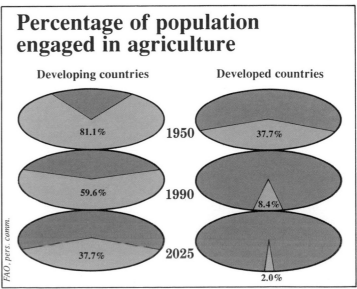

Developing countries   **Developed countries**

81.1%   1950   37.7%

59.6%   1990   8.4%

37.7%   2025   2.0%

FAO, pers. comm.

enough to prevent widespread hunger. Both the demand for imports and the inability to pay for enough of them will worsen over the next decades. Food aid is no answer, even if surplus countries are prepared to give it. Essential to relieve short-term famine, food aid undermines local production in more normal circumstances.

Falling harvests

Ominously, after four decades of growth, the global harvest began to falter in the second half of the 1980s. Per capita grain production, which grew from 246 kilograms to 345 kilograms between 1950 and 1984, fell back to 296 kilograms by 1988 – around the level that it had been in the mid-1970s. In 1988, for the first time in history, the United States produced less grain than it needed for its own people. In no less than four years between 1987 and 1992, the world as a whole failed to produce enough to satisfy demand.

World grain stocks fell from a record high in 1986 to approaching their lowest levels ever, and stood at minimum safety levels in early 1992. Prices rose by 48 per cent between 1986 and 1989, compounding the problems of countries and families that already could not afford to buy enough to eat.

Changing conditions

Bad weather accounts for part of the slump. Drought hit India in 1987, and the United States, Canada and China – the world's three biggest food producers – in 1988. And in 1992, the most serious drought of all gripped southern Africa; harvests fell by more than half in eight countries in the region. In June 1988, at the height of the drought in the American Midwest, Dr James Hansen, Director of NASA's Goddard Institute for Space Studies, told a Congressional hearing that he was "99 per cent certain" that the greenhouse effect was to blame, and scientists suspected the same factor at work four years later in southern Africa. No-one can yet be sure that the greenhouse effect has taken hold, or whether it is causing harvests to fail. But it does seem certain that it will take place before long, and have very similar effects.

Other causes include overuse, leading to erosion and desertification. Every year, the world's farmers lose about 26 billion metric tons of topsoil, about the same amount as covers the entire Australian wheatlands. At one stage, in the 1970s, American farmers lost 6 metric tons of soil for every metric ton of grain grown. The world has some spare capacity; 20 million hectares of US farmland were held in reserve in 1988, and bringing them back into production would increase the world's cropland by 2 per cent. But the Food and Agriculture

Organization of the United Nations (FAO) estimates that soil degradation could take 65 per cent of all the Third World's rainfed (non-irrigated) land out of production by the year 2000. And every year the world also loses 1.5 million hectares of irrigated fields to salinization.

Pollution is also thought to be cutting yields. A US Government survey suggests that just one pollutant, ozone, may have reduced American harvests by 5-10 per cent during the 1980s, and sulfur dioxide and other nitrous oxides will also have done damage. So does the depletion of the ozone layer in the stratosphere, where the poisonous gas protects life from harmful radiation.

Reducing hunger

Increasing per capita food production is a big task; making sure that it reaches the hungry is a gigantic one. The most important priority is to increase both production and consumption in developing countries. In the past, Third World governments have usually concentrated resources on the cities and on industry, neglecting farming and the countryside; food prices have been kept low to please city dwellers, to the ruination of agriculture.

Governments increasingly accept that this bias must end, but that will not in itself address the problem adequately. Increasing food prices may benefit the middleman rather than the farmer; concentrating attention on the richer landowners, as during the Green Revolution, will do little, or nothing, to help the poor or reduce hunger. It is much more effective to focus on small farmers, who both make up the bulk of the poor in many countries and have the greatest potential for raising production.

After independence in 1980, Zimbabwe switched attention away from the richer whites – who comprised only 1 per cent of the farmers but owned half the land, received 87 per cent of the credit, and got preferential prices for their produce – towards black subsistence farmers. Their maize production doubled by 1981 and more than trebled by 1985 – a bad year in the rest of Africa.

Where small farmers, particularly women, have been encouraged and given credit, harvests have increased and hunger has fallen. Land reform is particularly important. It splits up big estates, which are usually much less intensively farmed, and gives poor farmers and landless people the means to grow enough food to feed their families. The World Bank has estimated that such a Patchwork Revolution could increase yields even faster than the Green Revolution, with much more success in reducing hunger.

Drinking Water and Sanitation

Every day at least 25,000 people die from their everyday use of water, supposedly the very stuff of life itself. Dirty water is both the world's greatest killer, and its biggest single pollution problem.

The cost in lives

Diarrhea alone kills some 4 million young children a year. This is equivalent to wiping out every preschool child in the United Kingdom and Australia put together one year, killing every child under five in West Germany and Canada during the next 12 months, poisoning all their contemporaries in France, Norway, Sweden and Denmark in the third year, and so on. On average, every Third World child under five suffers three serious attacks of diarrhea a year. Children in rural Africa are commonly ill, from one cause or another, for 140 days in the year; many are sick nearly all the time. Few enjoy long enough breaks of good health to get properly better and catch up on growth. Repeated onslaughts of diarrhea not only threaten their lives, but cause malnutrition and so stunt their physical and mental development.

About 200 million people suffer from schistosomiasis, infected by contaminated water on their skin; in some areas, as around Lake Volta in Ghana, virtually every child is sick and blood in a young boy's urine, caused by the disease, is regarded as normal. Meanwhile some 500 million people have trachoma, one of the main causes of blindness, because they do not have enough clean water to wash in. Four fifths of all disease and a third of all deaths in the Third World have been linked to drinking dirty water; on average each person of a developing country loses a tenth of his productive time each year to water-related illness.

The World Bank estimated in 1992 that providing safe water and adequate sanitation to all the world's people – while not eliminating diarrhea entirely – would save the lives of 2 million children under five each year, cut the number of people suffering from schistosomiasis by three quarters, and allow 300 million more people to escape infection with roundworm. As Dr Halfdan Mahler, then Director-General of the World Health Organization, put it: "The number of water taps per 1,000 persons is a better indication of health than the number of hospital beds."

A rural scandal

Over most of the Third World, the number of taps is extremely small: over much of it the people never use a tap at all. In 1992 about 1.2 billion people did not have safe drinking water. Almost 1.8 billion had no sanitation – and inevitably their wastes got into the water. Some countries fare much worse: in the Central African Republic 88 per cent of the people lack clean water, in Ethiopia 81 per cent. Almost everywhere, the people living in villages and the countryside suffer more than those in towns and cities, and the poor are far less well served than the rich. In Ghana, for example, rural people are half as likely to have safe water as their urban counterparts, and the poorest fifth of the population are 12 times less likely to have taps in their houses than the richest fifth.

Water, of whatever quality, is used far less in the countryside. In rich countries people use between 350 and 1,000 liters of water each per day. In relatively well-off Third World urban areas, which have water piped into the houses but do not have dishwashers and washing machines, daily use is between 100 and 350 liters per person. In poorer urban areas, relying on public hydrants for their water, per capita consumption drops to between 20 and 70 liters a day. When there are no taps at all, as in much of the countryside, the people (in practice, the women) have to get water from where they can, from rivers, ponds or streams – or wells, if they are lucky – often miles away. In rural areas of Kenya, for example, per capita daily use often drops to between 2 and 5 liters, near to the absolute biological limit people need to stay alive.

In eastern Uttar Pradesh in India and in northern Burkina Faso in Africa, some women have to walk for four hours to collect water. In some parts of Africa, they can spend up to 12 hours a day at the task. It is a backbreaking, gruelling, daily grind that consumes scarce calories from their daily diet and takes up time that could be spent in growing food. Every filthy drop carried home is treated like gold. When one top United Nations official asked an African woman if she realized how important it was to get her children to wash their hands after defecating, she retorted: "I have to carry our water seven miles every day. If I caught anyone wasting water by washing their hands, I would kill them." Yet research by the International Diarrhoeal Diseases Research Centre in Bangladesh shows that handwashing cuts outbreaks of diarrhea by 40 per cent in children under five.

The economic cost

The economic cost is phenomenal. In India, where three young children die from dirty water every minute, waterborne diseases cost the country 73 million working days every year, and $600 million annually in health care and lost production. In only the first 10 weeks of Peru's cholera epidemic of 1991, the country lost $10 billion in reduced tourism and agricultural exports: this was three times as much as it had invested in water and sanitation in the 10 years of the 1980s. Typically, every sufferer from schistosomiasis loses 600 to 1,000

Drinking water

% of population without access to safe drinking water, 1985 – 88

- 70% and over
- 50 – 69%
- 25 – 49%
- 0 – 24%
- Insufficient data

Safe drinking water is taken for granted in the developed world. Yet for millions of people in developing countries safe water is a luxury.

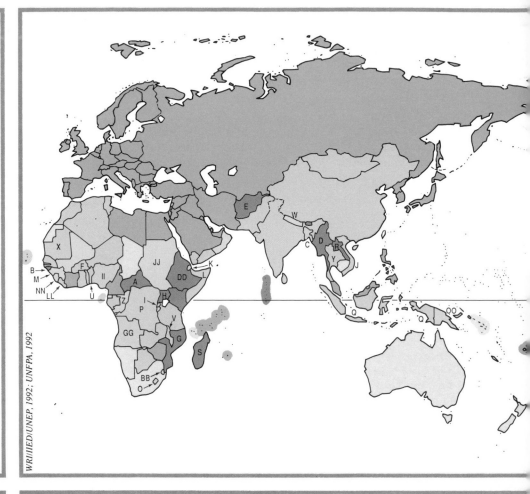

WRI/IIED/UNEP, 1992; UNFPA, 1992

Sanitation

% of population without access to sanitation services, 1985 – 88

- 70% and over
- 50 – 69%
- 25 – 49%
- 0 – 24%
- Insufficient data

Lack of sanitation facilities – toilets, latrines, proper drains – results in infection with debilitating diseases. Millions of Third World people, especially in rural areas, have no basic sanitation.

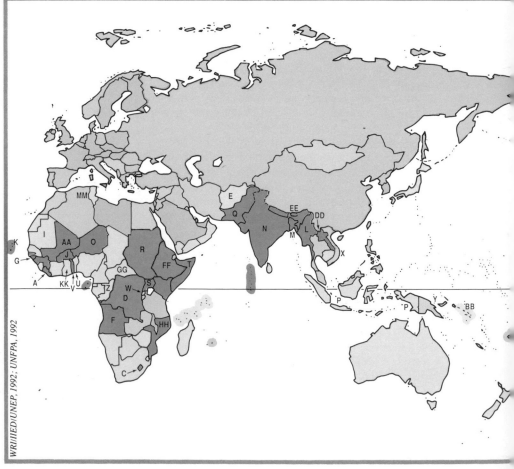

WRI/IIED/UNEP, 1992; UNFPA, 1992

Drinking water: the worst 25

Highest % of population without access to safe drinking water, 1988 (letters in brackets refer to top map).

Urban			*Rural*		
Central African Rep.	87%	(A)	Congo	98%	(Z)
Guinea-Bissau	82%	(B)	Uruguay	95%	(AA)
Bangladesh	63%	(C)	Paraguay	93%	(T)
Myanmar	62%	(D)	Swaziland	93%	(BB)
Afghanistan	61%	(E)	El Salvador	90%	(CC)
Burkina Faso	56%	(F)	Madagascar	90%	(S)
Mozambique	56%	(G)	Central African Rep.	89%	(A)
Uganda	55%	(H)	Ethiopia	89%	(DD)
Rwanda	54%	(I)	Uganda	88%	(H)
Vietnam	52%	(J)	Bolivia	85%	(EE)
Djibouti	50%	(K)	Afghanistan	83%	(E)
Somalia	50%	(L)	Argentina	83%	(FF)
Guinea	45%	(M)	Laos	83%	(R)
Haiti	45%	(N)	Mozambique	83%	(G)
Lesotho	41%	(O)	Zaïre	83%	(P)
Zaïre	41%	(P)	Angola	81%	(GG)
Indonesia	40%	(Q)	Nicaragua	81%	(HH)
Laos	39%	(R)	Nigeria	80%	(II)
Madagascar	38%	(S)	Sudan	80%	(JJ)
Paraguay	35%	(T)	Chile	79%	(KK)
Benin	34%	(U)	Djibouti	79%	(K)
Malawi	34%	(V)	Liberia	78%	(LL)
Nepal	34%	(W)	Peru	78%	(MM)
Mauritania	33%	(X)	Sierra Leone	78%	(NN)
Thailand	33%	(Y)	Papua New Guinea	77%	(OO)

WRI/IIED/UNEP, 1992; UNFPA, 1992

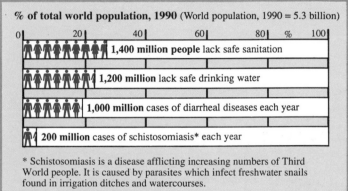

% of total world population, 1990 (World population, 1990 = 5.3 billion)

1,400 million people lack safe sanitation

1,200 million lack safe drinking water

1,000 million cases of diarrheal diseases each year

200 million cases of schistosomiasis* each year

* Schistosomiasis is a disease afflicting increasing numbers of Third World people. It is caused by parasites which infect freshwater snails found in irrigation ditches and watercourses.

Sanitation: the worst 25

Highest % of population without access to sanitation services, 1988 (letters in brackets refer to lower map).

Urban			*Rural*		
Liberia	96%	(A)	Congo	98%	(Z)
Jamaica	86%	(B)	Niger	97%	(O)
Lesotho	86%	(C)	Bangladesh	96%	(M)
Zaïre	86%	(D)	India	96%	(N)
Afghanistan	80%	(E)	Burkina Faso	95%	(J)
Angola	75%	(F)	Mali	95%	(AA)
Guinea-Bissau	70%	(G)	Solomon Islands	95%	(BB)
Nicaragua	68%	(H)	Somalia	95%	(T)
Mauritania	66%	(I)	Sudan	95%	(R)
Burkina Faso	65%	(J)	Chile	94%	(CC)
Cape Verde	65%	(K)	Laos	94%	(DD)
Myanmar	65%	(L)	Bhutan	93%	(EE)
Bangladesh	63%	(M)	Ethiopia	93%	(FF)
India	62%	(N)	Liberia	92%	(A)
Niger	61%	(O)	Pakistan	92%	(Q)
Indonesia	60%	(P)	Uganda	90%	(S)
Pakistan	60%	(Q)	Central African Rep.	89%	(GG)
Sudan	60%	(R)	Mozambique	89%	(HH)
Uganda	60%	(S)	Mexico	88%	(II)
Somalia	59%	(T)	Bolivia	87%	(JJ)
Benin	58%	(U)	Zaïre	86%	(D)
Togo	58%	(V)	Ghana	85%	(KK)
Rwanda	55%	(W)	Haiti	85%	(LL)
Vietnam	52%	(X)	Tunisia	85%	(MM)
Papua New Guinea	46%	(Y)	Togo	84%	(V)

WRI/IIED/UNEP, 1992; UNFPA, 1992

working days. In Venezuela, on the other hand, the Government has calculated that every dollar invested in providing clean water in the countryside is recouped more than five times over in increased production.

In 1980, the United Nations launched the International Drinking Water Supply and Sanitation Decade with the slogan "Clean water and adequate sanitation for all by the year 1990". During the decade, hundreds of millions of people were given safe water and proper sanitation for the first time. But – partly because of population growth, partly because pumps, wells and latrines installed during the decade broke down – there are roughly the same number of people without them in 1992 as there were in 1980.

India increased its funding six times over in the countryside and threefold in urban areas. Nepal, the Philippines, Sri Lanka, Sudan and Trinidad and Tobago all greatly increased their budgets. But investments in many other countries stayed low – and in some of the poorer countries they remained "negligible", to use the United Nations' description. Over the Third World as a whole the proportion of national budgets devoted to providing safe water and adequate sanitation remained much the same as before.

Nor was nearly enough money provided in overseas aid. Donors did increase the proportion of their budgets given for these purposes, but overseas aid as a whole stagnated during the 1980s, and the money was provided unevenly. Aid for water and sanitation in sub-Saharan Africa increased little during the decade. Africa received less than any other continent – even though it had the greatest needs. By the middle of the decade, almost all African countries had completed national action plans, or were preparing them. But even by 1985 many had had to be shelved for lack of funds. That year's devastating drought, and Africa's deepening economic crisis, diverted even more attention away from water and sanitation, and the rapid growth of population eroded such progress as was made.

An initiative which failed

The decade's target – to provide clean water and adequate sanitation for everyone on earth by 1990 – was always more of a dream than a realistic goal. No-one ever expected it to be achieved. Meeting the target would have cost $80 million for every day of the decade. By 1985, 650,000 people would have had to be provided with water and 1 million with sanitation every day for the remaining five years. This level of international commitment was, sadly, never in prospect, not in 1980, not in 1985, and not in 1990.

There is not much hope of it being achieved in the current decade either. Papers before the UN's International Conference on Water and the Environment in Dublin in January 1992 estimated that meeting the target by the year 2000 would cost the world $36 billion a year. It added, in a masterly understatement: "This is unlikely to be forthcoming in the current economic climate."

But new cheap technologies, pioneered during the International Drinking Water Supply Decade, offer some hope. In one UNICEF-supported program in Nigeria, for example, the average cost of a borehole and handpump fell by three quarters – from $20,000 to $4,000 between 1982 and 1990. Using such techniques, 80 per cent of the people who need sanitation and clean water could receive them in the 1990s at only one third of the total cost.

Providing these basic needs for everyone on earth remains more than a worthy aim. It is a minimum requirement for the human community, a yardstick by which to measure our failure to create a world fit for children to live in. Meanwhile, the daily toll of young lives goes on.

Health

A t any one time, one third of the world's people are ill. The vast majority of the sick are in developing countries, and an alarming number are children. Every year, over 14 million Third World children under five die of a handful of diseases; malnutrition is often a contributing factor. In poor countries, the majority of all deaths occur in children too young to go to school.

Two thirds of these deaths could be prevented by minimal health care. Most of the big Third World killers are diseases that have either been virtually eliminated in rich countries, or survive only as relatively minor problems. Every day in developing countries, nearly 11,000 children die from diarrhea, over 4,000 from measles, about 2,750 from malaria, another 2,250 from tetanus, around 1,500 from whooping cough and more than 6,000 from pneumonia.

Health and wealth

Wealth determines health, and health determines wealth. The poor get sick: their sickness ensures they remain poor, and weakens the economies of their countries.

The differences start at birth. In Canada, only two mothers in every 100,000 die in childbirth; in Burkina Faso, 1,680 of them perish. In Sierra Leone, Mali and Ethiopia more than 150 children in every 1,000 die before they reach their first birthday – around 20 times as many as in Scandinavia, France or the Netherlands. Within countries, too, death shadows poverty. In Kenya, children living in the relatively prosperous Central province are nearly three times more likely to see their fifth birthdays as children in the nearby, but poorer, Nyanza province. In Costa Rica, children of professional people are five times more likely to reach the age of two than those of agricultural workers.

Third World diseases

Four fifths of all illness in the Third World is caused by waterborne disease. Diarrhea is the leading cause of child deaths. Between 200 and 400 million children and adults get malaria every year; 5 million die. Some 200 million have the chronic and debilitating schistosomiasis. About 20 million suffer the agonies of guinea worm. Airborne diseases like pneumonia and bronchitis, responsible for between a quarter and a third of all child deaths, spread readily in cramped housing, and tuberculosis kills some 3 million people every year, 30,000 of them children.

In all, 40 per cent of deaths in Third World countries are caused by infectious and parasitic diseases, compared to only 8 per cent in industrialized countries. First World diseases are also affecting the Third World. Heart disease now kills one person in six in developing countries, and lung cancers are increasing as smoking spreads.

Over the next 10 years, the number of people with the HIV virus is expected to increase from 10 to 25 million – and half of the new cases will occur in the Third World. One third of all men aged 30-35 in Lusaka carry the HIV virus, as do 88 per cent of the female prostitutes in Nairobi. The death toll may be small compared to that of the big killers, but it can be particularly devastating because AIDS predominantly kills young and middle-aged adults in their most productive years. Many are young professionals or businesspeople, and the loss of large numbers could be destabilizing. The Harvard Institute for International Development has estimated that premature AIDS deaths could cost Zaïre 8 per cent of its GNP by 1995. The cost of treatment alone for each Zaïrean with AIDS can amount to 10 times the country's annual income per capita.

The scarce doctors

In developed countries there is one doctor for about every 500 people. In the poorest countries (excluding India and China), some 13,500 people have to share each doctor. In Africa as a whole, there is only one doctor for every 25,000 people; in Ethiopia, one for every 77,000. In East Africa there is, on paper, one doctor for every 17,500 people. But more than 80 per cent of them are based in the cities, while more than 80 per cent of the people live in the countryside. So in rural East Africa, the number of people per doctor is nearer 60,000. The situation is much the same in almost every Third World country, with hundreds of millions of people being deprived of medical care.

Life expectancy and health

In the absence of detailed medical statistics from developing countries, life expectancy figures are used as a rough guide to health. In many African countries life expectancy is below 50; in most Western countries it is around 70 or above. A child born in Sierra Leone or Afghanistan can expect to live to 43, but one born in Iceland or Japan can expect to reach 77 or 79. And yet, every country in the world has increased the life expectancy of its people since the 1960s; many developing countries have raised it by 10 years or more. The enormous toll of young children is predominantly responsible for the difference in life expectancy between rich and poor countries.

Saving the children

In 1960, 20 per cent of Third World babies died in their first year of life: by the end of the 1980s this had been

Population per physician
1990 estimate

- ◼ 20,000 and over*
- ◼ 10,000 – 19,999
- ◼ 5,000 – 9,999
- ◻ 2,000 – 4,999
- ◼ 1,000 – 1,999
- ◻ 500 – 999
- ◻ Under 500
- ◻ Insufficient data

*Three countries have more than 50,000 people per physician: Ethiopia (77,356), Rwanda (74,924) and Burkina Faso (57,359).

Access to health care

% of population with access to health care, selected countries, 1985 – 88

	Total	Urban	Rural
Mauritius	**100**	100	100
Saudi Arabia	**97**	100	88
Jordan	**97**	98	95
Iraq	**93**	97	78
Oman	**91**	100	90
Tunisia	**90**	100	80
Botswana	**88**	100	85
Algeria	**88**	100	80
Nicaragua	**83**	100	60
Congo	**83**	97	70
Iran	**80**	95	65
Tanzania	**76**	99	72
Honduras	**73**	85	65
Argentina	**71**	80	21
Morocco	**70**	100	50
Bolivia	**63**	90	36
Ghana	**60**	92	45
El Salvador	**56**	80	40
Pakistan	**55**	99	35
Cambodia	**53**	80	50
Sudan	**51**	90	40
Niger	**41**	99	30
Nigeria	**40**	75	30
Liberia	**39**	50	30
Guatemala	**34**	47	25
Myanmar	**33**	100	11
Côte d'Ivoire	**30**	61	11
Afghanistan	**29**	80	17
Rwanda	**27**	60	25
Somalia	**27**	50	15
Zaïre	**26**	40	17

UNICEF, 1991

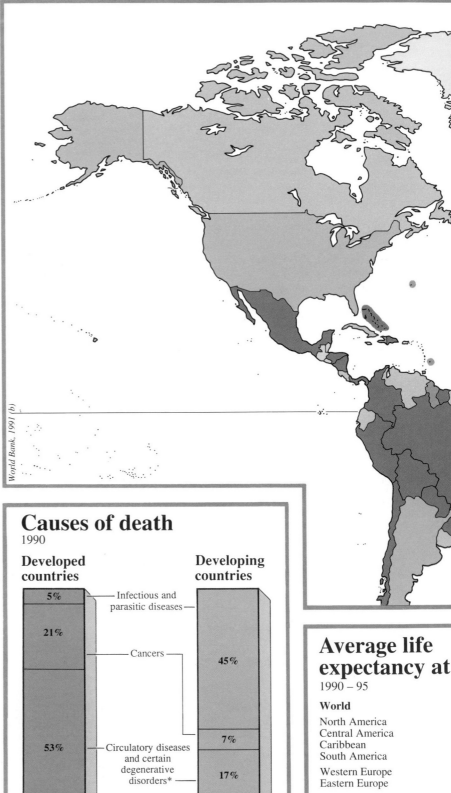

World Bank, 1991 (b)

Causes of death
1990

Developed countries / **Developing countries**

Cause	Developed	Developing
Infectious and parasitic diseases	5%	45%
Cancers	21%	7%
Circulatory diseases and certain degenerative disorders*	53%	17%
Childbirth-related	1%	10%
Injury and poisoning	7%	6%
Lung diseases	4%	6%
Other	9%	9%

*Includes diabetes, ulcers, liver complaints and nervous disorders

WHO, pers. comm.

Average life expectancy at birth
1990 – 95

World	66 years
North America	76
Central America	69
Caribbean	70
South America	68
Western Europe	76
Eastern Europe	72
North Africa	61
West Africa	51
Central Africa	52
East Africa	53
Southern Africa	62
USSR	71
West Asia	66
South Asia	59
East Asia	72
Southeast Asia	63
Australasia & Oceania	73

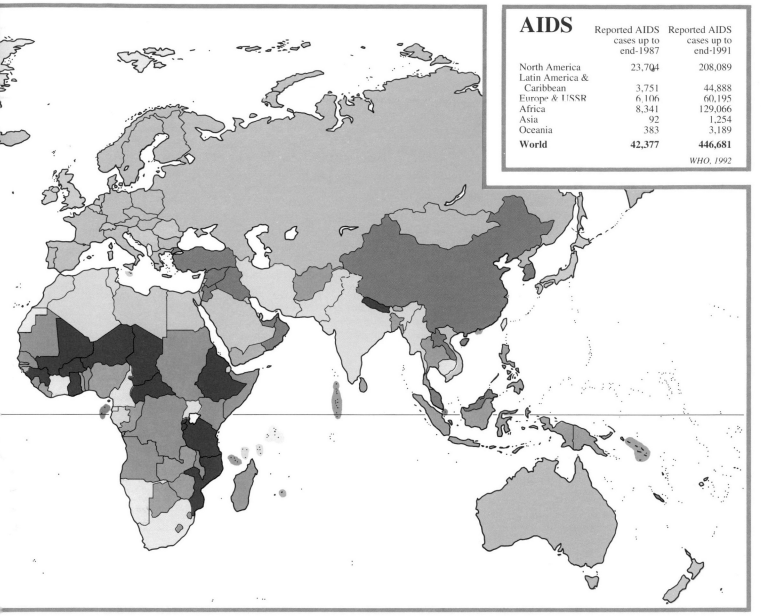

AIDS

	Reported AIDS cases up to end-1987	Reported AIDS cases up to end-1991
North America	23,704	208,089
Latin America & Caribbean	3,751	44,888
Europe & USSR	6,106	60,195
Africa	8,341	129,066
Asia	92	1,254
Oceania	383	3,189
World	**42,377**	**446,681**

WHO, 1992

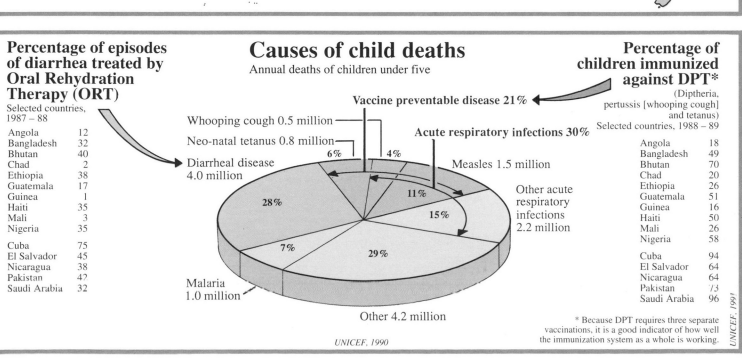

Percentage of episodes of diarrhea treated by Oral Rehydration Therapy (ORT)

Selected countries, 1987 – 88

Angola	12
Bangladesh	32
Bhutan	40
Chad	2
Ethiopia	38
Guatemala	17
Guinea	1
Haiti	35
Mali	3
Nigeria	35
Cuba	75
El Salvador	45
Nicaragua	38
Pakistan	42
Saudi Arabia	32

Causes of child deaths

Annual deaths of children under five

Vaccine preventable disease 21%

Whooping cough 0.5 million

Neo-natal tetanus 0.8 million

Diarrheal disease 4.0 million

Acute respiratory infections 30%

Measles 1.5 million

Other acute respiratory infections 2.2 million

6% 4% 11% 28% 15% 7% 29%

Malaria 1.0 million

Other 4.2 million

UNICEF, 1990

Percentage of children immunized against DPT*

(Diptheria, pertussis [whooping cough] and tetanus)
Selected countries, 1988 – 89

Angola	18
Bangladesh	49
Bhutan	70
Chad	20
Ethiopia	26
Guatemala	51
Guinea	16
Haiti	50
Mali	26
Nigeria	58
Cuba	94
El Salvador	64
Nicaragua	64
Pakistan	73
Saudi Arabia	96

* Because DPT requires three separate vaccinations, it is a good indicator of how well the immunization system as a whole is working.

UNICEF, 1991

reduced to 12 per cent. China cut its infant mortality rate to the level of some US cities. Moreover most developing countries managed to speed up their advance in the 1970s, despite economic difficulties.

They were helped by an international campaign for child health, promoting simple life-saving techniques. At the end of the 1970s, the World Health Organization set a target of immunizing 80 per cent of one-year-olds against diphtheria, tetanus, whooping cough, polio, tuberculosis and measles by 1990. At the time, only 10-20 per cent of children were protected. By 1988, immunization rates were approaching 70 per cent. Governments launched special campaigns: in 1985 El Salvador suspended its civil war for three "days of tranquillity" to vaccinate 400,000 children – an example followed in Afghanistan and the Sudan. The United Nations Children's Fund (UNICEF) estimates that the increased immunization is saving 2 million children's lives a year and has prevented 1.5 million more from being crippled by polio.

Meanwhile, UNICEF has been promoting the use of Oral Rehydration Therapy (ORT) – a simple mixture of salt and sugar, costing a few cents, dissolved in water. It prevents 90 per cent of deaths from diarrhea by allowing water to penetrate the gut of dehydrated victims. UNICEF says that it is already saving another million young lives every year.

Much remains to be done. In the world's poorest countries – and in Africa as a whole – the infant mortality rate is still one in five. Three million children still die each year because they have not been immunized. Only one family in every three is using ORT; if its use were universal, at least another 2.5 million lives could be saved annually. Breast feeding of babies still seems to be declining, and bottle-fed infants are up to 25 times as likely to die in childhood as those who are given the breast for their first six months of life. Every year 250,000 children go blind for the lack of a small supplement of vitamin A costing just 2 cents.

Barefoot health care

Simple though these solutions are, they cannot be spread unless someone takes them to the millions of villages where doctors will not tread. Nations have been agreed since the late 1970s on how to do this – by providing "primary health care" through a development of China's system of barefoot doctors. Selected villagers are given six months' training in basic health care and preventative medicine, and put in charge of a community of some 250 families: their training enables them to deal with most problems, and they can refer more difficult ones to qualified doctors in clinics and hospitals.

Most Third World countries have set up at least some primary health care on this pattern. Several, including Thailand, Zimbabwe and the Gambia, have had great success. But many large-scale schemes have failed in India, Indonesia, the Philippines and other countries. Even in Tanzania, which pioneered the concept in Africa, two thirds of the villages still have no health facilities.

Misdirected resources

Governments have usually starved the system of resources. The bulk of their money has gone to expensive hospitals and highly trained doctors to serve the elites in the cities. On average, developing countries spend three quarters of their health budgets on urban hospitals. Training full doctors at a cost of $60,000 a head is preferred to training paramedics at $500 each, even though paramedics will save far more lives and the doctors may well leave to practise in rich countries. The cost of operating the main hospital in Brazzaville equals the total Congolese expenditure on primary health care. Pakistan, by contrast, boosted its immunization coverage from 5 per cent of its children to 62 per cent, by postponing the building of a hospital for five years.

As the recession of the 1980s and early 1990s tightened its grip, primary health care systems were squeezed. Countries like Botswana and Jamaica suspended training community health workers, while Ecuador, Panama, Paraguay and Peru were unable even to afford vaccines in early 1989. Most governments attempted to shield expensive modern health services and let the cuts fall more heavily on primary health care. Hundreds of clinics were shut down. The World Bank reports that life expectancy began to fall again in nine African countries in the early 1980s, and since then ill-health has continued to advance over much of the Third World. Without a change in priorities, many of the achievements of the 1960s and 1970s could be lost.

The Education Gap

Education is the passport to modern life, and a precondition of national prosperity. But more than a quarter of the world's adults – 900 million – cannot read or write, and more than 100 million young children are deprived of even a primary school education. In most developing countries, after decades of educational expansion, spending on learning is falling.

The illiterate are virtually helpless in a world ruled by the written word. They are excluded from good jobs and handicapped in seeking even badly paid work. Notices and official papers can seem a mass of meaningless hieroglyphics. People who cannot decipher them are at the mercy of those who can; many, as a result, have been cheated of their rights or their land.

Studies show that people with even a basic education are healthier and eat better. They are more likely to plan their families and their children are more likely to survive. According to the World Bank, just four years of primary education enables farmers to increase productivity by 10 per cent, often the difference between hunger and sufficiency. National economic returns from education outstrip those from most other forms of investment.

Enrollment rise and fall

As they became independent, most developing countries enthusiastically embraced education. Two decades of astonishing expansion followed. Between 1960 and 1981 the world's 32 poorest countries (excluding India and China, which have long had good records) increased the proportion of their children enrolled in primary school from 38 to 72 per cent. The 38 next poorest achieved almost universal primary school enrollment by 1980; up from about two thirds in 1960. For some countries, the achievement was even greater. Ethiopia increased its enrollment more than sixfold, from 7 to 46 per cent. Tanzania and Laos achieved virtually universal enrollment by 1981 although only a quarter of their children were in school in 1960. Developing countries often spent a fifth of their national budgets on education. It seemed as if it would not be long before every child alive could be sure of going to school.

By the end of the 1980s that dream had turned to bitter disillusion. The decade brought economic disaster to developing countries. They slumped when rich nations went into recession at the beginning of the 1980s, the subsequent recovery passed them by and they were hit again by the renewed recession in the late 1980s and early 1990s. The educational expansion of the 1960s and 1970s first halted, then went into reverse.

Declining expenditure

The proportion of national expenditure going to education declined in more than half of developing countries over the 1980s. In two thirds, spending per pupil fell in real terms. In the world's 37 poorest countries the average expenditure per head on education dropped by a quarter. In Africa as a whole, says the World Bank, only $0.60 a year is spent on educational materials for each student: it estimates "minimum requirements" at $5. A United Nations Children's Fund (UNICEF) survey of schools in a rural area of Mozambique found that only half the classrooms had a blackboard, only a sixth had a desk for the teacher, only 5 per cent of first year pupils had a language book and only 13 per cent a mathematics book.

Building schools has virtually ceased in many countries and money is so scarce that 95 per cent of the educational budget is absorbed in teachers' pay. Salaries, low to begin with, have been cut. UNESCO (the United Nations Educational, Scientific and Cultural Organization) estimates that the salaries of almost two thirds of the Third World's teachers fell during the decade. Inevitably, the quality of education fell too.

By 1989, enrollment rates had dropped in one out of every five developing countries. In some African countries, the number of children in primary schools declined by a third between 1980 and 1985. Tanzania's universal primary school enrollment fell to less than 75 per cent. UNESCO's Director-General, Federico Mayor, warns that this threatens to "set back the countries of the South by a whole generation or even more."

Illiteracy and the poor

The United Nations General Assembly proclaimed 1990 as "International Literacy Year", the start of a decade in which the world should "eradicate illiteracy". No-one expects this target to be achieved. The proportion of the world's adults unable to read or write has fallen steadily; literacy rates in developing countries rose from 46 per cent to 66 per cent between 1960 and 1990. But, thanks to population growth, the number of illiterate has continued to grow, from about 700 million in 1960 to 900 million in 1990. This is more than a quarter of the world's adult population and 98 per cent of them live in the Third World. Rather more than a third of Asia's massive population cannot read or write. Latin America is better off, with a 17 per cent illiteracy rate. Africa, the poorest continent, fares far worse, with 54 per cent of its adults deprived of this most basic skill. Generally speaking, the poorer a country the higher the proportion of illiterate: more than two thirds of the adults in the very poorest countries cannot read or write.

In industrialized countries, absolute illiteracy was largely eradicated half a century ago: they contain only 2 per cent of the world's illiterate. But "functional illiteracy" remains. In Canada, the literacy of a quarter of all adults is seriously inadequate. In the United

Adult literacy rate

% of population aged 15 or over able to read and write, 1990

	95% and over
	80 – 94%
	60 – 79%
	40 – 59%
	Under 40%
	Insufficient data

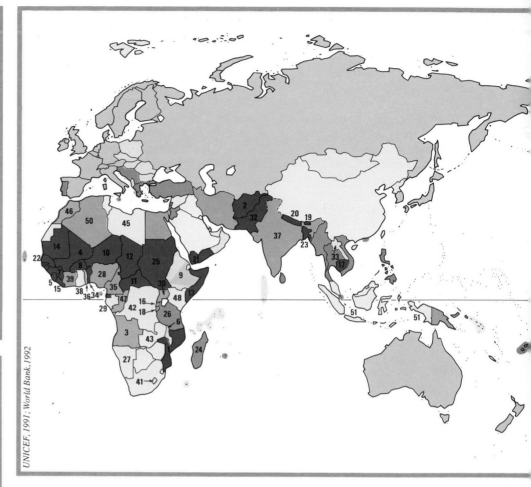

UNICEF, 1991; World Bank, 1992

Spending on education

As % of total government expenditure in selected developing countries

0	%	5	10	15	20	25	30

BOLIVIA **1986** **1972**
KENYA
BURKINA FASO
MOROCCO
TANZANIA
MALAWI
UGANDA
BANGLADESH
OMAN
PAKISTAN

UNICEF, 1989

School enrollment

% of 6-17 year-olds in school, 1990

	90% and over
	80 – 89%
	70 – 79%
	60 – 69%
	Under 60%
	Insufficient data

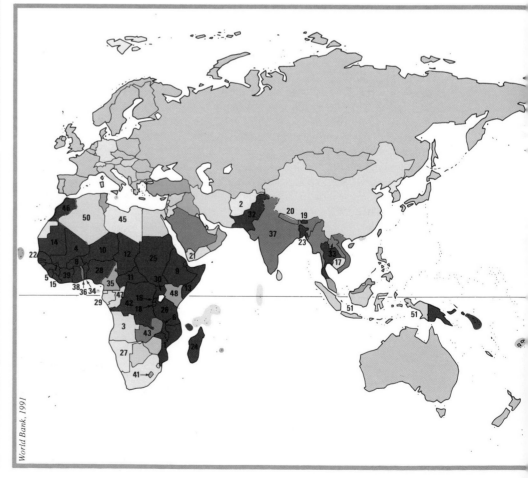

World Bank, 1991

38

Literacy and school enrollment: males and females compared

Ranked in order of countries' 1990 child mortality*.
Figures refer to both maps. na = not available.

— Adult literacy rate (%), 1990 —

Primary school enrollment (%), 1986 – 88 (gross)†

MALE			FEMALE	
45	76	1. MOZAMBIQUE	59	21
44	27	2. AFGHANISTAN	14	14
56	na	3. ANGOLA	na	29
41	29	4. MALI	17	24
31	68	5. SIERRA LEONE	48	11
31‡	73	6. MALAWI	59	15‡
35	41	7. GUINEA	18	13
28	41	8. BURKINA FASO	24	9
na	46	9. ETHIOPIA	28	na
40	37	10. NIGER	20	17
52	82	11. CENT. AFRICAN REP.	51	25
42	73	12. CHAD	29	18
27	26	13. SOMALIA▲	13	9
47	61	14. MAURITANIA	42	21
50	82	15. LIBERIA	50	29
64	69	16. RWANDA	66	37
48	na	17. CAMBODIA	na	22
61	68	18. BURUNDI	50	40
51	31	19. BHUTAN	20	25
38	104	20. NEPAL	47	13
47	132	21. YEMEN	39	21
52	71	22. SENEGAL	49	25
47	76	23. BANGLADESH	64	22
88	97	24. MADAGASCAR	92	73
43	59	25. SUDAN	41	12
93	67	26. TANZANIA	66	88
na	na	27. NAMIBIA	na	na
62	103	28. NIGERIA▲	81	40
74	127	29. GABON▲	125	49
62	76	30. UGANDA	63	35
85	97	31. BOLIVIA	85	71
47	51	32. PAKISTAN	28	21
na	102	33. LAOS	85	na
31	84	34. BENIN	43	16
66	119	35. CAMEROON	100	43
56	124	36. TOGO	78	31
62	113	37. INDIA	81	34
70	78	38. GHANA	63	51
67	92	39. CÔTE D'IVOIRE▲	65	40
59	83	40. HAITI	72	47
na	102	41. LESOTHO	127	na
84	84	42. ZAÏRE	68	61
81	102	43. ZAMBIA	92	65
92	125	44. PERU	120	79
75	na	45. LIBYA	na	50
61	85	46. MOROCCO	56	38
70	na	47. CONGO	na	44
80	98	48. KENYA	93	59
76	104	49. HONDURAS	108	71
70	105	50. ALGERIA	87	46
84	120	51. INDONESIA	115	62
63	82	52. GUATEMALA	70	47

* Mozambique has the world's highest child mortality rate, Afghanistan the second highest, and so on. ▲ 1984-86. ‡ 1985.

† The gross enrollment ratio is the total number of children enrolled at primary school level, expressed as a percentage of all children of primary school age. This figure sometimes exceeds 100 where children older or younger than primary school age are being educated at that level.

UNICEF, 1991 - UNFPA, 1992

States, estimates range from 5 to 25 per cent; in France, the total numbers range from 2 to 8 million people, depending on the study. Most are among the poorest members of their societies.

In the Third World, too, the poor suffer most. The poorer a child's family, the less likely he (or, particularly, she) is to start school, the more likely those who do start are to drop out. Even during the 1970s boom, the differences were dramatic. The child of an official in Burkina Faso, for example, was 60 times more likely to reach secondary school than a farmer's child; in Kenya, a civil servant's child was 1,000 times more likely than a farmer's to get to university.

The disadvantaged countryside

More poor people in the Third World live in the countryside, where schools and teachers are always scarcer. In Bolivia, two thirds of rural people over 25 have never attended school, compared with less than a quarter of city dwellers. From El Salvador to Indonesia, from Botswana to Brazil, city adults are twice as likely to have had some schooling as country people. But even in the cities, the poor miss out: in Calcutta, over 60 per cent of children do not attend school because they have to work to help keep the family going, or look after younger siblings to enable their mothers to work.

Most poor families struggle to put at least one child through school in the hope that he will get a good job and be able to look after them. But, as times get harder, more and more parents are pulling their children out. Of the 100 million six-year-olds who first went to school in Third World countries in 1990, UNICEF estimates that 40 million will drop out without even completing primary school education. Almost all will be illiterate for life.

Two thirds of the children who either never start school or drop out early are girls. Two thirds of the world's illiterate are women. Yet women's education is particularly important. The World Bank identifies it as "perhaps the single most important determinant of family health and nutrition", and its research shows that infant mortality rates fall steadily, and dramatically, for every year women spend at school. But tradition, prejudices and the burden of work to be done at home ensure that daughters are pulled out of school first. In the first grade of Kampala's primary schools the sexes are evenly balanced: by the seventh grade there are more than twice as many boys as girls.

Primary education: the productive dollar

Every dollar invested in primary education, according to another World Bank study, is 50 per cent more productive than one invested in secondary schooling, and gives twice as much as one spent on universities. Yet, throughout the Third World, these spending priorities are reversed. Early schooling is squeezed to pay for higher education, even though 100 children could complete primary school for the same sum it takes to produce a single university graduate. Even during the educational boom, only one Third World child in every three went to secondary school, only one in every 20 made it to higher education. Yet young people with secondary education are twice as likely to be unemployed as those who left school at the primary stage.

A few countries have started to change their priorities, emphasizing primary education. Zimbabwe doubled its number of primary schools in its first five years of independence; the proportion of its budget spent on education is the fifth highest in the world, and the curriculum has been reorientated to meet local needs. Jamaica has cut its illiteracy rate from 50 per cent to 18 per cent over the last 15 years. Bangladesh has opened more than 2,500 basic village primary schools with appropriate syllabuses since 1985 at an annual cost of just $15 per pupil: only 1.5 per cent of the children drop out, compared to 60 per cent of their peers in the ordinary primary schools – and 95 per cent of pupils, the majority girls, continue their education after leaving.

But all these countries are under harsh economic pressure. There is little hope for the children of Third World countries, even if their governments do change their priorities, unless their countries are enabled to develop.

GNP and the Debt Squeeze

This century has witnessed a phenomenal economic boom. By 1990, the world was producing 20 times as much in goods and services as in 1900, while industrial production had grown 50-fold. But most of the wealth has gone to relatively few nations and people.

Per capita GNP: the highs and lows

Three countries – the United States, West Germany and Japan – together accounted for more than half of the world's entire economic output. The 10 million people of Belgium shared about the same economic output as the whole of sub-Saharan Africa, a continent of 450 million.

The easiest way to compare national wealth is by per capita GNP (Gross National Product). By this measure, Switzerland is the world's richest nation with $32,790 of economic output for each of its people in 1990. Six other countries – Luxembourg, Finland, Japan, Norway, Sweden and Denmark – all have per capita GNPs over $22,000, with the United States just lower, at $21,700.

By contrast, 47 countries, mainly in Africa and Asia – containing 3 billion people – have per capita GNPs of $500 or less. Mozambique is the poorest in the world, with a per capita GNP of just $80. Most have been officially designated by the United Nations as "Least Developed Countries" (LDCs) on the grounds of low per capita income, very low literacy and little industrialization. (China and India, which have per capita GNPs of $370 and $350 respectively but do not qualify under other criteria, are among those excluded, as is Indonesia, the world's fifth most populous country.) Their average per capita income is about $200 a year, roughly a quarter of that for developing countries as a whole. In the words of the United Nations Conference on Trade and Development: "They share in common totally inadequate levels of living for most people, reflected in acute hunger and malnutrition, high levels of infant mortality and disease, illiteracy and other stigma of mass poverty."

The growth of inequality

Over the last decades, as the world economy has grown, the rich countries have got richer, but the poorest have got poorer still. Studies at the University of Pennsylvania suggest that while average per capita income has doubled since 1950, citizens of wealthy countries got three times richer, while per capita incomes in the poorest nations have, at best, remained the same. In 1960 the richest fifth of the world's people enjoyed 20 times the income of the poorest fifth; by 1990 they were 60 times wealthier.

Developing-country economies grew until the mid-1970s. Some stagnated after the 1973 oil shock; many slumped during the 1980s and early 1990s. Three quarters of all Latin American countries suffered a 10 per cent drop in per capita income over the 1980s; three quarters of all African countries sustained at least a 15 per cent fall. And per capita income continued to fall in Africa and the Middle East in both 1990 and 1991. Thirteen African countries, with a third of the continent's population, are now poorer, in these terms, than they were at independence. Worldwide, the number of LDCs grew from 31 in 1980 to 42 in 1989.

There are exceptions: countries like China, India, Indonesia, Botswana, Cameroon, the Congo, Mauritius, Pakistan, Thailand and the newly industrializing countries of East Asia raised per capita incomes appreciably in the 1980s. But, in general, the picture is bleak. The Worldwatch Institute reported in 1989: "For the poor, particularly in Africa and Latin America, the 1980s have been an unmitigated disaster, a time of falling earnings and rising debt, of falling food supplies and rising death rates."

The poorest suffered most. Figures for average per capita incomes conceal the plight of the majority; in most countries between 60 and 70 per cent of the people earn less than the average. In countries as diverse as Malawi and Mexico, Côte d'Ivoire and Costa Rica, Malaysia and Kenya, the poorest two fifths of the people only earn a quarter to a third of the average. In Peru, they earn less than a sixth.

Income differentials: robbing the poor

There are huge inequalities, though some countries are more unequal than others. In China, the richest fifth of the population earn three times as much as the poorest fifth; in Bangladesh they earn seven times as much; in India and the Philippines 10 times, in Côte d'Ivoire 25 times, and in Brazil 28 times as much. Income distribution often does more to determine the plight of the poorest than national wealth. Sri Lanka's average per capita income is about a fifth of Brazil's but, because it is shared more equally, its poorest people are actually better off.

Studies in Africa, Asia and Latin America show that the poor got even smaller shares during the 1980s as national incomes fell. The Food and Agriculture Organization of the United Nations (FAO) found that the proportion of national income that reached the poorest tenth of Brazil's rural population declined by 14 per cent in the early 1980s – while the richest tenth increased their share by a third.

China, on the other hand, reduced the numbers of people living in absolute poverty from 150 million to 70 million during the 1980s, even with an increasing population. Indonesia has also cut poverty, and there are

Per capita GNP US$, 1990

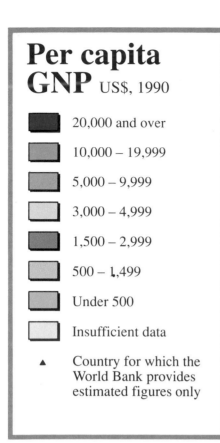

- ■ 20,000 and over
- ▨ 10,000 – 19,999
- ▨ 5,000 – 9,999
- ▧ 3,000 – 4,999
- ▨ 1,500 – 2,999
- ▨ 500 – 1,499
- ▨ Under 500
- ▨ Insufficient data
- ▲ Country for which the World Bank provides estimated figures only

A country's ability to develop can be hampered by the severity of its indebtedness. The World Bank uses the ratio of a country's external debt total (EDT) to its production (GNP) as one of the indicators of severe indebtedness. Its critical level is where EDT exceeds 50% of GNP.

Per capita external debt total US$, 1990

(World Bank developing countries only)

- ■ 2,000 and over
- ▨ 1,500 – 1,999
- ▨ 1,000 – 1,499
- ▧ 500 – 999
- ▨ 200 – 499
- ▨ Under 200
- ▨ Not reported by World Bank
- • Country which the World Bank considers to be severely indebted

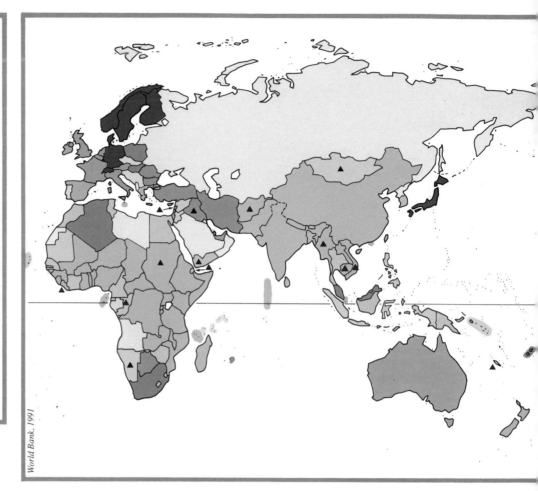

World Bank, 1991

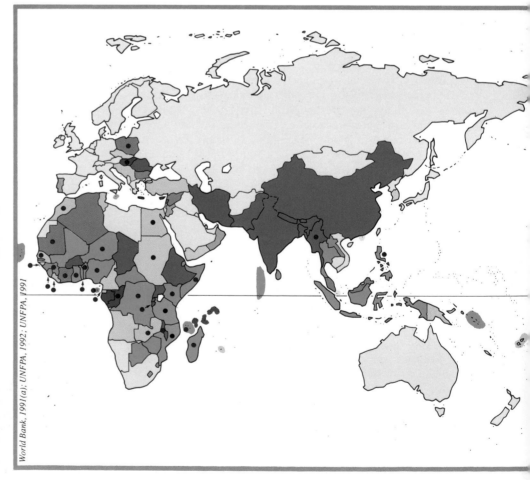

World Bank, 1991(a); UNFPA, 1992; UNFPA, 1991

42

GNP: the top and bottom 35

US$ million, 1990

	The top 35			*The lowest 35*	
1	United States	5,445,825	1	São Tomé & Príncipe	47
2	Japan	3,140,948	2	Kiribati	54
3	West Germany	1,411,346	3	Maldives	96
4	France	1,099,750	4	Tonga	100
5	Italy	970,619	5	Western Samoa	121
6	United Kingdom	923,959	6	St Christopher & Nevis	133
7	Canada	542,774	7	Equatorial Guinea	136
8	Spain	429,404	8	Dominica	160
9	China	415,884	9	Vanuatu	167
10	Brazil	402,788	10	Guinea-Bissau	176
11	India	294,816	11	St Vincent & Grenadines	184
12	Australia	290,522	12	Solomon Islands	187
13	Netherlands	258,804	13	Grenada	199
14	Republic of Korea	231,132	14	Comoros	227
15	Switzerland	219,337	15	The Gambia	229
16	Mexico	214,500	16	Bhutan	273
17	Sweden	202,498	17	St Lucia	286
18	Belgium	154,688	18	Guyana	293
19	Austria	147,016	19	Seychelles	318
20	Iran	139,120	20	Cape Verde	331
21	Finland	129,823	21	Antigua & Barbuda	363
22	Denmark	113,515	22	Belize	373
23	Indonesia	101,151	23	Swaziland	645
24	Norway	98,079	24	Lesotho	832
25	Turkey	91,742	25	Laos	848
26	South Africa	90,410	26	Somalia	946
27	Thailand	79,044	27	Sierra Leone	981
28	Argentina	76,491	28	Mauritania	987
29	Yugoslavia	72,860	29	Chad	1,074
30	Hong Kong	66,666	30	Burundi	1,151
31	Poland	64,480	31	Central African Republic	1,194
32	Greece	60,245	32	Mozambique	1,208
33	Algeria	51,585	33	Fiji	1,326
34	Israel	50,866	34	Surinam	1,365
35	Portugal	50,692	35	Togo	1,474

For location of the above countries, please see endpapers.

World Bank, 1991

Heavily indebted countries

External debt total compared with foreign exchange earnings, 1990

* 1987 † 1988 ‡ 1989

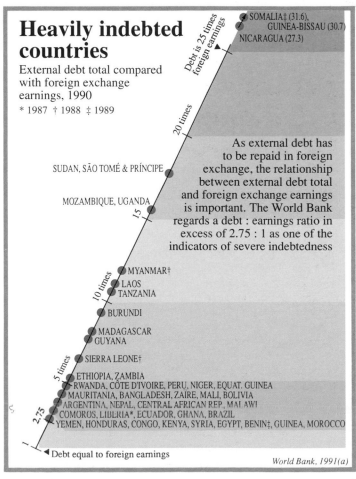

Debt is 25 times foreign earnings

SOMALIA‡ (31.6), GUINEA-BISSAU (30.7)
NICARAGUA (27.3)

20 times

SUDAN, SÃO TOMÉ & PRÍNCIPE

MOZAMBIQUE, UGANDA

15

As external debt has to be repaid in foreign exchange, the relationship between external debt total and foreign exchange earnings is important. The World Bank regards a debt : earnings ratio in excess of 2.75 : 1 as one of the indicators of severe indebtedness

MYANMAR†
LAOS
TANZANIA
10 times

BURUNDI

MADAGASCAR
GUYANA

SIERRA LEONE†
5 times

ETHIOPIA, ZAMBIA
RWANDA, CÔTE D'IVOIRE, PERU, NIGER, EQUAT. GUINEA
MAURITANIA, BANGLADESH, ZAIRE, MALI, BOLIVIA
ARGENTINA, NEPAL, CENTRAL AFRICAN REP, MALAWI
2.75
COMOROS, LIBERIA*, ECUADOR, GHANA, BRAZIL
YEMEN, HONDURAS, CONGO, KENYA, SYRIA, EGYPT, BENIN‡, GUINEA, MOROCCO

1

◄ Debt equal to foreign earnings

World Bank, 1991(a)

indications that India and Pakistan have done the same. But despite this, the numbers of the poor worldwide have grown even faster than population. About 1.2 billion people – close to a quarter of the population of the world – now live in absolute poverty, unable to satisfy their basic needs for food, shelter and clothing, and the numbers are expected to continue to grow throughout the 1990s. About a third of the absolute poor are literally wasting away from hunger.

Poverty has also grown in some rich countries, even though national incomes have risen. One fifth of American children live in poverty; so do a fifth of the people of the former Soviet Union. Income distribution is often as unequal as in the Third World; in the UK the richest fifth earn six times as much, in the US 12 times, and in France 13 times as much as the poorest fifth.

The tragedy of the Third World

But the greatest tragedy of the 1980s was the halting – and reversing – of Third World development. The prices of the raw materials most developing countries export slumped while the cost of the manufactured goods they had to import increased. Aid stagnated and debt grew, dominating many economies.

By 1990, the Third World owed $1,180 billion – almost half of its collective GNP – to rich-country governments and banks. Sub-Saharan Africa's debt was approaching its entire collective GNP, and 3.5 times its export earnings. Individual countries are far worse off: several owe more than $2,000 a head. Brazil is the developing world's largest debtor, owing $117 billion in 1990, followed by Mexico with some $96 billion, neither far short of the entire African debt. But they are better placed than many much poorer countries who owe far less. Brazil's debt comes to less than two fifths of its GNP. The Philippines' much smaller debt, around $30 billion, represents 87 per cent of its GNP; Zambia owes just $7.5 billion, but this amounts to more than three times its economic output.

The crisis started with the oil price rises of the 1970s. Oil exporting countries deposited their new wealth with rich-country banks, who wanted to earn money by lending it out again. Developing countries were anxious to borrow, since they were having to pay out to import newly expensive oil. Gigantic loans were easily agreed, often to finance wasteful projects, without either party paying much attention to the consequences. When interest rates rose at the beginning of the 1980s, the growing burden became unsustainable; for every percentage point that interest rates go up, the biggest debtors have to pay an extra $5 billion a year.

Between 1982 and 1988, developing countries repaid $830 billion in interest and principal – far more than they originally owed – and still their debt grew. Sub-Saharan Africa has struggled to pay off its debts, which are mainly to rich-country governments. More than a quarter of all its earnings from exports between 1985 and 1987 went to service debts – but even so it met only about 60 per cent of its obligations. More than 40 countries worldwide have debts more than 2.75 times as great as earnings from abroad. With few exceptions, notably Mexico, the greatest relative burdens are borne by the poorest countries.

A grotesque reversal

Mounting debt and stagnating aid have led to a grotesque reversal of financial flows between rich and poor nations. At the beginning of the 1980s, poor countries received a surplus of about $40 billion a year from rich countries. By the early 1990s, they were paying back many billions of dollars a year more to service their debts than they received. This will certainly get worse. Banks are reluctant to throw good money after bad to provide new loans, and the low interest rates of the 1950s and 1960s are unlikely ever to return. Unless there is a massive international bid to solve the crisis, the poor will be locked into financing the rich for the indefinite future, with devastating effects on their economies.

Even the rich suffer, because the poor are spending so much money servicing loans that they cannot afford to import goods. Six million jobs in Europe, North America and Australasia are estimated to have been lost in the 1980s as a result of the debt crisis. But piecemeal attempts to resolve it have come too late, are too little and have mainly been directed at saving rich-country banks from collapse rather than solving the mounting world crisis.

Such measures as have been taken have made things worse for the poor. The World Bank and the International Monetary Fund provided money to bail out poor countries – on condition that they adopt harsh "adjustment" policies, which include heavy cuts in spending. Health, education and living standards usually suffer: the World Bank itself has calculated that "hard core" poverty almost doubled in Côte d'Ivoire as a result of adjustment. And another World Bank paper admitted that the austerity measures were not even achieving their goal of giving Third World economies a beneficial short, sharp shock: they were, it concluded, "taking far longer and proving far more arduous than originally expected."

Industrialization and Development

For hundreds of millions of people, the 1980s were a lost decade. Development stagnated, and often regressed, in much of the Third World. A third of the entire population of the world live in countries which experienced either zero growth or actual decline during the decade. More than 40 Third World countries left the 1980s with lower per capita incomes than when they entered them, and most of them fell even lower in 1990 and 1991. Sub-Saharan Africa's share of world trade in 1990 was only a quarter of what it had been in 1960.

The prices of the raw materials that poor countries sell to survive fell to record low levels, while the cost of the manufactured goods that they import went on rising. They became shackled with ever-increasing debt, paying about a quarter of their earnings to service it. Aid stagnated, and bank lending to developing nations fell by almost two thirds. And if the first years of the 1990s are anything to go by, this decade does not promise to be any better.

Third World exports

A glance at the map of many Third World countries shows how the railways and main roads run to the capital and the ports, like rivers draining the hinterland to the sea. This reflects an economic pattern developed in the 19th century, and still in operation: Third World countries overwhelmingly produce raw materials for export to rich countries.

Eighty eight per cent of sub-Saharan Africa's export earnings come from the sale of these raw materials. More than half of all Third World countries rely on just one or two crops or minerals for over half their foreign exchange: as the 1980s began, for example, the export earnings of the Sudan were 65 per cent dependent on cotton, of Mauritius 68 per cent reliant on sugar, and of Burundi 93 per cent dependent on coffee.

This ties entire Third World economies to the prices they can get for just a few commodities, and most of these have been falling since the mid-1970s. By 1985 they had plummeted to their lowest levels since records were first compiled in 1957 – and they went on falling. By 1987, the prices of 33 commodities monitored by the World Bank (they excluded oil) stood at about half the level they had been in 1960. They fell by about 40 per cent in the 1980s alone.

Falling demand: falling prices

The prices are mainly determined by demand in the industrial world. Since the mid-1970s, recession and slow growth in rich countries – and particularly in the industries that use raw materials from developing countries – have kept demand and prices low. Most agricultural commodities are produced by dozens of developing nations – 62 grow coffee – who compete for this same sluggish demand. And if prices should rise, rich countries can cut consumption. Many of the commodities, like bananas or coffee, are luxuries that people can do without. Many more are being replaced by substitutes developed in rich countries: optical fibers are replacing copper wire in telecommunications, causing havoc for Zaïre and Zambia, while sugar substitutes in soft drinks have cut demand for sugar cane by about a quarter. These trends will continue – advances in biotechnology will bring a whole range of new substitutes – and there is little chance that prices will recover in the foreseeable future.

Little of what is ultimately paid for Third World commodities actually goes to developing countries. Only about 15 cents in every dollar spent by consumers on products from these raw materials goes back to the countries they came from. The rest goes to banks, shops, traders and processing industries in the rich world.

Industry is concentrated in rich countries. The Third World accounts for only 14.2 per cent of the world's industry, and 60 per cent of this is in just nine countries, mainly in Asia and Latin America: least developed countries share just 0.21 per cent of world industry between them. The prices of the manufactured products poor countries bought from the rich rose by a quarter between 1980 and 1988. As commodity prices slumped and debt repayments rose, most Third World countries have had to cut back their imports. In 1988 Latin America imported only three quarters the volume of goods it bought from overseas in 1980; Africa cut its imports to two thirds over this period.

If more developing countries industrialized, they could earn more from their commodities by processing them and turning them into finished products, and cut their import bills at the same time. But they face formidable obstacles. Most countries, lacking big home markets, have to try to sell as much as possible to the industrialized world – and, indeed, need to earn scarce foreign exchange. In doing so, they have the advantages of cheap labor and locally-available raw materials. But they have to face well-established rivals with markets and connections already set up, the best technology and science to hand, long production runs and mass production techniques. Even if the market for their goods were free, it would not be easy to penetrate. And it is not free at all.

Tariffs and barriers

Poor exporters generally have to jump over a whole series of tariffs and other barriers before they can reach the well-protected markets of the rich. Generally speaking, semi-manufactured goods like metal or cloth

Income group

World Bank classification, 1992

- Low income
- Lower middle income
- Upper middle income
- High income
- Not classified

Share of industry and services in GDP (%)

1990 or latest available figures

- ● 90% and over
- ◒ 80 – 89%
- ◓ 60 – 79%
- ○ Under 60%

No symbol: Insufficient data

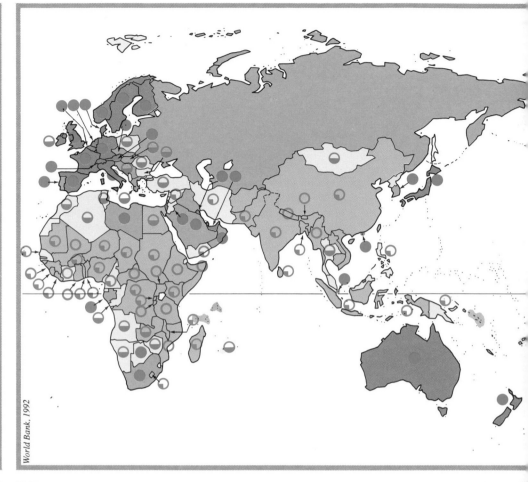

World Bank, 1992

% share of merchandise exports 1990

- Over 85% manufactured goods
- 50 – 84% manufactured goods

- 50 – 84% primary commodities
- Over 85% primary commodities

- Insufficient data

On this map primary commodities include fuels, minerals, metals, oils, foods, live animals and so on. Manufactured goods include machinery and plant, transport equipment, textiles and clothing and the full range of finished manufactured goods.

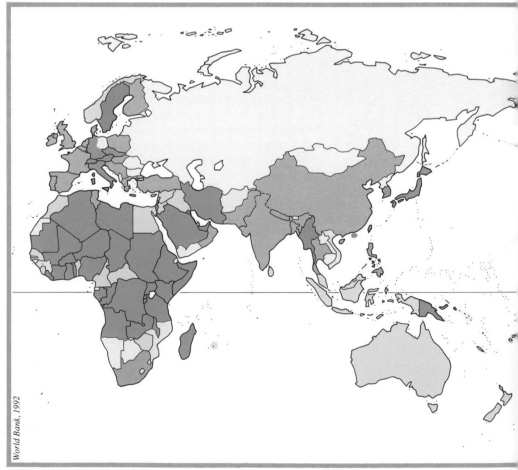

World Bank, 1992

Gross domestic product (GDP)

Selected countries, 1990, US$ million

Low income countries

China —————— 364,900	Indonesia — 107,290	Mozambique 1,320
India ————— 254,540	Togo ————1,620	Bhutan ———— 280

Lower middle income countries

Turkey ———— 96,500	Ecuador 10,880	Pap. New Guinea - 3,270
Poland ———— 63,590	Angola — 7,700	Botswana ——— 2,700

Upper middle income countries

Brazil ————— 414,060	Greece ——— 57,900	Oman ———— 7,700
Republic of Korea 236,400	Uruguay ——— 8,220	Gabon ———— 4,720

High income countries

United States — 5,392,200	Australia —— 296,300	United Arab
Japan———— 2,942,890	Switzerland 224,850	Emirates – 28,270
United Kingdom— 975,150	Ireland ——— 42,500	Kuwait —— 23,540

World Bank, 1992

% share of total national household income in selected countries, 1980s

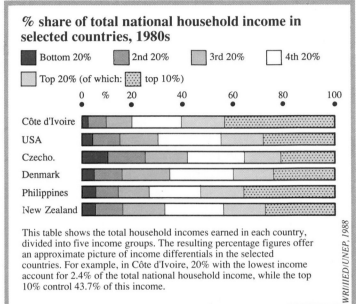

■ Bottom 20%   ▨ 2nd 20%   ▨ 3rd 20%   □ 4th 20%

□ Top 20% (of which: ▨ top 10%)

This table shows the total household incomes earned in each country, divided into five income groups. The resulting percentage figures offer an approximate picture of income differentials in the selected countries. For example, in Côte d'Ivoire, 20% with the lowest income account for 2.4% of the total national household income, while the top 10% control 43.7% of this income.

WRI/IIED/UNEP, 1988

Commodity prices

1972=100. 1980 prices (US$)

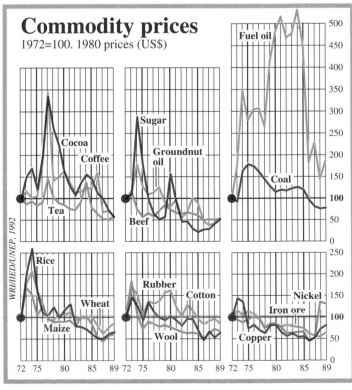

WRI/IIED/UNEP, 1992

47

face tariffs double those for raw materials like metal ores or cotton – while finished goods like metal products or clothes are penalized twice as highly again. Non-tariff barriers, such as quotas, are even stricter. Together, they are biased against developing countries; cloth imported into the European Community from poor countries faces tariffs four times as high as those imposed on cloth from other rich nations. The World Bank estimates that such trade barriers cost Third World countries between $50 and $100 billion a year by preventing them selling goods and reducing prices.

Faced with these obstacles, most developing countries have been unable to industrialize. Africa, the worst affected continent, increased the contribution of manufacturing industry to its collective Gross Domestic Product (GDP) from just 9 per cent in 1965 to 11 per cent in 1987.

In the long run the barriers do not even benefit the rich. They are erected to protect jobs in sensitive industries, like textiles. But study after study shows that reducing and eliminating them would increase employment, because the resulting increase in world trade would create many more jobs than were lost. But the vulnerable industries, naturally, lobby for protection, and they are often concentrated in politically sensitive areas. Rather than embark on programs to help industries diversify and to retrain workers, governments find it easier to increase protection. Despite several international agreements designed to make access easier for Third World products, 20 of the world's 24 industrialized countries were more protectionist in 1992 than they were 10 years before.

The Third World's economic double-bind – caught between falling commodity prices and the rising cost of manufactures, but unable to industrialize – is at the heart of the debt crisis. Most developing countries are unable to pay off what they owe, and increasing their commodity exports does not necessarily help; between 1981 and 1983 Sudan tripled its cotton exports to help meet debt payments, but its revenue increased only slightly because prices fell. Indeed, raising production helps to drive prices down. It can also do grave environmental damage, by using the best land to grow cash crops – pushing subsistence farmers onto marginal land – and by increasing indiscriminate logging of tropical rainforests. Indonesia's Environment Minister, Emil Salim, points out: "Today we have to export three times as much timber to buy one tractor as we did in the 1970s."

Stagnant aid

In all, the United Nations Development Programme estimated in 1992, the bias of the world's economy against developing countries costs them $500 billion a year – around a fifth of their combined Gross National Product (GNP). This is nearly 10 times the amount they receive in aid. For 20 years aid remained at about a third of 1 per cent of the rich world's combined GNP – less than half the UN target of 0.7 per cent. Only four countries of the Organization for Economic Cooperation and Development (Norway, the Netherlands, Denmark and Sweden) exceed this level; the UK, Switzerland and the US are among those that do not even meet half of it, while Saudi Arabia and Kuwait provided between 3 and 6 per cent of their GNPs in aid between 1976 and 1986.

But even if aid were generous, commodity prices were rising rather than falling, Third World countries could industrialize and the debt crisis did not exist, the world's poorest people would probably still be destitute. For most developing countries practise the same kind of economic bias against the poor within their borders as they suffer from in international markets. Most governments have long rigged the internal terms of trade against their primary producers, the farmers, in favor of their wealthy elites and city dwellers. Prices paid to farmers, the majority of the population, have been kept low, and the money has been lavished on grandiose schemes and modern industries that employ few people.

While one quarter of the world's population is so poor that it is effectively excluded from the market place, neither national economies nor the international economic system will work efficiently. Economically enfranchising the poor would benefit Third World economies; poor people tend to spend money on food and goods produced locally, helping to build the economy, while the rich tend to spend it on imported luxuries. It is also a social imperative; while the poor and the countryside are neglected, people will continue to migrate in their millions to the cities, swelling the urban slums and increasing political instability.

There are signs that some Third World countries are beginning to correct this bias, though progress remains limited. But talks between rich and poor nations on reforming the world economic system have been stalemated for more than a decade. Getting them started again would be the first step on a long road to tackling the escalating crisis in the international economy.

Indigenous People and Refugees

Hundreds of millions of people are being forced off the land they have owned for centuries, to swell the ever-mounting numbers of the dispossessed. Indigenous cultures are under assault as their lands are invaded, and desperately poor people are being forced to try to cultivate marginal areas, or to migrate to the cities. Often these two groups come into conflict.

Indigenous peoples

About 250 million people – one in every 20 on earth – belong to indigenous cultures. The original inhabitants of their lands, they uniquely know how to live in harmony with their environment. But everywhere they are now ruled by newcomers. Their communities are often divided by the frontiers of modern nation states. Their problems go back for centuries. As lands were "discovered" and "civilization" advanced, they were massacred, perished from diseases to which they had no immunity, or were absorbed, partly or wholly, into the alien cultures, as the poorest of the poor.

In the century after the Spanish Conquistadores, led by Cortez, landed in Central and South America, the population of the Aztec empire was reduced by two thirds; 20 million men, women and children perished. Of the 6 to 9 million Indians who originally lived in the Amazonian rainforest, only about 200,000 now survive. There were 300,000 Aborigines in Australia when "The First Fleet" landed in Botany Bay; a century later 60,000 remained. Every single Carib Indian on the island of Hispaniola was killed or deported by the Spanish colonialists, to be replaced by slaves from Africa. The people of the 600 Indian nations of North America were reduced by half to two thirds by 1860. Treaties were made and broken, and the people increasingly chased from the land that had been their own.

Today, remaining indigenous peoples are scattered over the face of the earth in around 70 countries. Among them, well over 150 million live in Asia; two thirds in China and India. There are at least 3 million in North America, and at least 30 million in Central and South America. There are about 250,000 Australian Aborigines, 300,000 New Zealand Maoris, and some 100,000 Inuit in the Arctic countries. In Sweden the Lapps make up less than 0.1 per cent of the population, and Amazonian Indians comprise the same proportion of the people of Brazil. But Inuit are 90 per cent of Greenland's population, and Indians make up two thirds of the people of Bolivia.

As the Independent Commission on International Humanitarian Issues (ICIHI) noted in 1987, indigenous peoples are always "the poorest of the poor" in both developed and developing countries. The Mayan Indians that make up the majority of Guatemala's people can expect to live 11 years less than the European minority that rules them; Australian Aborigines, on average, live 20 years less than whites and are nine times more likely to die of infectious diseases. Infant mortality among North American Indians is double the rate for Canada and the US as a whole.

The unemployment rate among indigenous people in New Zealand is four times the national average; in Australia seven times. And when they are in work they usually are left with the worst jobs. In Latin America some still work as bonded laborers, virtual slaves. In the US, Indian children are twice as likely to drop out of school as those of other races. In Australia, Aborigines are 45 times more likely than whites to be sent to jail.

"Since the Second World War," adds the ICIHI, "the number of incursions into indigenous peoples' lands has escalated worldwide." For example, nearly half of all the mineral production in India is carried out on indigenous lands; in the US, 40 per cent of all uranium reserves are on Indian lands. The indigenous populations usually receive only derisory rents and royalties – if they get anything at all. Gold miners have invaded Indian lands in the Amazon, poisoning their rivers with mercury. Hydro-electric dams flood the land of indigenous people in the US, Brazil, India and elsewhere. And Australian Aborigines and Pacific Islanders have, in the past, been affected by nuclear tests.

Most serious of all are the invasions of subsistence farmers. They are increasingly in conflict with nomads in the world's drylands, and with the rainforest peoples. They almost always replace a sustainable system with a destructive one. Indigenous people have learned to live in the most hostile environments, from the Arctic to the equator, from the desert to the rainforest. And they have evolved a host of sophisticated strategies for doing so.

The clash of interests

Rainforest Indians, such as the Kuna of Panama and the Kenyah of central Kalimantan in Indonesia, have managed forest ecosystems sustainably for centuries. The Kuna Indians live on barren sand and coral islands off the Caribbean coast. Unable to grow food at home, they visit the mainland to carve garden plots out of jungle to grow a wide variety of food crops. They carefully recycle nutrients and abandon the plots when they begin to wear out after five to 10 years, starting new ones in a different part of the forest. Through this centuries-old land management system, the Kuna have been able to provide for all their needs, without destroying the forest, upon which their very lives depend.

Refugees

Refugee populations, 1990

![]	2,000,000 and over
![]	500,000 – 1,999,999
![]	100,000 – 499,999
![]	10,000 – 99,999
![]	500 – 9,999
![]	Under 500

Refugee populations are as defined by the United Nations High Commissioner for Refugees.
Palestinian and Cambodian refugees under the mandates of other UN organizations are excluded, as are populations fleeing environmental degradation and others in a refugee-like predicament.

●   Selected cities with major squatter settlements
(See chart on right)

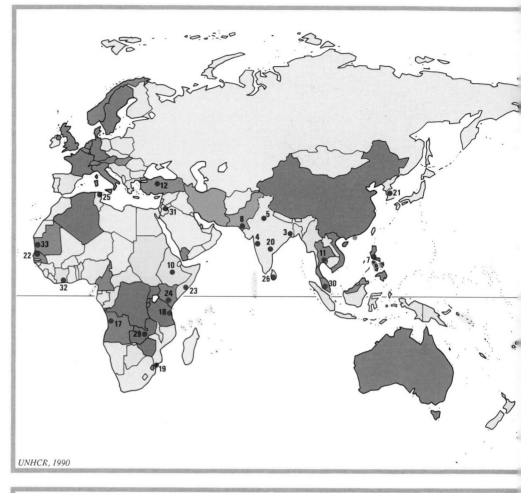

UNHCR, 1990

Indigenous peoples

▲   Hunters

●   Gatherers

❖   Shifting cultivators

◆   Pastoral nomads

★   Settled farmers

✳   Industrial workers

Arrows indicate general, not precise, locations.

Peoples under threat:
(See list on right)

■   Country locations

☐   General locations

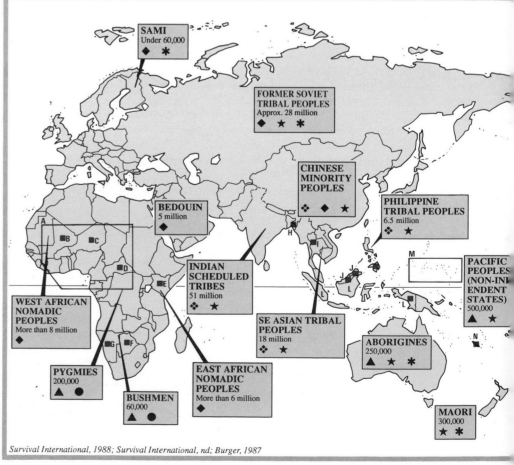

Survival International, 1988; Survival International, nd; Burger, 1987

Informal (squatter) settlements

Number of people living in squatter settlements in selected major cities, 1980s (figures in thousands)

Over 1 million squatters

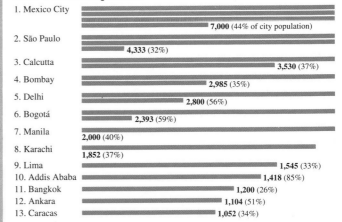

1. Mexico City	**7,000** (44% of city population)
2. São Paulo	**4,333** (32%)
3. Calcutta	**3,530** (37%)
4. Bombay	**2,985** (35%)
5. Delhi	**2,800** (56%)
6. Bogotá	**2,393** (59%)
7. Manila	**2,000** (40%)
8. Karachi	**1,852** (37%)
9. Lima	**1,545** (33%)
10. Addis Ababa	**1,418** (85%)
11. Bangkok	**1,200** (26%)
12. Ankara	**1,104** (51%)
13. Caracas	**1,052** (34%)

Under 1 million squatters

14. Guayaquil	**976** (75%)	21. Seoul	**508** (6%)	28. Port au Prince	**270** (50%)
15. Santiago	**935** (25%)	22. Dakar	**490** (50%)	29. Lusaka	**265** (50%)
16. Medellín	**805** (33%)	23. Mogadishu	**480** (80%)	30. Kuala Lumpur	**242** (25%)
17. Luanda	**671** (70%)	24. Nairobi	**421** (33%)	31. Amman/Zarqa	**186** (25%)
18. Dar es Salaam	**645** (60%)	25. Tunis	**396** (50%)	32. Abidjan	**185** (27%)
19. Maputo	**620** (79%)	26. Colombo	**336** (57%)	33. Nouakchott	**94** (70%)
20. Hyderabad	**603** (23%)	27. Panama City	**307** (73%)		

It must be stressed that these estimates are little more than educated guesses. No reliable official figures exist or are being gathered. The populations of most of these cities have increased since these "guesstimates" were made. Even assuming that the percentage of squatters has remained steady, this would mean a significant increase in their absolute numbers. However, the percentages have probably also increased.

Hardoy & Satterthwaite, 1989; UNHCR, 1989: WHO/UNEP, 1988

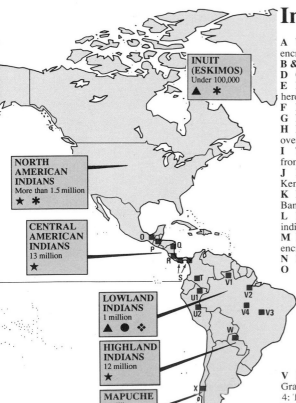

Indigenous peoples under threat

A WESTERN & CENTRAL AFRICA Pygmy cultures threatened by tropical deforestation and encroachment of plantations.
B & C MALI & NIGER Tuareg nomads threatened by enforced settlement in refugee camps.
D CENTRAL AFRICAN REPUBLIC Bororo peoples subjected to expulsions and human rights abuse.
E KENYA Masai under threat from land alienation, Degodia herders subjected to massacres, Somali herdsmen subjected to human rights abuse and massacre.
F BOTSWANA Bushmen (San) threatened by economic development and exploitation.
G NAMIBIA Bushmen (San) threatened by expulsion from national park.
H BANGLADESH Hill tribes of Chittagong threatened by resettlement programs which have brought over 100,000 settlers to the area.
I THAILAND Karen peoples threatened by tribespeople displaced by Indo-China war. Further threats from opening up of the area for economic exploitation.
J MALAYSIA Logging in Sabah and Sarawak threatens many indigenous peoples including the Punan. Kenyah and Kayan peoples under threat from a massive hydro-electric scheme.
K PHILIPPINES Manobo peoples threatened with land alienation by sugar cane and fruit production. Bangsa Moro peoples of southern Philippines threatened by policies initiated by Marcos regime.
L INDONESIA Government policy of transmigration, oil exploitation and palm plantations threaten indigenous peoples in many parts of Indonesia. Policy of forced resettlement threatens people of East Timor.
M MICRONESIA Indigenous peoples threatened by US and French weapons testing and military encroachment.
N NEW CALEDONIA Kanak peoples threatened by economic development and military encroachment.
O MEXICO Lacandon Maya Indians under threat from cattle ranchers and deforestation.
P GUATEMALA Guatemalan Maya threatened by forcible resettlement and military encroachment.
Q NICARAGUA Miskito Indians threatened as a result of civil war.
R COSTA RICA Indian reserves under threat from settlement and development.
S PANAMA 80,000 Guaymi Indians threatened by state and private mining interests. East of the canal, Kuna Indians under threat from developers and tropical deforestation.
T COLOMBIA Paez and Guambiano Indians subject to land alienation.
U PERU 1: Yagua Indians threatened by economic slavery. 2: Amuesha Indians threatened by Pichis-Palcazu road scheme.
V BRAZIL 1: Yanomami peoples threatened by mining interests. 2: Indigenous peoples threatened by Grand Carajas hydro-electric scheme. 3: Apinaye Indians under threat from landless settlers. 4: Txukarramae subjected to forcible expulsion from land.
W PARAGUAY Aché Indians victims of genocide.
X CHILE Mapuche Indians threatened by 1980s Government policy aimed at the "liquidation of the Indian communities".

Similarly, the Kenyah Indians have lived in dense tropical forests for centuries, by making clearings on more fertile soils, on moderate slopes, and in secondary forest that is eight to 20 years old. The original primary forest is never cut down. Instead reserves are maintained close to the villages and used for medicines, pharmaceuticals and supplementary food. Once their garden soils begin to wear out, the Kenyah leave the area fallow for up to 20 years and move on to another secondary forest area, which is again cleared and planted. So secondary forest is used again and again for gardens, while nearby areas of primary forest are left intact.

Like forest peoples everywhere, the Kuna and Kenyah are under threat. The Kuna are engaged in a struggle with Panamanian farmers and cattle ranchers who want to clear more of the lush jungle along the Caribbean coast and open it up to more "productive" uses. On Kalimantan, the Kenyah are fighting for their lives as thousands of poor migrants encroach on their ancestral lands, seeking livelihoods from the forest. More hopefully, the Colombian Government has returned half of its part of Amazonia to 70,000 Indians from 50 tribes, while in 1991 Brazil gave the Yanomami people rights to 90,000 square kilometers of land.

The biggest threat comes from another group of the dispossessed: poor people with no land on which to grow food for their families. Brazil opened up Amazonia to defuse demand for land reform from poor and landless people in the northeast of the country. Denied a share of the best land, they fell the trees to grow crops in soil that is exhausted within a few years, forcing them to move on ever deeper into the forests. Worldwide, some 150 million people are felling the rainforest simply to scratch a living.

Environmental and economic refugees

Others of these new dispossessed have to leave farming altogether. In the early 1980s, drought and desertification drove at least 10 million people in Africa off their land and into refugee camps or city slums. One in every six of Haiti's people have fled, partly to escape political repression and economic hardship, but partly because about a third of the nation's land is now virtually uncultivatable as a result of deforestation and erosion.

The numbers of such environmental refugees will grow as land is increasingly made useless by overuse, and particularly as the greenhouse effect warms up the world climate. Dr Pier Vellinga, of the Inter-governmental Panel on Climatic Change, estimates that a 1-meter rise – expected before the end of the next century – in sea level will make 300 million people homeless. Tens or hundreds of millions more will be displaced by droughts and harvest failures as rainfall patterns change.

Economic refugees already crowd the cities of the Third World. Two thirds of Bombay's pavement dwellers came to the city because they had no land; half had nothing, except the clothes they wore, when they arrived. Most migrants, if they are fortunate to find somewhere better than the street to live, end up in illegal squatter settlements on land no-one else wants, in swamps or dustbowls, on steep slopes or next to stinking rivers, beside – or on top of – garbage dumps.

Over 3 million poor squatters have built their make-shift shacks from discarded tin, wood and cardboard on the steep hills surrounding Rio de Janeiro. At the end of February 1988, torrential rains tore apart the denuded hills, sending mud avalanches into the valleys, killing 277 people – mostly women and children – and leaving 20,000 without any shelter.

Political refugees

By the mid-1980s there were some 13 million political refugees around the world, including 2 million Palestinians, 2 million Indochinese and 1.5 million Central Americans. Almost all are taking refuge in countries with little ability to cope with the influx; the 20 countries with the highest proportion of refugees in their populations all have per capita incomes of less than $700 a year.

The Convention Relating to the Status of Refugees, agreed in 1951 and now ratified by about 100 countries, lays down minimum standards of treatment for them and protects them from being returned to a country where they would be at risk. But its definition of refugees – "persons who are outside their country because of a well-founded fear of persecution for reasons of race, religion, nationality, membership of a particular social group, or political opinion" – is limited. It does not cover people who are forced to flee their homes – for example, through civil war – but remain within their own countries: there are thought to be far more of these "internally displaced people" worldwide than there are official refugees. And it excludes the far greater number of economic and environmental refugees. The definition will almost certainly have to be re-thought in the face of the growing refugee problems of the next decades.

Deserts and Desertification

The world's deserts are spreading, extending over more and more land. Desertification threatens about one third of the world's land surface and affects the lives of some 850 million people.

Desert landscapes vary from the hot, arid wastelands of the Middle East and the sea of sand around the Gulf of Arabia and parts of the Sahara, to the high, cold deserts of the Andes Mountains in South America. Even the Antarctic is often called a frozen desert because it gets only a few centimeters of rain a year. Whether too high and cold, too hot and sandy, or too rocky and impassable, deserts all have one thing in common: their poor soils and harsh climates make animal husbandry or agriculture impossible, without massive (and often prohibitively expensive) imports of water, fertilizers, fodder and labor.

Every continent is affected by desertification. "Sometimes thousands of kilometers away from the margins of the Sahara, Gobi, Atacama and the other so-called true deserts, desertification is taking place," explains Dr Mostafa Tolba, Executive Director of the United Nations Environment Programme (UNEP). "The situation has been accurately likened to a skin disease in which existing eruptions worsen and coalesce with new outbreaks of the disease. And, as with any disease, treating the symptoms is secondary to tackling the causes."

What causes desertification?

There are five main causes: over-cultivation of poor soils; over-grazing by sheep, goats, cattle and camels on fragile rangelands; excessive cutting of fuelwood; deforestation, especially of upland watersheds; and inappropriate irrigation practices which result in the salinization or alkalization of agricultural land.

People have been creating deserts since the beginning of settled agriculture, 10,000 years ago. Nearly the whole of Mesopotamia – the once fertile crescent lying between the Tigris and Euphrates rivers – is now desert. Yet 4,000 years ago, it was the cradle of civilization; an area that supported tens of thousands of people and nurtured the world's first true cities, outside China. Centuries of over-use, combined with poor irrigation techniques, sterilized the land and was the main cause of the civilization's collapse.

UNEP estimates that some 3.3 billion hectares of the world's rangeland, rainfed cropland and irrigated land – an area about the size of North and South America combined – is under assault by desertification. About 63 per cent of all rangelands – 2.6 billion hectares – are affected to one degree or another. Likewise, just over 60 per cent of the world's rainfed croplands – 347 million hectares – are subject to desertification, along with 30 per cent (roughly 40 million hectares) of the world's irrigated lands. Some 26 billion metric tons of topsoil are eroded away every year: over the last two decades the world has lost some 500 million metric tons – the same amount as covers the entire cropland of the United States.

Every year 6 million hectares of productive land are lost altogether at a cost by one 1992 UN estimate of some $42 billion a year. Another 21 million hectares – an area the size of the State of Kansas – are so impoverished that they are no longer worth farming or grazing. If present trends continue, by the year 2000 desertification could threaten the livelihoods of 1.2 billion people worldwide. In 1992 the Food and Agriculture Organization of the United Nations (FAO) warned that the world's soils were degrading so fast that in little more than 30 years it might no longer be able to feed its growing population.

The rapid advance of deserts

The process of desertification appears to be accelerating in sub-Saharan Africa, Mediterranean Africa, the Middle East, Western Asia (from Iran to Bangladesh), South Asia, the western United States, and parts of South America and Mexico. Over the last half century, some 65 million hectares of Africa south of the Sahara have been turned to desert. In Mali, the desert has spread 350 kilometers south in just 20 years, in the Sudan it has advanced 100 kilometers in 17. About 78 million people now have to cope with severely degraded land.

Africa has particularly fragile soils. Four fifths of them have inherent limitations on their fertility and, generally speaking, they are low in organic matter and clay and so erode easily. The harsh climate, including long periods of drought, interspersed with precipitate rains, makes the continent even more susceptible to desertification – two thirds of it is drought-prone, and even in "normal years" the rains are highly erratic. Over the past 15 years, average annual rainfall in the West African countries of Mali, Senegal, Burkina Faso and Niger has been significantly less than over the half century between 1934 and 1984 as a whole.

Traditionally, African farmers were able to live with irregular rains and poor soils; they rotated their crops and rested their most vulnerable land, letting it lie fallow for up to 20 years at a time. But the "carrying capacity" of the land is coming under increasing strain; a World Bank study in West Africa in 1980 found that the northernmost parts of the region were already unable to support the demands placed upon them for fuelwood and agriculture, and that seven entire countries were approaching the limits of what they could stand.

Farmers have been forced onto marginal land by increasing population and decreasing rainfall and, in many countries, by the appropriation of the better land

Spreading deserts

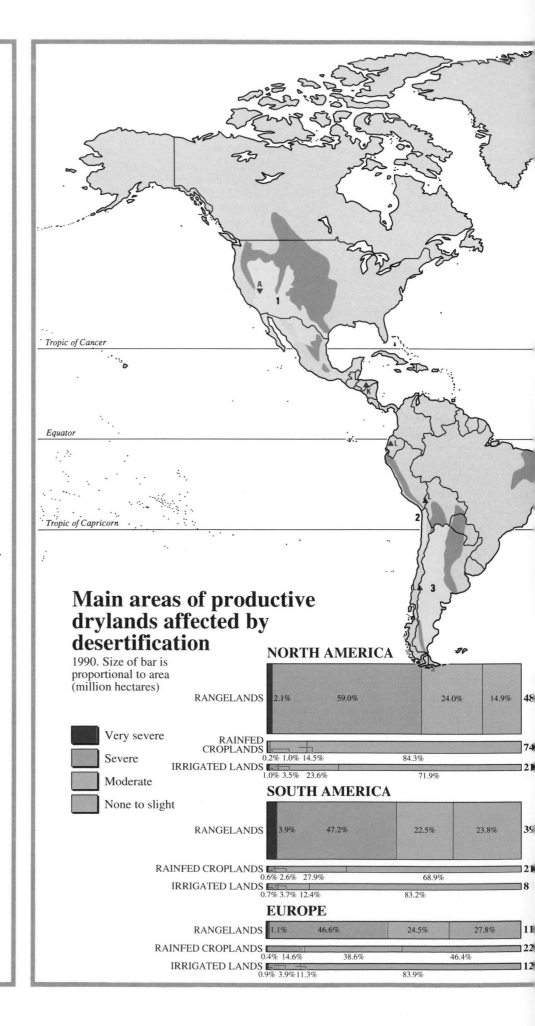

Main areas of productive drylands affected by desertification

1990. Size of bar is proportional to area (million hectares)

■ Very severe
▨ Severe
▨ Moderate
☐ None to slight

NORTH AMERICA

RANGELANDS   2.1%   59.0%   24.0%   14.9%   48

RAINFED CROPLANDS   0.2% 1.0% 14.5%   84.3%   74

IRRIGATED LANDS   1.0% 3.5% 23.6%   71.9%   21

SOUTH AMERICA

RANGELANDS   3.9%   47.2%   22.5%   23.8%   39

RAINFED CROPLANDS   0.6% 2.6% 27.9%   68.9%   21

IRRIGATED LANDS   0.7% 3.7% 12.4%   83.2%   8

EUROPE

RANGELANDS   1.1%   46.6%   24.5%   27.8%   11

RAINFED CROPLANDS   0.4% 14.6%   38.6%   46.4%   22

IRRIGATED LANDS   0.9% 3.9% 11.3%   83.9%   12

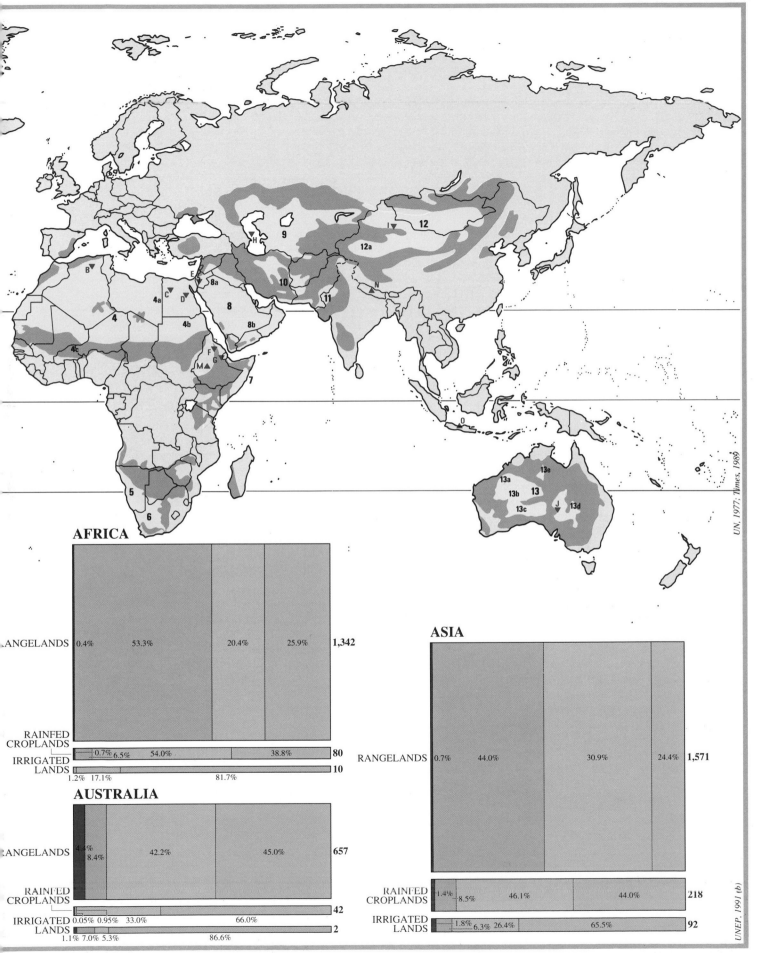

AFRICA

RANGELANDS | 0.4% | 53.3% | 20.4% | 25.9% | **1,342**

RAINFED CROPLANDS | 0.7% 6.5% | 54.0% | 38.8% | **80**

IRRIGATED LANDS | 1.2% 17.1% | 81.7% | **10**

ASIA

RANGELANDS | 0.7% | 44.0% | 30.9% | 24.4% | **1,571**

RAINFED CROPLANDS | 1.4% 8.5% | 46.1% | 44.0% | **218**

IRRIGATED LANDS | 1.8% 6.3% 26.4% | 65.5% | **92**

AUSTRALIA

RANGELANDS | 4.4% 8.4% | 42.2% | 45.0% | **657**

RAINFED CROPLANDS | | **42**

IRRIGATED LANDS | 0.05% 0.95% | 33.0% | 66.0% | **2**

1.1% 7.0% 5.3% | 86.6%

UN, 1977; Times, 1986

UNEP, 1991 (b)

55

by wealthy landowners. They are compelled to sacrifice their futures in order to grow enough food to meet their families' immediate needs. Fallow periods are cut and more land is cleared to grow more food, reducing fuelwood supplies and grazing area. As the vegetation cover, which stitches the soils together, is stripped away, soils unravel into dust. They blow away in the wind, or are gouged out by torrential rains. At the height of the Sahelian drought of the early 1970s, three times as much topsoil as normal was blown as dust across the Atlantic to Barbados, 4,700 kilometers away.

Already poor farmers are made entirely destitute as their land erodes away. Ethiopia's highlands, the scene of repeated famines, used to be rich and fertile; they supported agrarian-based societies for millennia. Since 1900, more than 90 per cent of the forests of the highlands have been cut down. Every year almost a billion metric tons of topsoil are washed away. Already 20,000 square kilometers of land have lost so much that they can no longer grow crops.

Desertification is not just a Third World phenomenon. Over three quarters of North American rangelands have been affected by desertification. In Canada soil erosion has been costing farmers $1 billion a year. Some 152 million hectares of the former Soviet Union have lost fertility through erosion: 1.5 billion metric tons of topsoil are lost each year at a cost of up to $35 billion. And when it opened up the "new lands" with irrigation, salinization and alkalization sterilized more than 7 million hectares, rendering them unfit for food production. If irrigated soils are not drained properly, salts build up in the root zones of the crops, killing them and ruining the soil. Rehabilitating can cost billions of dollars.

Deserts can be turned back

Deserts can be turned back by mobilizing local communities and national resources. Vast sums of money are not always necessary – but political commitment is indispensable. The UN claims that at least $4.5 billion is needed annually for the next 20 years to halt desertification and reclaim degraded land; but much has been done with little money.

A "great green wall", rivalling the Great Wall built to keep out rampaging Mongols, has been planted in northern China over the last decade in an effort to hold off the advancing desert and to stabilize badly eroded uplands. This San Bei forest belt will eventually cover 3.5 million square kilometers, stretching in a wide arc across the dry hills. So far, local communities, assisted by government experts and supervisors, have planted a forest belt 7,000 kilometers long and 400-1,700 kilometers wide protecting nearly 8 million hectares of cropland. Annual grain harvests have increased by 13 per cent.

Although total forest cover in northern China has risen from 4 to 6 per cent, the amount of money spent by the Chinese Government was a mere $600,000. Despite setbacks in some areas, where seedlings were not tended properly and died, the project is an example of what political commitment can accomplish in partnership with local communities.

Political will plus local action

In Rajasthan, India, imported acacia trees from the Middle East were used to stabilize 60,000 hectares of sand dunes. And in parts of West Africa, the kad tree (Acacia albida) was used to revitalize exhausted cropland and pasture. Not only does the tree have leaves during the dry season, giving shade when it is most needed, but it also buffers the wind, fixes nitrogen from the air (increasing the yield of crops around it) and provides fuelwood. Its pods and seeds provide protein-rich fodder for cattle and goats.

Ranchers in the arid southwest US are trying out a new management concept of heavy, but transient livestock grazing adapted from Africa. Animals are moved on once an area has been grazed, giving the rangeland a chance to recover: fodder plants have doubled their productivity.

In Kenya, women planted trees to reduce wind erosion. In Haiti, 35 million trees have been planted. Simple lines of stones placed in fields in Burkina Faso have trapped soil that would otherwise have been washed away, raising crop yields by 50 per cent.

The Earth Summit in June 1992 agreed, at the instigation of African countries, to negotiate an international treaty to combat desertification.

Freshwater: Abundance and Scarcity

Seen from space, the earth is the water planet. Water, indeed, is the very essence of life on it. Yet although 70 per cent of the globe's surface is covered by ocean, less than 3 per cent of the world's water is fresh, and more than three quarters of that is frozen, mainly at the poles. Ninety eight per cent of the remaining freshwater lies underground. Only about a hundredth of a per cent of the world's total water is easily available to terrestrial life, including man.

Nevertheless, there is plenty of it – or there would be if it were more evenly distributed. The "great pump" of the world's hydrologic system, which makes the planet habitable, causes over 113,000 billion cubic meters of freshwater to fall on the land each year as rain and snow – enough to flood the continents 80 centimeters deep, and more than enough, in theory, to meet all foreseeable needs.

Unequal distribution

The world, however, is divided into water "haves", areas which receive enough for vegetation to thrive, and "have nots", which do not. On this basis, most of Africa and the Middle East, much of western United States and northwest Mexico, parts of Chile and Argentina, and nearly all of Australia suffer severe water shortages.

People, too, receive grossly unequal amounts of freshwater: Iceland, for example, gets enough rain and snow to provide every one of its 250,000 people with 674,600 cubic meters of freshwater every year; Kuwait, with seven times the people, gets scarcely a drop to share between them. In Israel and Saudi Arabia, roughly 4,000 people compete for every 1 million cubic meters of water. In France and India, the same amount is shared between 350 people; in Sweden and Malaysia, it is split among about 100. And, of course, there are vast differences in rainfall within individual countries.

About 2 billion people in 80 countries around the world live in areas suffering from chronic water shortage and, as human and animal populations grow, the crisis will get worse. By the end of the 1990s, Egypt will have only two thirds as much water for each of its inhabitants as today, Kenya only half as much. By then, six of East Africa's seven countries and all five of the nations of the south Mediterranean coast will face severe shortages. Poland, Israel – and arid parts of the United States – are also approaching a critical situation.

Changes in the global climate, brought about by the greenhouse effect, are likely to cause great disruption, because rainfall patterns will change as the earth heats up. It is extremely hard to predict what will happen, but one study suggests that the entire western US – much of which is already short of water – could find its supplies cut to 40-76 per cent of present levels.

Groundwater

When surface water supplies are insufficient, humanity has traditionally mined some of the vast resources naturally stored underground. But groundwater is also unequally distributed, only a little of it is economically exploitable and tapping it consumes liquid capital. The vast Ogallala aquifer, which lies under eight of the states of the US Great Plains, is being so depleted that water tables are falling by up to a meter a year. A quarter of all the US irrigated cropland is only kept watered at the cost of depleting groundwater supplies, and some areas have already had to be abandoned.

The same is happening in the Third World. Bangkok's water table has plunged 25 meters since the late 1950s, saltwater has penetrated its wells and the city itself is sinking fast. The pumping of water for agriculture in Tamil Nadu, south India, has caused its water table to fall by the same amount in a decade; in the north of the country, the number of villages short of water in the state of Uttar Pradesh has risen from 17,000 to 70,000 in two decades; out of 2,700 water supplies provided by the Government, 2,300 have simply dried up. In China, 50 cities are threatened by shortages; the water table beneath Beijing is sinking by up to 2 meters a year and a third of its wells are thought to be waterless.

Water quality in peril

Water is not only becoming scarcer, its quality is also being degraded. The river Vistula in Poland is so polluted with industrial and municipal wastes that its water is unusable, even for industrial purposes, along most of its 1,068-kilometer length. The industrialized countries have dumped billions of tons of pollutants into their rivers, estuaries and coastal waters. Aquifers are also becoming contaminated with chemical poisons. In China, 41 large cities use polluted groundwater. Nitrates from chemical fertilizers contaminate water beneath farmland; 24,000 wells out of 124,000 samples in the US were found to be seriously contaminated.

Global patterns of use

Since 1950, the world's use of water has increased three and a half times over and per capita use has almost trebled. The amounts vary widely among countries. Guyana uses the most water per head – about 7,600 cubic meters a year, followed by Iraq with about 4,600. Americans each use 2,160 cubic meters, but the US' huge population makes it the biggest total consumer. Countries as diverse as Bulgaria, Madagascar and Chile use about 1,600 cubic meters a head – while Britain, Sri Lanka, Honduras and Morocco use about 500 cubic meters and Bhutan, Guinea Bissau and the Solomon Islands use less than 20 cubic meters per person each year.

Global water surplus and deficiency

Millimeters per year

SURPLUS:

 1000 and over

 0 – 999

DEFICIENCY:

 0 – -999

-1000 and below

Surplus areas have enough water, without irrigation, to support a wide variety of vegetation, including crops.
Deficiency areas do not. In addition, dry areas under long periods of stress may have trouble maintaining native vegetation.

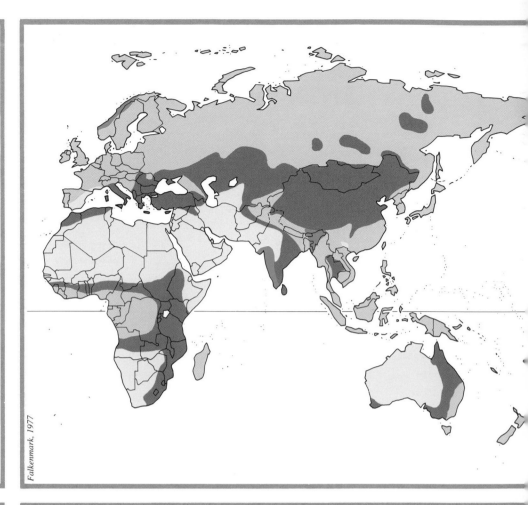

Falkenmark, 1977

Water availability

Internal renewable water resources per capita. Thousand cubic meters per year, 1990

 50 and over

 10 – 49.9

 5 – 9.9

 2 – 4.9

Under 2

Insufficient data

Countries with under 2,000 cubic meters per person per year are considered to be chronically short of water. Those with 2,000 – 4,900 may also experience acute water shortages during all or part of the year.

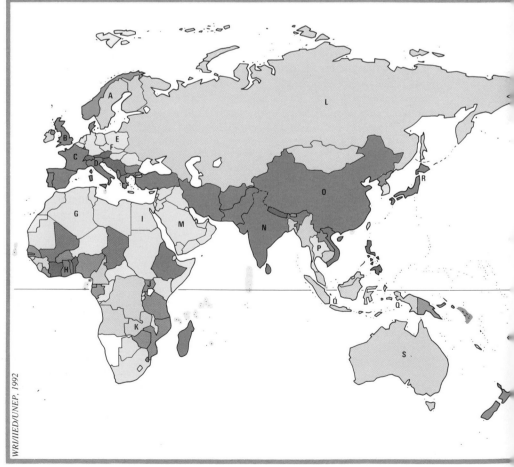

WRI/IIED/UNEP, 1992

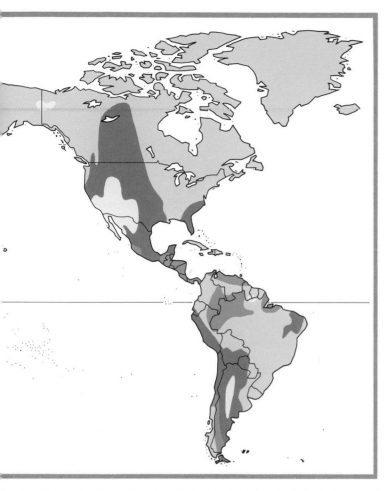

Distribution of the world's water

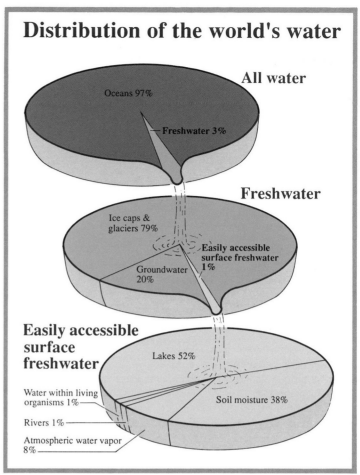

All water

Oceans 97%

Freshwater 3%

Freshwater

Ice caps & glaciers 79%

Easily accessible surface freshwater 1%

Groundwater 20%

Easily accessible surface freshwater

Lakes 52%

Soil moisture 38%

Water within living organisms 1%

Rivers 1%

Atmospheric water vapor 8%

Use of water

Water withdrawal by sector, selected countries, 1980s.
Letters refer to lower map

WRI/IIED/ UNEP, 1992

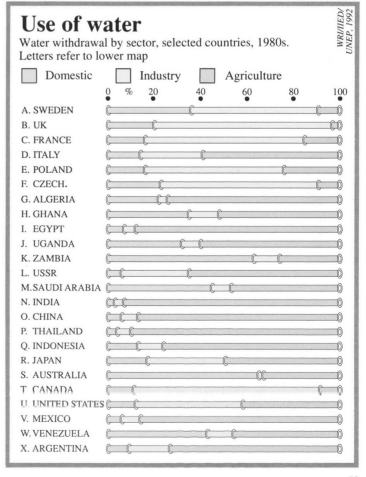

☐ Domestic ☐ Industry ☐ Agriculture

	0 % 20 40 60 80 100
A. SWEDEN	
B. UK	
C. FRANCE	
D. ITALY	
E. POLAND	
F. CZECH.	
G. ALGERIA	
H. GHANA	
I. EGYPT	
J. UGANDA	
K. ZAMBIA	
L. USSR	
M. SAUDI ARABIA	
N. INDIA	
O. CHINA	
P. THAILAND	
Q. INDONESIA	
R. JAPAN	
S. AUSTRALIA	
T. CANADA	
U. UNITED STATES	
V. MEXICO	
W. VENEZUELA	
X. ARGENTINA	

Surprisingly, little of the water is drunk; only about 5 per cent of worldwide consumption is for all domestic uses. About 75 per cent is used for irrigation, and the remaining 20 per cent or so in industry. But the uses differ between nations as much as the amounts consumed; countries which use little water usually consume high proportions for domestic purposes while dry countries often use great amounts for irrigation. In developed countries as a whole, two fifths of the freshwater is consumed by industry; in the US, half is used by factories and power plants, in Eastern Europe up to 80 per cent. In the Third World, 85 per cent is used to grow food; in India 93 per cent goes to irrigation.

New sources

Freshwater consumption is bound to grow with rising populations and affluence. There are two ways of trying to meet it: finding new resources and using water more efficiently.

Only about 3,500 billion cubic meters of the 113,000 billion that fall as rain are used. But about two thirds of the rainfall evaporates back off the land, leaving some 41,000 billion cubic meters to flow down to the sea. Most of this runs off the land after storms and floods, leaving only 14,000 billion cubic meters as a steady flow that is reasonably easy to exploit. And more than two thirds of this runs through uninhabited or sparsely populated land.

That leaves 9,000 billion cubic meters that could, theoretically, be exploited. There is scope for using more of it, particularly in developing countries. Building small dams, which avoid the manifold drawbacks of large projects is a productive way of trapping more water for irrigation; China built 90,000 of them between 1950 and 1980. But major increases in exploitation are frustrated by the failure of the rain to fall where it is most needed. Transporting water for long distances, even within countries, is difficult and expensive. The US stopped building big dams, long aqueducts and canals, partly because they waste both water and money, and the USSR abandoned plans to reverse the flow of rivers discharging into the Arctic and Pacific Oceans to water the arid Central Asian republics. Desalination of seawater is becoming cheaper, but is still too expensive for most Third World countries.

Saving water

Reducing waste offers much more potential. Only a little over a third of the water used in irrigation actually helps the crops to grow; the rest is wasted. If such waste around the river Indus in Pakistan could be cut by just a tenth, another 2 million hectares of cropland could be irrigated.

Improving the flow of irrigation channels can cut wastage by 40 per cent. Using sprinklers to spray the water over the crops can reduce it by 70 per cent. Special sprinklers that deliver the water directly to the crop roots, instead of spraying it around, can cut wastage by a further quarter. These are being used in Israel, the US, the former USSR, France, Italy and several other countries, but still cover only 1 per cent of the world's entire irrigated land.

Israel is increasingly using its urban wastewater to irrigate crops. More than a third of it is already reused, and the country plans to recycle four fifths of it by the end of the decade. It has also cut its use of water in industry by 70 per cent for every unit of production. Other countries have also made great savings in industry. Between 1965 and 1974, Japan increased the proportion of recycled water in its industries from a third to two thirds. By 1975, US factories used water an average of 2.2 times before discharging it, and between 1978 and 1983 the country cut its water intake for industry by nearly a quarter, even though the number of its factories grew. Even greater savings are expected in the 1990s.

Pricing properly

Charging realistic prices for water would aid conservation. Water is usually heavily subsidized, particularly for irrigation. In many developing countries, the farmers who benefit from irrigation systems pay back only a tenth to a fifth of their costs. In California's giant Central Valley irrigation project, the farmers have only paid back 5 per cent of the costs over the last 40 years – and, because it is so cheap, one third of the water is used on inappropriate crops. Farmers have little incentive to save water, particularly when, as in many countries, they are charged according to the amount of land they irrigate, regardless of how much water they use.

Using water resources more efficiently – and proper pricing – will be needed to deal with the water crisis, which is likely to grow worse in an increasingly thirsty world.

Freshwater Pollution

Everywhere, it seems, surface waters are being polluted with a frightening assortment of municipal, industrial and agricultural wastes. Even in industrialized countries, where water quality legislation has taken hold, pollution is still a nagging problem. And for much of the developing world, rivers and lakes are often clogged with a virulent mixture of industrial toxins, untreated sewage and agricultural chemicals. By the turn of the century, the world will be generating about 2,300 cubic kilometers of wastewater a year.

Sewage treatment plants and more stringent controls on industrial polluters have reduced effluents in rivers and lakes in North America and Western Europe. Oxygen levels in many rivers and streams may have improved considerably over the past three decades, but problems remain. Only 60 per cent of the total population of the 21 OECD countries (Organization for Economic Cooperation and Development) are served by sewage treatment plants. Coverage ranges from nearly 100 per cent in Sweden and Denmark to less than 1 per cent in Turkey, 11 per cent in Portugal and 39 per cent in Japan.

The river Rhine, which drains one of the most highly industrialized regions in the world, was so polluted with chemical wastes in the 1970s that one concerned German partially developed a role of film in it. Today, fish have returned to parts of the river where they were absent for decades and concentrations of heavy metals have declined, thanks to improved sewage and industrial wastewater treatment plants. But the Rhine is still polluted with toxic chemicals and mineral wastes. Its biggest pollution problem now comes from chlorides (salt) washed out of agricultural land and from mining operations. It has also suffered serious damage from spills of chemicals. In a sense the Rhine typifies many rivers in the developed world. Whereas some pollutants have been controlled and reduced, others continue to pose serious threats to the human health and ecology of the region.

Water pollution in the developed world

The industrialized OECD nations have registered some success in controlling pollution from "point sources" – that is from measurable discharge points. But they are still plagued with diffuse pollution – such as run-off from agricultural lands packed with pesticides and fertilizers and stormwater flowing from cities and towns which can contain sewage, heavy metals, oils, hydrocarbons, garbage, chemicals from road de-icing, organic wastes from animals, and dust, among other things. These sources are difficult to control and are causing increased pollution of rivers and lakes throughout Europe and North America.

Lakes, in particular, are becoming eutrophic – overdosed with nutrients such as fertilizers and animal wastes. Since the 1950s, concentrations of phosphates and nitrates in freshwater rivers and lakes throughout Europe and North America have increased significantly. Nutrients from fertilizers and animal wastes were found to be the main contaminants in nearly 60 per cent of the polluted lakes surveyed in the United States. Switzerland's Lake Geneva and the lake districts of Friesland and Zuid-Holland in the Netherlands are victims of high amounts of agricultural chemicals percolating into their waters from surrounding farmland. Massive fish kills and degraded water quality result.

Increasing concentrations of nitrates

"Nitrates in drinking water," says *World Resources 1988-89*, "may cause blood poisoning in infants, hypertension in children, gastric cancers in adults and fetal malformations. The combination of high nitrates with pesticides, a common phenomenon, may form nitrosamines, which are both carcinogenic [cancer-forming] and mutagenic [causing birth defects]."

Nitrate concentrations have been increasing in European rivers and lakes since the 1960s. In some areas, like the Rhine, average nitrate levels are approaching the World Health Organization's recommended limit of less than 10 milligrams per liter of water. In 1990, at least 1.7 million Britons were still drinking water above this limit, in breach of a European Community directive which took effect in 1985.

Urban run-off is a major threat to freshwater in some areas. The polluted run-off from city streets in Florida is thought to account for more than half of the state's water pollution and nearly 85 per cent of all heavy metals found in its rivers and swamps.

Poisoned rivers worldwide

In Eastern Europe, the former Soviet Union and much of the Third World, however, the problems are more varied and more severe. Water courses are fouled with all manner of wastes and debris. Many countries lack basic water quality legislation. But even where legislation exists, as in much of Eastern Europe, enforcement is slack or non-existent. Pollution control takes a back seat to industrial protection and economic growth.

The river Vistula, which flows through the heartland of Poland, is so bloated with pollution from factories and municipalities that its waters are unsuitable even for industrial use along most of its length. Once the Vistula empties into the Baltic Sea, its contaminated plume stretches all the way to Sweden. Around 80 per cent of water samples taken from 200 major rivers in the former

Polluted rivers

Thick lines indicate severely polluted stretches of river. Thin lines are shown only to connect polluted stretches of the same river or to connect with the sea.

1	Skeena	80	Don
2	Fraser	81	Kuban
3	Columbia	82	Kura
4	Sacramento	83	Terek
5	San Joaquin	84	Kuma
6	Nelson	85	Volga
7	Severn	86	Kama
8	Moose	87	Pechora
9	St John	88	Ob
10	Hudson	89	Tobol
11	Delaware	90	Ishim
12	Roanoake	91	Irtysh
13	Savannah	92	Syr Darya
14	Alabama	93	Amu Darya
15	Platte / N. Platte	94	Yenisey
16	James	95	Lena
17	Missouri	96	Amur
18	Mississippi	97	Oum er Rbia
19	Ohio	98	Sénégal
20	Arkansas	99	Gambia
21	Red	100	Corubal
22	Brazos	101	Cess
23	Río Grande	102	Sassandra
24	Río Grande de Santiago	103	Komoé
		104	Volta
25	Río Balsas	105	Black Volta
26	Mezcalapa	106	White Volta
27	Cauca	107	Niger
28	Magdalena	108	Benue
29	Orinoco / Apure	109	Ogooué
30	Negro / Amazon	110	Zaïre
31	Ucayali	111	Cuanza
32	São Francisco	112	Cunene
33	Paraná	113	Orange
34	Paraguay	114	Vaal
35	Salado	115	Gourits
36	Carcaraña	116	Great Fish
37	Río de la Plata	117	Limpopo
38	Curaco	118	Save
39	Colorado	119	Zambezi
40	Bio Bio	120	Betsiboka
41	Elqui	121	Rovuma
42	Loa	122	Ruaha / Rufiji
43	Tyne	123	Tana
44	Tees	124	Juba
45	Mersey	125	Shibeli
46	Severn	126	Nile
47	Thames	127	Euphrates
48	Dal	128	Tigris
49	Bug	129	Indus
50	Vistula	130	Luni
51	Oder	131	Narmada
52	Elbe	132	Kaveri
53	Weser	133	Tungabhadra
54	Main	134	Krishna
55	Rhine / IJssel	135	Godavari
56	Meuse / Vesdre	136	Mahanadi
57	Schelde	137	Ganges
58	Seine	138	Brahmaputra
59	Loire	139	Irrawaddy
60	Garonne	140	Perok
61	Douro	141	Tulangbawang
62	Tagus	142	Ci Manuk
63	Guadiana	143	Mas / Surabaya
64	Guadalquivir	144	Chao Phraya / Mae Nam Yom
65	Segura		
66	Ebro	145	Mekong
67	Rhône / Saône	146	Xi Jiang / Pearl
68	Drac	147	Min Jiang
69	Po	148	Huangpujian
70	Inn	149	Yuan Jiang
71	Danube	150	Yangtze
72	Neva	151	Yalong
73	Dvina	152	Han Shui
74	Prut	153	Huang He
75	Dnestr	154	Wei He
76	Ingulets	155	Liao Hi
77	Dnepr	156	Fly
78	Desna	157	Murchison
79	Sev. Donets	158	Mackenzie

The rivers shown on the map and listed left are considered generally severely polluted, or exceed GEMS* median values in more than one category of pollutant.

* The UN-sponsored Global Environmental Monitoring System.

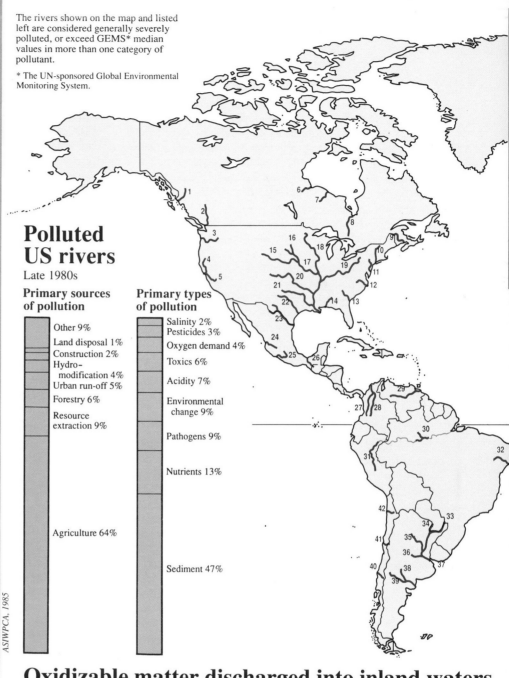

Polluted US rivers

Late 1980s

Primary sources of pollution

Other 9%
Land disposal 1%
Construction 2%
Hydro-modification 4%
Urban run-off 5%
Forestry 6%
Resource extraction 9%

Agriculture 64%

Primary types of pollution

Salinity 2%
Pesticides 3%
Oxygen demand 4%
Toxics 6%
Acidity 7%
Environmental change 9%
Pathogens 9%
Nutrients 13%

Sediment 47%

ASIWPCA, 1985

Oxidizable matter discharged into inland waters

Selected countries, late 1980s, kilograms per person

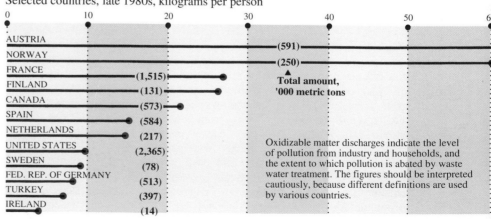

AUSTRIA	(591)
NORWAY	(250)
FRANCE	(1,515)
FINLAND	(131)
CANADA	(573)
SPAIN	(584)
NETHERLANDS	(217)
UNITED STATES	(2,365)
SWEDEN	(78)
FED. REP. OF GERMANY	(513)
TURKEY	(397)
IRELAND	(14)

▲ Total amount, '000 metric tons

Oxidizable matter discharges indicate the level of pollution from industry and households, and the extent to which pollution is abated by waste water treatment. The figures should be interpreted cautiously, because different definitions are used by various countries.

OECD, 1991

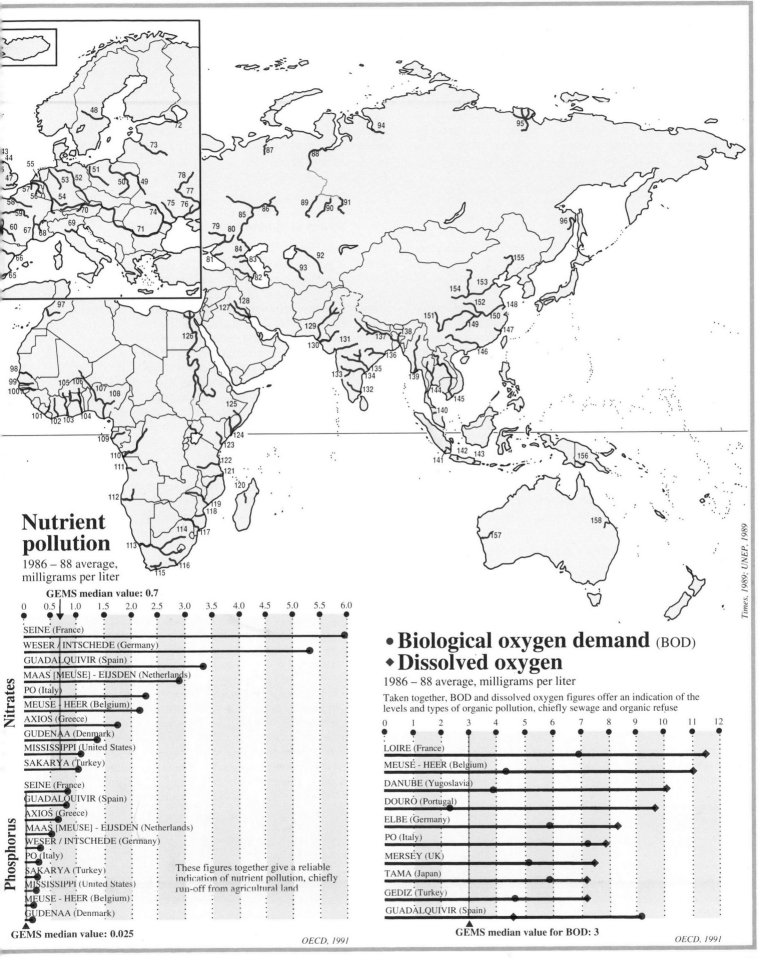

Nutrient pollution

1986 – 88 average,
milligrams per liter

GEMS median value: 0.7

0	0.5	1.0	1.5	2.0	2.5	3.0	3.5	4.0	4.5	5.0	5.5	6.0

Nitrates

SEINE (France)
WESER / INTSCHEDE (Germany)
GUADALQUIVIR (Spain)
MAAS [MEUSE] - EIJSDEN (Netherlands)
PO (Italy)
MEUSE - HEER (Belgium)
AXIOS (Greece)
GUDENAA (Denmark)
MISSISSIPPI (United States)
SAKARYA (Turkey)

Phosphorus

SEINE (France)
GUADALQUIVIR (Spain)
AXIOS (Greece)
MAAS [MEUSE] - EIJSDEN (Netherlands)
WESER / INTSCHEDE (Germany)
PO (Italy)
SAKARYA (Turkey)
MISSISSIPPI (United States)
MEUSE - HEER (Belgium)
GUDENAA (Denmark)

These figures together give a reliable indication of nutrient pollution, chiefly run-off from agricultural land

GEMS median value: 0.025

OECD, 1991

Times, 1989; UNEP, 1989

• Biological oxygen demand (BOD)
◆ Dissolved oxygen

1986 – 88 average, milligrams per liter

Taken together, BOD and dissolved oxygen figures offer an indication of the levels and types of organic pollution, chiefly sewage and organic refuse

0	1	2	3	4	5	6	7	8	9	10	11	12

LOIRE (France)
MEUSE - HEER (Belgium)
DANUBE (Yugoslavia)
DOURO (Portugal)
ELBE (Germany)
PO (Italy)
MERSEY (UK)
TAMA (Japan)
GEDIZ (Turkey)
GUADALQUIVIR (Spain)

GEMS median value for BOD: 3

OECD, 1991

63

Soviet Union showed dangerously high levels of bacterial and viral agents. Effluents from industries along the river Volga make up 10 per cent of the river's average flow at Volgograd: three quarters of them have received no treatment whatsoever.

India's rivers are little more than open sewers, carrying untreated wastes from urban and rural areas and factories to the sea. About 70 per cent of the country's surface waters are polluted; out of some 3,119 towns and cities, only 217 have partial or complete sewage treatment facilities.

The Ganges, sacred to Hindus and one of the most celebrated rivers in the world, exemplifies the problems. Almost 600 of its 2,525 kilometers are dangerously polluted with human and animal wastes, and with increasing amounts of toxic and hazardous effluents from industries and agriculture. Only 12 out of 132 industrial plants pumping effluents directly into the river have waste treatment plants, in working order. Many of the cities along the river, like Varanasi, have no sewage treatment plants at all.

The Government of India has allocated $195 million for a five-year clean-up program. In an effort to bring the river back to life, sewage treatment plants will be upgraded and new facilities built, animal wastes will be collected in urban areas and fertilizers and pesticides will be regulated more carefully in the river's catchment area.

Asia's rivers most degraded

Asia's rivers are, perhaps, the most degraded in the world. Increasing population pressures account for much of the region's poor water quality. As cities and industries expand, without proper waste treatment plants, rivers and streams are used increasingly as receptacles for their wastes.

Out of 78 rivers monitored in China, 54 are seriously polluted with untreated sewage and industrial wastes. Between a quarter and a sixth of the flow of the river Huangpujian, a major source of drinking water for Shanghai, is untreated waste. Less than 1 per cent of Taiwan's sewage gets even minimal treatment – and the island has the world's record levels of hepatitis B.

More than 40 of Malaysia's major rivers are so polluted with industrial and agricultural wastes, including oil palm and rubber processing residues, sewage and industrial chemicals, that they are said to be biologically dead, while the rivers that run through Manila are so full of untreated industrial wastes and raw sewage that they are almost devoid of life.

Integrated management is needed

The only way most countries, developed and developing alike, can hope to cope with mounting pollution of their freshwater resources is to evolve and implement integrated management strategies, which not only help clean up water courses but prevent pollution in the first place. Few countries have managed to do this. Sweden, which has one of the best pollution control records of any industrialized country, comes close to integrated management through its complicated permit system.

In essence, Swedish environmental law requires that 37 types of polluting industries – involving more than 800 separate companies – apply for an operating permit with the National Franchise Board for Environment Protection. Through this permit system, companies must give detailed reports regarding the effluents or emissions to land, water and air that result from their production processes. The permit is reviewed by the Franchise Board, which functions almost like an environmental court of law. Once the permit is issued, industries must comply with its stipulations or face stiff fines, even closure.

"The permit is a kind of contract between the plant and society," observes Mr Olle Aslander, Deputy Director of the National Environment Protection Board's Technical Department. "We have a long tradition of using a consensus approach to controlling emissions from factories. The idea is to allow the company to operate, while at the same time ensuring that environmental demands will be met."

Unfortunately, real integration of water management, pollution control and land management has not proved possible in most countries. Even the Swedes continue to struggle with their system, which some criticize as being too bureaucratic and cumbersome. Despite the problems, it is clear that in order to safeguard remaining water resources, integrated management of land and water will have to become the norm, not the exception.

Tropical Forest Destruction

Over the past five decades, a tragedy of untold proportions has been unfolding in the world's tropical forests. These unique ecosystems, the richest and oldest on earth, are being destroyed at unprecedented rates. All of humanity is affected.

Tropical forests regulate waterflow and protect watersheds for farmers who grow food for over a billion people; they regulate climate and produce oxygen, provide hardwood timber and fuelwood, are home to indigenous people, and harbor untapped genetic resources worth countless billions of dollars.

The falling forests

Only about half of the mature tropical forests that once graced the planet still stand. The latest estimates suggest that between 750 to 800 million hectares of the original 1.5 to 1.6 billion hectares have been felled. Most of what remains is in the Amazon Basin, where the forest covered more than 600 million hectares, an area nearly two thirds the size of the United States.

The forests are being destroyed at an ever-quickening pace. Until recently, the best estimates, based on a 1980 survey by the Food and Agriculture Organization of the United Nations (FAO), suggested that some 11.4 million hectares of tropical forest were being felled each year. But a 1990 survey, in *World Resources 1990-91* – which for the first time used satellite observations to build up a global picture – suggested that the rate of destruction had increased to between 16.4 and 20.4 million hectares annually. The higher figure is an area over twice the size of Austria. The figures represent areas that have been permanently cleared for other uses; many millions of hectares more are severely degraded each year.

Individual countries showed an even more rapid increase in the rate of destruction. In Myanmar there was a more than sixfold increase, from 105,000 hectares in 1980 to 677,000 in the 1990 report. And in India there is a tenfold difference in the two figures, rising from 147,000 hectares of tropical forest destroyed annually in 1980 to 1.5 million in 1990. However, after increasing the destruction for most of the 1980s, Brazil reversed the trend at the end of the decade: between 1988 and 1990, deforestation fell from 3.5 million hectares to about 1.3. Partly as a result of international pressure, partly through a change of government, Brazil ended subsidies for deforestation, clamped down on illegal felling and established new reserves.

Many other countries, including Indonesia, Thailand, Malaysia, Bangladesh, China, Sri Lanka, Laos, Nigeria, Liberia, Guinea, Ghana and Côte d'Ivoire have already lost large areas of their rainforest. Eighty per cent of the forests of the Philippine archipelago have already been cut down. In 1960 Central America still had four fifths of its original forest; now it is left with only two fifths of it. Half of the Brazilian state of Rondonia's 24.3 million hectares have been destroyed or severely degraded in recent years. Here, as in rainforests all over the world, indigenous people are being driven from the land they have lived in – and managed sustainably – for thousands of years.

Tropical forests come in two varieties: wet and dry. Both are under threat. The worst destruction is occurring in the wet ones. Latin America has lost 37 per cent, Asia 42 per cent and Africa 52 per cent of their original tropical moist forests.

Three steps to destruction

Indiscriminate logging, long thought to be the main reason for deforestation, now takes second place to shifting cultivation by landless forest farmers – estimated to be about 150 million worldwide – and the conversion of forest to agricultural plots, plantations and pastureland.

Tropical rainforest eradication is often a three-stage process. Logging companies carve out concessions and bulldoze access roads into pristine rainforest to extract timber. Peasant families follow the roads into the jungle in a desperate search for land and livelihood. They clear the forest to grow subsistence crops, cutting down all the trees and burning them and using the ashes as fertilizer. After just three or four harvests, insect plagues, weeds and soil impoverishment force them to move on and repeat the cycle in undisturbed areas. Some seed the plots with grass and sell them to cattle ranchers who then complete the third and final stage of destruction. Some two thirds of Central America's forest has been destroyed to provide cheap beef, often for hamburgers. By one estimate, two trees are destroyed to provide each hamburger.

In some cases, the forest is cleared away to make room for large plantations to supply rich countries with strawberries, eggplants, peppers, pineapples, bananas, sugar, groundnuts, palm kernels and cotton. Barbados, for example, replaced forests with sugarcane, and in Malaysia tropical forests are disappearing at the rate of 255,000 hectares a year, to make room for new rubber and oil palm plantations.

The green deserts

Tropical rainforests have been called "green deserts". Their soils are poor in nutrients, so the profusion of vegetation has evolved ingenious life-support systems. Trees and plants send out shallow roots, like tentacles, in all directions, soaking up nutrients from the forest's waste products – fallen leaves, dead trees and rotting organic matter.

Tropical forest destruction

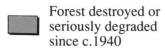
Extent of tropical forests, late 1980s

Forest destroyed or seriously degraded since c.1940

(Most, but not all of these current and former areas of tropical forest consist of closed forest).

- Tropical forests contain 155,000 of the 250,000 known plant species. Less than 1.5 million of the world's estimated 5 – 30 million species have been scientifically described.

- 700 tree species (the equivalent of Canada and continental US) were found in 10 1-hectare plots in Borneo.

- One-fifth of all bird species are found in Amazon forests.

- 90% of all primates are found only in tropical forest regions of Latin America, Africa and Asia.

- Scientists believe that 10% of the world's species could become extinct by 2000 and 25% by 2009.

- 90 different Amazonian tribes are thought to have disappeared this century.

- Genes from the wild are used to improve resistance of crops to pests or disease.

Medical benefits from tropical forest plants

Forest Indians of Northwest Amazonia use over 1,300 plant species as medicines.

Worldwide over 3,000 different species are used by indigenous peoples to control fertility.

120 pure chemical substances used in medicine are derived from less than 90 higher plant species. Less than 1% of 250,000 known tropical rainforest plants have been screened for use in life-saving drugs. 40% of the world's drugs come from the wild. The trade is worth $40 billion per year worldwide.

Diosgenin from the wild yam *Dioscorea spp* from Mexico and Guatemala enabled the contraceptive pill to be developed. In China and India, wild yams are still processed to make oral contraceptives.

Tubocurarine made from curare from the Amazonian liana *Chondodendron tomentosum* is used as a muscle relaxant during surgery.

Industry

Tropical forests produce far more than timber:
- Rubber
- Gum
- Latexes
- Resins
- Tannins
- Steroids
- Waxes
- Rattans
- Bamboo
- Essential and edible oils
- Pesticides
- Nuts and fruits
- Lubricants
- Flavorings and dyestuffs

In 1986 rubber exports earned $3 billion for tropical countries. Rattan earned $4 billion worldwide.

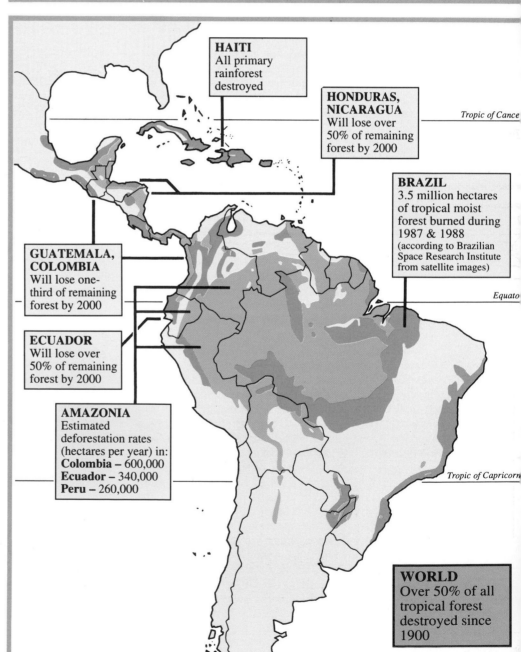

Worldwide distribution of tropical forests

HAITI
All primary rainforest destroyed

HONDURAS, NICARAGUA
Will lose over 50% of remaining forest by 2000

Tropic of Cance

BRAZIL
3.5 million hectares of tropical moist forest burned during 1987 & 1988 (according to Brazilian Space Research Institute from satellite images)

Equato

GUATEMALA, COLOMBIA
Will lose one-third of remaining forest by 2000

ECUADOR
Will lose over 50% of remaining forest by 2000

AMAZONIA
Estimated deforestation rates (hectares per year) in:
Colombia – 600,000
Ecuador – 340,000
Peru – 260,000

Tropic of Capricorn

WORLD
Over 50% of all tropical forest destroyed since 1900

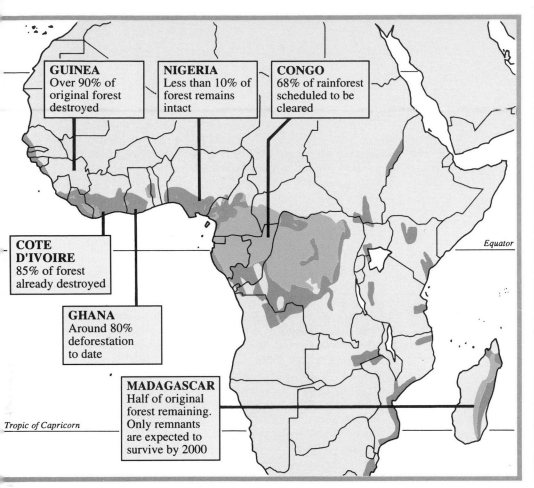

Total tree cover* and main causes of destruction

Latin America
(23 tropical countries: total tree cover 895.7 million hectares)
Felling and burning to create small farms and cattle ranches; opening up areas for colonization and development schemes.

Africa
(37 tropical countries: total tree cover 703.1 million hectares)
Fuelwood cutting, small-scale and shifting agriculture.

Asia
(16 tropical countries: total tree cover 336.5 million hectares)
Logging, fuelwood gathering, agricultural expansion and plantations.

*Early 1980s, closed and open forest

GUINEA
Over 90% of original forest destroyed

NIGERIA
Less than 10% of forest remains intact

CONGO
68% of rainforest scheduled to be cleared

COTE D'IVOIRE
85% of forest already destroyed

GHANA
Around 80% deforestation to date

MADAGASCAR
Half of original forest remaining. Only remnants are expected to survive by 2000

Equator

Tropic of Capricorn

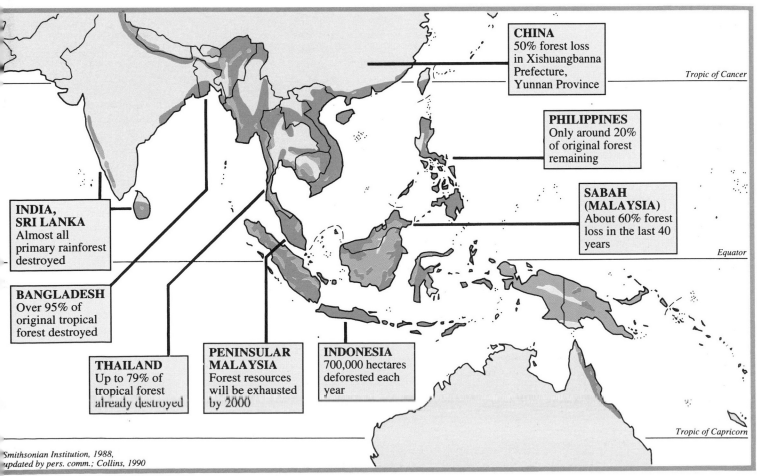

CHINA
50% forest loss in Xishuangbanna Prefecture, Yunnan Province

Tropic of Cancer

PHILIPPINES
Only around 20% of original forest remaining

SABAH (MALAYSIA)
About 60% forest loss in the last 40 years

Equator

INDIA, SRI LANKA
Almost all primary rainforest destroyed

BANGLADESH
Over 95% of original tropical forest destroyed

THAILAND
Up to 79% of tropical forest already destroyed

PENINSULAR MALAYSIA
Forest resources will be exhausted by 2000

INDONESIA
700,000 hectares deforested each year

Tropic of Capricorn

Smithsonian Institution, 1988, updated by pers. comm.; Collins, 1990

Strip away the trees and the exposed soils deteriorate rapidly, eroded by the torrential rains that can deluge tropical forests with over 2.5 centimeters of rain in 30 minutes. A single storm can remove up to 185 metric tons of topsoil from one treeless hectare. After the rains cease, the sun bakes the earth into a hard ocher-colored mass. Such "laterization" often renders the exposed soil incapable of supporting any kind of vegetation.

Genetic diversity under threat

Plants and animals vanish with the forests. A typical 1,000-hectare patch of tropical moist forest contains as many as 1,500 species of flowering plants, up to 750 species of tree, 400 bird species, 150 kinds of butterflies, 100 different types of reptile and 60 species of amphibians; the insects are too numerous to count. Panama has as many plant species as the whole of Europe. Peninsular Malaysia has 7,900 species of flowering plants; the UK, which is twice the size, contains only 1,430. A single volcano in the Philippines, Mount Makiliang, is home to more types of woody plant than the entire United States.

Mature tropical rainforests cover only about 7 per cent of the earth's surface, but harbor perhaps half of all its species, most of them as yet undiscovered. Clearing them may drive a million or more species to extinction by the year 2000.

The loss of even one species diminishes the whole of humanity, for it is a storehouse of genetic resources. All civilizations have been built on the diversity of nature, since crops and livestock were first developed from the wild, and we are still dependent on it for food, medicines and industrial raw materials. Crop breeders increasingly rely on wild strains to improve domesticated varieties and safeguard them against disease: a wild coffee from Ethiopia's fast-disappearing forests, for example, was used to save plantations throughout Latin America from devastation and several national economies from disaster. Half of all the medicine prescribed worldwide is originally derived from wild products, and the US National Cancer Institute has identified more than 2,000 tropical rainforest plants with the potential to fight cancer. Miracle substances wait in the rainforest to be discovered, and are being destroyed as they wait.

The way forward

In recent years, saving what remains of the world's rainforests has become an international cause.

Governments, international organizations and citizens groups are paying it increasing attention. Many initiatives have been launched including action plans, and debt-swap deals under which countries protect particular rainforest areas in return for alleviation of some of their foreign debt. But no plan will succeed if it simply seeks to preserve the rainforest. Third World nations, deep in financial crisis, will have no interest in saving the forest unless they can be shown that it pays.

Recent studies have shown that traditional non-destructive uses of the rainforest, like tapping rubber, harvesting fruit, oils and medicinal plants and practising agroforestry, achieve much higher economic returns than logging, slash-and-burn agriculture or cattle ranching. One, in the Peruvian Andes, showed that these uses produced seven times as much income per hectare over 50 years as intensive logging; another, by the Institute of Economic Botany, showed that fruit and latex from tropical forests were worth nine times as much as its timber. The Lacandon Maya Indians of Chiapas, Mexico, practise a highly efficient form of traditional agroforestry, using a multi-layered cropping system which permits them to cultivate up to 75 crop species on single hectare plots for up to seven consecutive years. With skilled husbandry of the forest's resources, one Lacandon farmer will clear no more than 10 hectares of rainforest during his lifetime.

Yanesha Indians in Peru's tropical forests along the border with Brazil are now managing their forest by cutting trees in narrow strips, leaving wide sections of forest intact, and allowing it to recolonize the cleared areas. They export wood to Europe and the US where it sells at about 15 per cent more than destructively logged hardwoods from tropical forests.

Many other peoples, including the Chagga of Mount Kilimanjaro and the Lua of Thailand, practise similar systems; there is traditional agroforestry almost everywhere there are rainforests. It is much more productive than the cultivation that destroys the forests, and it allows farmers to stay on their land without having to move on. So it abates the land hunger which forces people into the forest and offers those who do settle there a sustainable way to farm in harmony with their environment.

The Earth Summit in Rio de Janeiro in June 1992 failed to agree either on an international convention to cover the world's tropical and temperate forests or even to work towards one, and settled merely for a weak statement of principles.

The Tropical Timber Trade

Tropical timber is one of the five most valuable exports from Third World countries. A single rainforest tree can be worth $1,000 and developing countries make a total of $8 billion a year from selling hardwoods overseas. Demand for the beautiful timber is growing fast – rich countries import 16 times more than they did in 1950 – and tropical forest nations are naturally anxious to earn the foreign exchange. But present methods of exploitation are wasteful and help to cause the destruction of the world's forests.

Demand for tropical timber

Ninety per cent of the trees of tropical rainforests are hardwoods. Exotically colored and grained, supremely versatile in their uses, they have long been specially valued. Nearly 2,000 years ago, Pliny described them being shipped long distances to Rome from the East. A major trade emerged with the 19th-century colonial empires, and economic powers of the 20th century are now the main markets. Japan receives more than 53 per cent of the world's tropical hardwood trade: wood is its biggest import after oil. The United States accounts for 15 per cent: it imports wood worth over $1 billion each year, and its demand is expected to double by the end of the century. The UK, by contrast, imported only a third as much in the 1980s as 50 years before: Europe as a whole now receives 32 per cent of the trade.

Tropical hardwoods are in demand because they have a relatively uniform structure and the large tree dimensions make it possible to obtain large planks of high quality. They are then used in the form of sawnwood, plywood and veneers for building construction and furniture (which account for about 50 per cent of total consumption), vehicles, ships, household goods, games, toys, musical instruments, kitchen utensils and coffins. Much of the woodwork in Western homes originates in the tropics – everything from window and door frames to furniture and the veneer on radio, stereo and TV sets. In heavy construction, tropical hardwoods are used for railway sleepers, harbor pilings, wharves, jetties, mooring posts, gangway planks and for fencing and railing waterways. Tropical forests are the source of about 30 per cent of all the world's log exports, 10 per cent of all sawnwood exports and 60 per cent of all its veneer and plywood exports. Hardwoods are also turned into humble pulp: one Japanese company is transforming a rich rainforest in Papua New Guinea into packaging for cameras, calculators and other electrical goods.

Meeting the demand

Exploitation of rainforest by logging companies is accelerating, especially in Southeast Asia, the source of about three quarters of exported tropical timber. Africa accounts for most of the rest, though its exports dropped from 8.3 to 4.8 million cubic meters a year between 1973 and 1983, partly because its forests were increasingly logged out and partly because more timber was used for domestic consumption. Latin America's exports, always much smaller, fell from 525,000 cubic meters to 47,000 cubic meters over the same period because almost all of the timber is used at home. In 1980, Brazil produced over 156 million cubic meters of hardwood timber, but exported only 7,000 cubic meters of it: the rest was used domestically for building and in paper products.

In fact, a few countries account for almost all the world's exports. More than 70 per cent of all tropical hardwoods for export are produced by just six countries: Indonesia, Malaysia, the Philippines, Papua New Guinea, Brazil and Côte d'Ivoire.

The Food and Agriculture Organization of the United Nations (FAO) predicts that by the turn of the century production of tropical hardwoods will have more than doubled from 1980 levels, and that developing countries will consume two fifths of it.

The effects of over-exploitation

Logging has already led to the virtual extinction of some trees. Afromarsia, a leguminous tree from tropical Africa widely used in furniture and veneers, is now hard to find – less than 40 years after first entering commercial trade. Loggers only want the best wood, and so concentrate on relatively few valuable species: only about a dozen of Southeast Asia's 630 hardwood species make up the bulk of the exports. These species are so widely scattered that there are often only two or three commercially important trees in every hectare of rainforest. Felling these causes much more widespread damage: as each tree comes crashing down and is dragged away by heavy machinery it damages its neighbors, leaving them easy prey to disease. Studies show that extracting just a few trees per hectare often injures a third to two thirds of those that remain.

Increasingly, the wealth of the forest is being cut away. Nigeria now imports 100 times as much as it sells overseas. All accessible hardwood forests in the Philippines are expected to be logged out in the early 1990s. Nearly half of Peninsular Malaysia's rainforests have been logged over the past 30 years. Timber extractors are cutting away the Côte d'Ivoire's forests at the rate of 500,000 hectares a year, so that all exploitable areas could be exhausted by the end of the century. Virtually all of Indonesia's easily accessible lowland forests, including some areas designated as "protected forest", have been let as timber concessions. Indonesia now requires logging companies to undertake

Total tropical hardwood production

Million cubic meters, 1989

■	100 and over
■	20 – 99.9
■	5 – 19.9
■	2 – 4.9
■	Under 2
■	Does not produce tropical hardwood

Tropical hardwood production figures include not only timber that is traded internationally but also timber used locally, including that used for fuelwood and charcoal production.

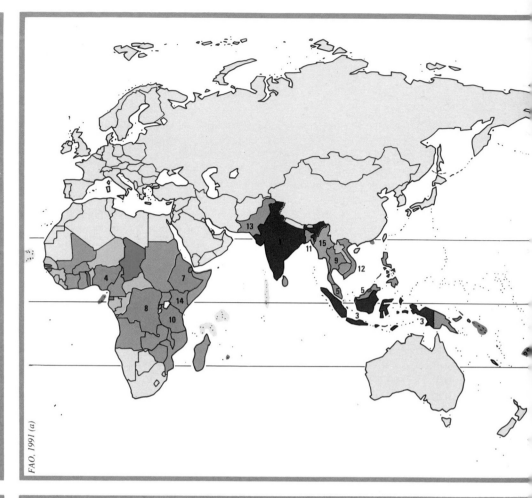

FAO, 1991 (a)

Tropical hardwood exports

US$ million, 1989

■	1,000 and over*
■	100 – 999
■	10 – 99
■	Under 10
■	Does not export tropical timber

*Only two countries fall into this category: Indonesia (exports worth $3,493 million), and Malaysia ($3,029 million)

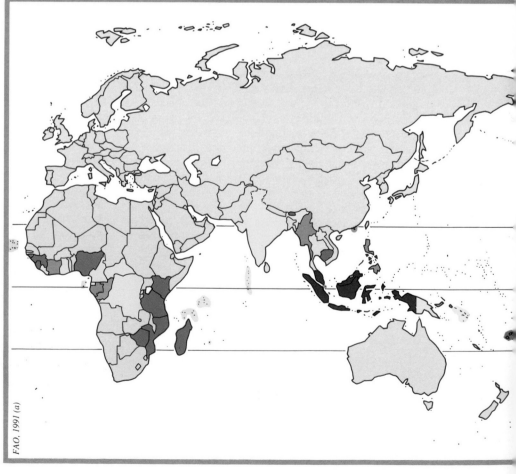

FAO, 1991 (a)

Tropical timber: the 15 largest producers

Million cubic meters, 1989. Figures refer to top map

1. **INDIA (Total: 248.1,** of which fuelwood: 226.6)

Fuelwood

▲ **Total tropical timber production**

2. **BRAZIL (175.3,** 133.2)

3. **INDONESIA (174.3,** 135.3)

4. **NIGERIA (99.3,** 91.4)

5. **MALAYSIA (48.3,** 6.0)

6. **PHILIPPINES (38.0,** 32.6)

7. **ETHIOPIA (35.7,** 34.0)

8. **ZAIRE (35.3,** 32.6)

9. **THAILAND (35.1,** 30.5)

10. **TANZANIA (31.7,** 30.1)

11. **BANGLADESH (30.1,** 29.3)

12. **VIETNAM (26.9,** 23.8)

13. **PAKISTAN (23.0,** 22.1)

14. **KENYA (22.7,** 21.6)

15. **MYANMAR (22.3,** 17.4)

Fuelwood, which includes charcoal production, is produced mainly for domestic consumption, whereas non-fuelwood may be largely exported (eg Malaysia exports over 75% of its non-fuelwood) or largely used within the country (eg Brazil exports under 2% of its non-fuelwood).

FAO, 1991 (a)

The tropical timber trade

Selected major importers of tropical timber, trade flows in thousands of cubic meters, 1987

☐ Logs   ☐ Sawnwood   ☐ Plywood

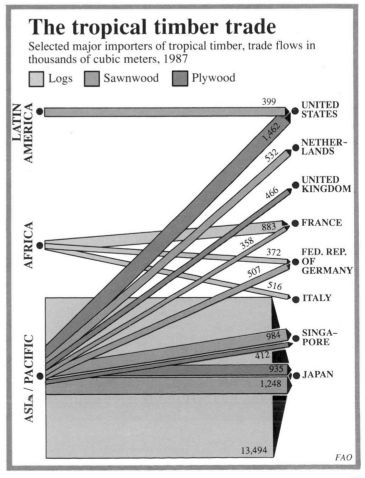

LATIN AMERICA

AFRICA

ASIA / PACIFIC

399 — **UNITED STATES**
1,462 — **NETHER-LANDS**
532 — **UNITED KINGDOM**
466 — **FRANCE**
883 — **FED. REP. OF GERMANY**
358
372
507
516 — **ITALY**
984 — **SINGA-PORE**
412
935 — **JAPAN**
1,248
13,494

FAO

reforestation programs, but a recent survey indicated that few comply. The World Bank estimates that more than two thirds of the countries which now export tropical timber will have run out of wood by the year 2000.

Latin American exports are expected to grow as Southeast Asian forests are increasingly denuded and as rising home consumption uses up what remains. Experts are predicting that Asia's share of exports could drop to as little as 10 per cent early next century.

Action needed

Outrage at the devastation caused by unrestrained logging has led to calls for developed countries to boycott tropical timber, and this has put pressure on timber companies to improve their operation. The German building ministry has announced a ban, and one out of every six Britons say they will not buy tropical wood products. Many experts believe, however, that boycotts might be counterproductive in the long run because they would drive down the price of timber, and exporters might decide to clear-fell the forests for pulp, accelerating the destruction in order to maintain their incomes. They could also increase pressure on equally valuable temperate forests.

The best solution would be for exporting countries to grow valuable trees on special plantations, using forest land already ruined by logging. Zambia has established enough plantations to meet all its needs for industrial timber until the end of the century, but it is very much the exception. Only 2 per cent of all exports comes from managed plantations and, because it takes 30 years for tropical hardwoods to grow, a campaign to establish new ones would not affect timber supplies until 2020.

Exporting countries might be encouraged to look after their forests if they got better value from them. They would get at least five times as much revenue from each tree if they processed the wood themselves, rather than merely exporting logs to be turned into veneer and furniture in industrialized countries. Similarly, countries could get better value from each hectare of forest by harvesting the other potentially valuable trees that are smashed in the process of extracting the few fashionable species. Malaysia has developed markets for them. But in Thailand and Sarawak, timber worth about $150 million is needlessly burned each year.

Getting better value from the forests could reduce the rate of destruction and give countries a bigger incentive to ensure that they survived to go on producing timber. On present trends everyone stands to lose: estimates suggest that demand for tropical hardwoods could exceed supply by some 400 million cubic meters a year by the year 2000, and by 3 billion cubic meters by 2025. Controlling unplanned deforestation, allowing previously logged forests to regenerate, protecting national parks, establishing plantations and setting up sustainable management programs would both conserve remaining forests and ensure the survival of the tropical timber trade.

The International Tropical Timber Agreement

Concern about future supplies of tropical timber led to the establishment of the International Tropical Timber Agreement (ITTA). One of the prime movers behind the Agreement was Japan, the world's greatest importer. After nine years of negotiations, ITTA was adopted in 1983 and finally ratified by the requisite number of countries in 1985.

It has the contradictory aim of promoting the tropical timber trade while encouraging conservation and the sustainable use of forests. In 1990 it laid down the objective of ensuring that the entire tropical timber trade should come from sustainable sources – as opposed to less than one per cent of it today. But its achievements have been insignificant. Several countries, meanwhile, have decided only to import wood from sustainable sources by 1995: they include Switzerland and the Netherlands. Austria has gone further, already banning imports of unsustainably harvested timber. Unless the timber trade makes a rapid transition from simply mining the forests to really sustainable management, bans and boycotts will increase.

The Fuelwood Crisis

Two billion people are caught in the "poor man's energy crisis" – the shortage of fuelwood. Nearly half of humanity has not yet entered the age of fossil fuels and depends on wood for warmth, light and cooking. But as trees are cut down, fuelwood is becoming scarce over large parts of sub-Saharan Africa, the Middle East and Asia.

Fuelwood accounts for a large proportion of all energy consumption in many developing countries. More than 90 per cent of energy use in such nations as Burkina Faso, Tanzania, Nepal and Ethiopia comes from wood. It even supplies more than four fifths of oil-rich Nigeria's needs, and many other countries get half to three quarters of their energy from wood. Even in countries with large industrial sectors, which use a great deal of fossil fuels, fuelwood dominates the life of the country people; wood accounts for only a third of India's total energy consumption, but – together with dung and crop wastes – meets over 90 per cent of rural dwellers' energy needs.

An escalating shortage

Already 100 million people in the Third World cannot get sufficient fuelwood to meet their minimum energy needs, and close to 1.3 billion are consuming fuelwood resources faster than they are being replenished. On average, consumption outpaces supply by 30 per cent in sub-Saharan Africa as a whole; by 70 per cent in the Sudan and India; by 150 per cent in Ethiopia; and by 200 per cent in Niger. If present trends continue, the Food and Agriculture Organization of the United Nations (FAO) predicts that, by the turn of the century, another 1 billion people will be faced with chronic fuelwood shortages.

Thus within a decade, more than half the Third World's population would not be able to meet their minimum needs for energy or will be forced to consume wood faster than it can be grown. There is little chance that they could find any other source of heat and light. Usually there are no viable alternative fuels and, where there are alternatives, they cost too much. Already even a few sticks of wood can be prohibitively expensive in areas affected by severe shortages. It costs some rural families in India and Pakistan as much to heat the evening dinner bowl as it does to fill it. Health and nutrition are affected. Disease and death spread as boiling water becomes an unaffordable luxury. Mothers are forced to feed their children on cereals, which cook quickly, rather than slower-cooking and more nutritious foods, like beans. Nearly two thirds of Rwanda's people now only cook one meal a day, most of the rest even less frequently.

Even in wood-rich areas like the Amazon, some towns are feeling the energy pinch. In Manaus, Brazil, for example, the majority of the poorer residents depend on charcoal for preparing meals. As the jungle recedes further away every year, due to expanding agriculture and logging, the cost of trucking in fuel rises. A week's supply of charcoal for an average family costs around $2 – a considerable amount of money in a town where the average wage is $2 a day. In Ouagadougou, Burkina Faso, and Port au Prince, Haiti, families spend nearly a third of their income on fuel.

As supplies of fuelwood dwindle, villagers – mainly women and children – have to spend more time searching for wood. Women in the village of Kalsaka in Burkina Faso walk three and a half hours through searing heat to collect wood that used to be available a stone's throw away from their huts. Families in the uplands of Nepal spend 230 person-days a year on fuelwood collection and, in some parts of Tanzania, 250-300 days are required to meet family needs. Inevitably, less time can be spent in growing food. Nepalese women, for example, now spend nearly half as much time tending crops and vegetable gardens as they used to, because of the search for firewood. And fetching wood and water often costs African rural women up to 500 calories a day.

Complex causes of the crisis

Ostensibly, the fuelwood shortage has been attributed to the over-exploitation of forests and woodlots by the rural poor as human numbers and energy needs increase. This is an important factor. Yet the roots of the fuelwood crisis are more intricate: rampant logging, in combination with animal foraging and slash-and-burn cultivation, contributes to the fuelwood deficit. Deforestation is followed by erosion, desert encroachment, loss of biomass productivity and reduced water-retention capacity of the soil.

Another consideration, often overlooked, is the increasing consumption of wood and charcoal by urban dwellers. Urbanization concentrates people and puts increasing stress on basic necessities such as food, energy, drinking water and shelter, exacting a heavy toll on the surrounding countryside. A study by the Beijer Institute of Stockholm, made for the Kenyan Government, found that proportionally one of the major contributors to deforestation was not rural fuelwood use, which was found to be mostly sustainable, but the wholesale conversion of wood to charcoal for sale to people living in towns and cities.

Charcoal is used to meet city needs because it is so much lighter than the original wood, and therefore much cheaper to transport. But converting wood to charcoal in traditional earthen pits consumes more than half its energy. So each town dweller uses twice as much wood for a given amount of energy as a country person who

The fuelwood crisis

Areas of fuelwood deficit in developing countries, 1980s

- Prospective deficit
- Deficit
  (fuelwood resources below requirements)
- Acute scarcity
  (unable to provide a minimum supply)
- Desert
- Not known to have deficit
  (including some sub-desert areas)

Fuelwood and charcoal producers in the developing world, 1989

Thousand cubic meters

The 30 largest producers

1	India	245,127
2	Brazil	182,806
3	China	177,610
4	Indonesia	136,079
5	Nigeria	100,430
6	Ethiopia	37,884
7	Thailand	34,115
8	Kenya	33,884
9	Philippines	33,075
10	Zaïre	32,557
11	Tanzania	31,114
12	Bangladesh	29,272
13	Vietnam	23,776
14	Pakistan	23,226
15	Sudan	20,112
16	Myanmar	17,407
17	Nepal	17,244
18	Ghana	16,068
19	Mexico	15,204
20	Colombia	15,086
21	Mozambique	15,022
22	Uganda	12,507
23	Zambia	11,565
24	Cameroon	10,142
25	Côte d'Ivoire	9,830
26	Turkey	9,721
27	Sri Lanka	8,302
28	Malaysia	8,258
29	Burkina Faso	8,141
30	Peru	7,669

Other selected countries

A	Guatemala	7,490
B	Malawi	7,275
C	Madagascar	7,049
D	Somalia	6,896
E	Ecuador	6,642
F	Chile	6,540
G	Zimbabwe	6,269
H	Rwanda	5,602
I	Papua New Guinea	5,533
J	Haiti	5,488
K	Paraguay	5,288
L	Cambodia	5,236
M	Honduras	5,172
N	Mali	5,163

FAO, 1991(a)

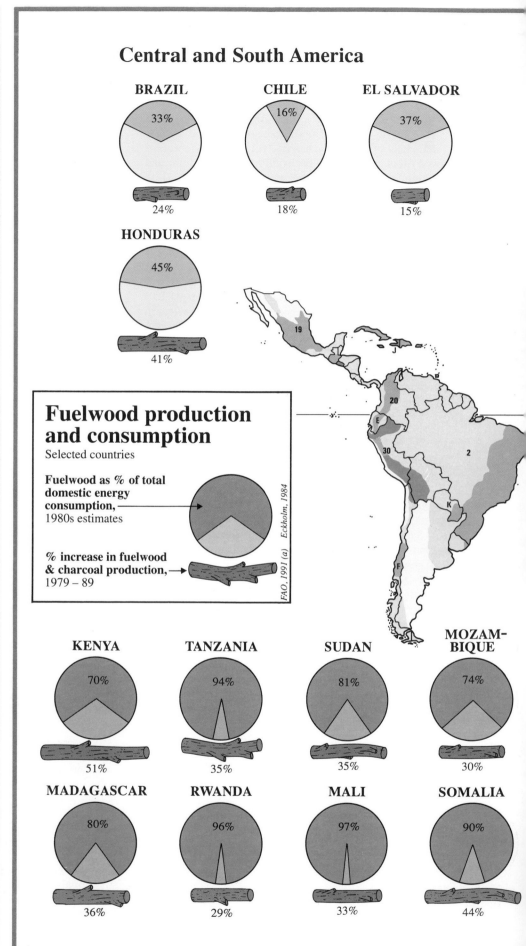

Central and South America

BRAZIL 33% / 24%
CHILE 16% / 18%
EL SALVADOR 37% / 15%
HONDURAS 45% / 41%

Fuelwood production and consumption

Selected countries

Fuelwood as % of total domestic energy consumption, 1980s estimates

% increase in fuelwood & charcoal production, 1979–89

FAO, 1991 (a) Eckholm, 1984

KENYA 70% / 51%
TANZANIA 94% / 35%
SUDAN 81% / 35%
MOZAMBIQUE 74% / 30%
MADAGASCAR 80% / 36%
RWANDA 96% / 29%
MALI 97% / 33%
SOMALIA 90% / 44%

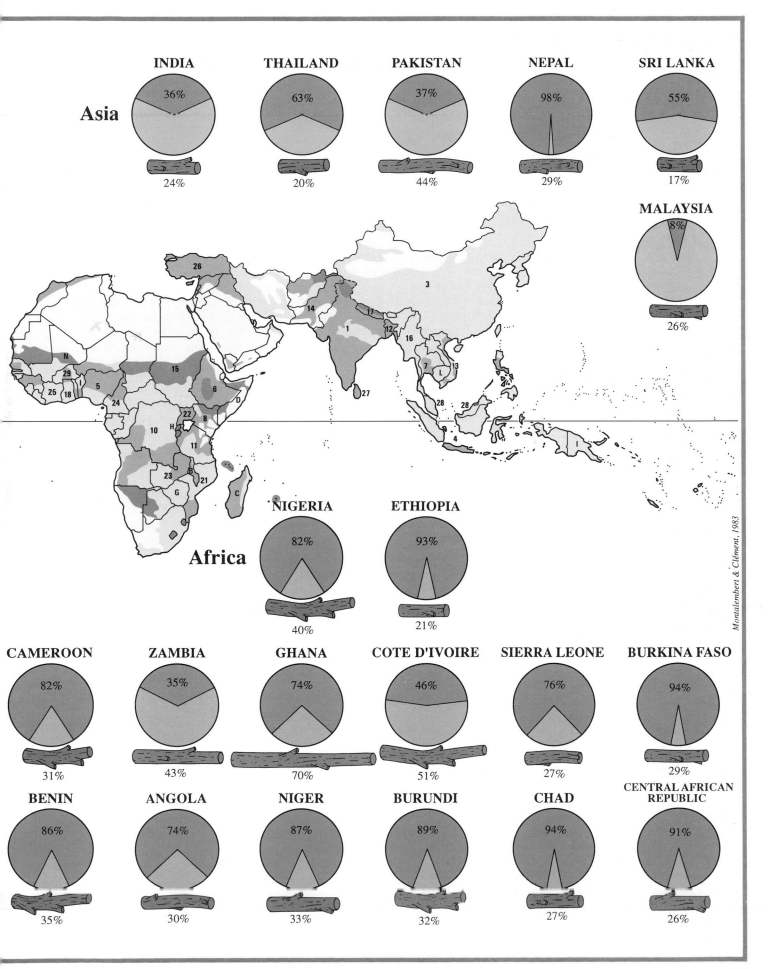

Asia

INDIA
36%
24%

THAILAND
63%
20%

PAKISTAN
37%
44%

NEPAL
98%
29%

SRI LANKA
55%
17%

MALAYSIA
8%
26%

Africa

NIGERIA
82%
40%

ETHIOPIA
93%
21%

CAMEROON
82%
31%

ZAMBIA
35%
43%

GHANA
74%
70%

COTE D'IVOIRE
46%
51%

SIERRA LEONE
76%
27%

BURKINA FASO
94%
29%

BENIN
86%
35%

ANGOLA
74%
30%

NIGER
87%
33%

BURUNDI
89%
32%

CHAD
94%
27%

CENTRAL AFRICAN REPUBLIC
91%
26%

Montalembert & Clément, 1983

75

sticks to the original wood.

An expanding "ring of desolation" surrounds many Third World cities – such as Ouagadougou in Burkina Faso, Niamey in Niger and Dakar in Senegal – as a result of urban fuelwood demands. By the year 2000, estimates the World Bank, half to three quarters of all West Africa's fuelwood consumption will be burned in its towns and cities.

In several areas, the fuelwood crisis is beginning to hit agricultural productivity, especially where drylands have been deforested to make way for irrigated agriculture or grazing land. Once fuelwood becomes so scarce that people have to burn animal dung or crop residues to cook their meals, the search for fuel takes on new and terrible dimensions. Burning dung and crop residues, instead of returning them to the fields, robs the soil of fertilizers, reducing food output. Worldwide some 800 million people have to resort to these fuels. In Asia and Africa, at least 400 million metric tons of animal dung are burned each year so cannot be used to enrich the soil. If this natural fertilizer had been used on croplands instead, an extra 20 million metric tons of grain could have been produced. In Nepal, the use of dung and crop residues cuts grain yields by 15 per cent.

Planting the right trees

The World Bank estimates that meeting the fuelwood crisis will require planting 55 million hectares with fast-growing trees at a rate of 2.7 million hectares a year, five times the present annual rate of 555,000 hectares. But simply planting more trees, even fast-growing ones, is only part of the solution. The right species have to be selected for each particular environment, taking into account growth rates, water and mineral needs, and adaptability, among other things. Above all, the trees have to be planted where they will be used and by the people who will use them. This implies an emphasis on community woodlots, not the creation of huge fuelwood plantations filled with monocultures, located too far from where the wood is needed.

Some species show great promise. The *Leucaena* or *ipil ipil*, a native of Mexico, is one of the world's fastest growing trees. It can reach heights of 20 meters in six years. A *Leucaena* plantation can provide up to 50 metric tons of wood per hectare per year, or five times the average for cultivated pines in temperate regions. Its nitrogen-fixing roots also replenish the soil.

Similarly, *Euphorbia*, a leafless shrub capable of prodigious growth, may become one of Kenya's most valuable energy resources. Field studies report yields of 80 metric tons per hectare without fertilizers or irrigation and it thrives in arid and semi-arid savanna, where few crops can grow.

Burning fuel more efficiently

At the same time, more efficient stoves must be developed to allow the wood to burn longer, extracting more energy from each stick. The traditional three-stone fire is a very inefficient way of burning wood; its conversion efficiency may only reach 6 per cent of the fuel value of the wood. Unfortunately, many improved stoves have proved unworkable, because they have technical problems, or are made with material too expensive or too hard to obtain. Improved stoves, however, must respond to local needs. The materials for building them have to be locally available, easily accessible and relatively cheap. A good example is the Lorena stove from Guatemala, which cuts fuelwood consumption in half. Molded from mud and sand, and fitted with a simple metal damper and pipe, it costs the equivalent of $5. With proper maintenance, the Lorena stove will last for years. The World Bank estimates that the use of more efficient stoves – together with fuel substitution where practicable – could reduce fuelwood consumption by a quarter by the year 2000.

A plan to resolve the crisis

According to the United Nations, the following measures need to be taken, if the fuelwood crisis is to be resolved:

- A fivefold increase in current levels of tree planting for fuel; estimated to cost $1 billion a year for the next 20 years.

- Improved fuelwood distribution networks.

- The adoption of better conversion technologies.

- Use of more efficient wood-burning stoves.

- Encouraging the proliferation of family or community woodlots, using appropriate fast-growing species adapted to local conditions and needs.

The threat of massive fuelwood shortages in the next century should prompt governments and international agencies to commit money and manpower to solving the "poor man's energy crisis" now, before it turns into a human tragedy.

Damaged Watersheds

The world's mountain ranges – among the most critical areas on earth – are being devastated by continuous assault from human activities. In tropical developing countries alone, an estimated 160 million hectares of these upland watersheds – an area larger than the United Kingdom, France, Germany, Italy, Switzerland, Belgium, Luxembourg and the Netherlands all put together – has been seriously degraded in the last three decades.

The function of watersheds

At least 2 billion people depend on the stability of these uplands. When they are healthy and covered with trees they regulate water supplies, absorbing the rains and releasing them to percolate into groundwater and flow down rivers. They help to regulate the climate and they provide habitats for a wide variety of plants and animals found nowhere else.

Trees help bind the soil to the ground. As they go, the soil is left exposed and is washed off the hills by the rains. Crop yields fall. Vital water supplies are usually disrupted. Instead of replenishing water sources, rains rush off slopes which have been stripped of trees and denuded by overgrazing or overcultivation; both floods and droughts result.

Everywhere the forests are being destroyed. In developing countries, they are felled by logging companies and, above all, by poor subsistence farmers who have been forced off better land by population growth and unequal land holdings. In developed countries they are often destroyed by the *Waldsterben* (forest death) syndrome triggered by pollution that has swept across Europe and parts of North America.

A litany of disaster

The once fertile Ethiopian highlands – where 70 per cent of the country's people live – supported agrarian societies for millennia, but are now synonymous with disaster. Over the last 50 years they have lost 90 per cent of their trees. Every year at least a billion metric tons of topsoil are washed away. Already 20,000 square kilometers of land have lost so much soil that they can no longer sustain crops. Yields are declining on the rest and, within 25 years, 100,000 square kilometers – an area the size of the former East Germany – are expected to be useless for agriculture; by then the already pitifully low incomes of the people will have declined by a third.

Half the trees in the Alps are dying of *Waldsterben*. Pollution is brought in on the winds from across Europe and rises from the exhausts of the vehicles that increasingly throng the area; roads through the Alps now carry 20 per cent of all passengers and 25 per cent of all goods carried in the European Community. Holidaymakers using the 40,000 ski runs in the mountains add to the erosion. Avalanches and landslides are increasing. Nearly 60 plant species have disappeared from France's Haute Savoie region so far this century; more than half of Europe's threatened species of fauna and flora live in the mountain region.

China's Yellow river (Huang He) actually flows 3 to 10 meters above the surrounding land as it approaches the sea. Over the years it has carried so much silt that this has raised the river bed, and it now runs in a raised channel, enclosed by dikes. More than a billion metric tons of soil are carried down the river every year, half of it scoured from the already badly eroded Loess Plateau in its middle reaches. Some 430,000 square kilometers of the plateau have been reduced to bare hillsides, slit with huge erosion scars. On average, every hectare loses 65 metric tons of soil a year. In this part of its course the Yellow river consists of equal parts of water and sediment; if the soil content gets any higher the river will reach the level officially classified as liquid mud.

The natural instability of the Andes in Ecuador and Peru is aggravated by tropical storms. Half of the land has now been stripped of trees and more than 15 metric tons of soil are being lost per hectare each year; in severe cases as much as 200 metric tons per hectare can be lost. The crisis is expected to get worse.

Forty years ago most of the Central American isthmus was covered in forest; now only a few major areas remain. As a result, 40 per cent of the land on the Pacific side of the isthmus, where most of the people live, is being eaten away by erosion. The Panama canal is silting up with soil stripped from local watersheds; big ships may not be able to use it from the year 2000.

The destruction of the Himalayas, the world's greatest mountain chain, affects the largest number of people. Over the last 40 years at least 40 per cent of its forests have been cut down. By 1992 China had cut and taken away $54 billion worth of timber from Tibet since occupation while Nepal's mountains have lost at least half their trees since the conquest of Everest in 1953. Between 35 and 75 metric tons of soil are stripped from each hectare of its bare slopes every year. Hill people keep watch day and night for landslides during the monsoons; up to 20,000 have been recorded in a single day, often sweeping away terraced fields laboriously carved into the mountain sides. In many villages three quarters of all the farmers have lost part of their land; tragically often whole villages are buried. One Nepalese official laments, "Our biggest export to India is now soil."

Damaged watersheds

Areas of major
damaged watersheds
A. Andes; **B**. Alps; **C**. Ethiopian
Highlands; **D**. Himalayas; **E**. West
& South Ghats; **F**. Loess Plateau,
China

Rivers carrying heavy sediment loads:

(Main streams and tributaries only)

1	Yukon	40	Nile
2	Mackenzie	41	Tigris
3	Columbia	42	Indus
4	Hudson	43	Krishna
5	Delaware	44	Ganges
6	Ohio	45	Brahmaputra
7	Missouri	46	Irrawaddy
8	Río Grande	47	Mekong
9	Colorado	48	Song Hong
10	Río Balsas	49	Xi Jiang
11	Magdalen	50	Yangtze
12	Orinoco	51	Huang He
13	Amazon	52	Amur
14	Río de la Plata	53	Kapuas
15	Colorado	54	Mahakan
16	Río Negro	55	Fly
17	Bio Bio	56	Murray
18	Vistula		
19	Oder		
20	Elbe		
21	Rhine		
22	Douro		
23	Tagus		
24	Guadiana		
25	Rhône		
26	Drac		
27	Po		
28	Drin		
29	Danube		
30	Don		
31	Volga		
32	Niger		
33	Sanaga		
34	Zaïre		
35	Zambezi		
36	Rufiji		
37	Tana		
38	Juba		
39	Shibeli		

WRI/IIED/UNEP, 1988, updated by editors

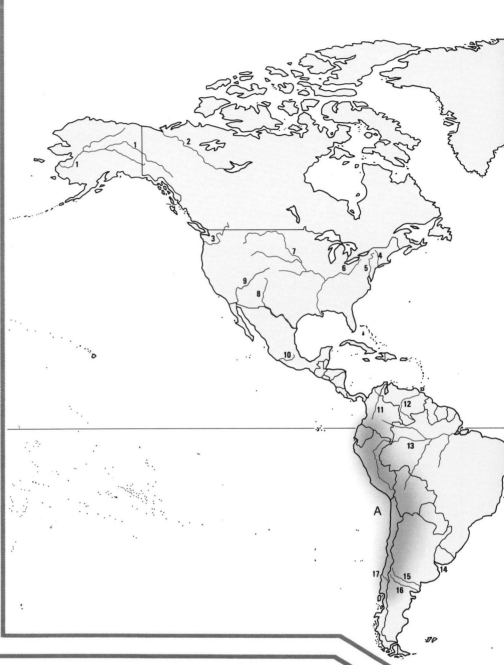

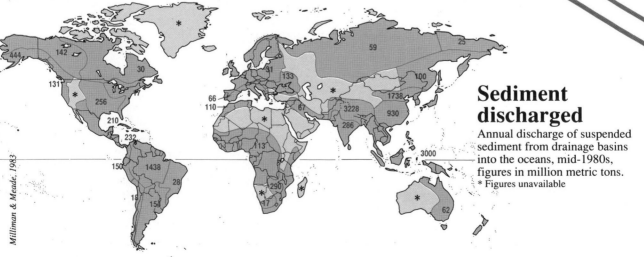

Milliman & Meade, 1983

Sediment discharged

Annual discharge of suspended
sediment from drainage basins
into the oceans, mid-1980s,
figures in million metric tons.
* Figures unavailable

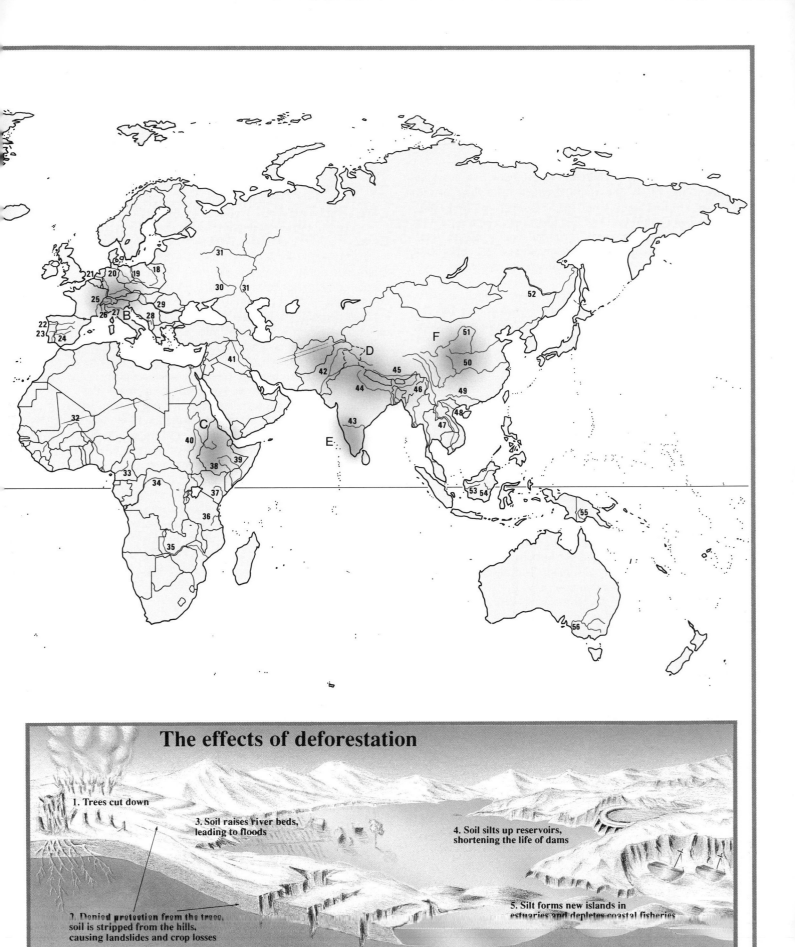

The effects of deforestation

1. Trees cut down

3. Soil raises river beds, leading to floods

4. Soil silts up reservoirs, shortening the life of dams

2. Denied protection from the trees, soil is stripped from the hills, causing landslides and crop losses

5. Silt forms new islands in estuaries and depletes coastal fisheries

Consequences of deforestation

As the topsoil goes, harvests fall. In just five years in the early 1980s rice yields dropped by a third in the most affected areas, while maize harvests fell by a fifth. The average hill farmer could grow only enough food to feed his family properly for eight months of the year. Sixty per cent of the children are stunted by hunger, and as the erosion continues malnutrition increases even more.

The Ganges and Brahmaputra rivers together carry more than 3 billion metric tons of soil to the Bay of Bengal every year, spreading it over 3 million square kilometers of the seabed. Whole new islands have surfaced in the bay, formed of Himalayan soil. Much of the sediment never reaches the sea, but coats the river beds, raising them by more than 15 centimeters a year. As the rains rush ever more rapidly off increasingly bare mountain sides into ever shallower river channels, there are more and more floods.

A Himalayan tragedy

The Himalayas are the youngest mountain range on earth, as well as the highest and longest, and are still inching upwards from the original impact of India with the rest of Asia. Their youth makes them particularly prone to erosion: even if they remained forested they would be losing soil, and the ferocity of the monsoons would cause floods downstream. But deforestation has escalated the problem immensely.

In the past, when the Himalayas were covered with trees, Bangladesh used to suffer from overwhelming floods about once every half century. By the 1970s one was occurring every four years, and the pace of destruction continued into the 1980s with a series of devastating disasters; one in 1988 covered 85 per cent of the country. India's flood-prone area increased from 25 million hectares in the late 1960s to 59 million hectares in the late 1980s. Meanwhile, regular water supplies are drying up. In Uttar Pradesh, the number of villages short of water has risen from 17,000 to 70,000 in 20 years; deforestation has been cited as a cause.

A Himalayan hope

There is, however, one bright spot in the Himalayas: the small mountain Kingdom of Bhutan, which proudly proclaims itself as the "last green patch" left in the mountains. Here the hills are still cloaked with magnificent forests and, as Jigme Singye Wangchuck, the King, says with pride: "The rivers run clear even in the rainy season."

The country remained isolated, by deliberate policy, for centuries, and only began to open up to the rest of the world in the late 1960s. When it was realized what was happening to its neighbors, the Government became determined not to develop in the same way, and adopted a strict conservationist policy. In 1974 it stipulated that 60 per cent of the country should remain under forest cover in perpetuity; no small target as about 35 per cent of its land is above the tree line. In 1979 it stopped private contractors cutting timber – national production fell by 87 per cent in two years. Felling by shifting cultivators was also banned, though less successfully. A tree planting program was launched to increase the forest area, while another law cut the population of goats – one of the great destroyers of the land – by four fifths and restricted them to stalls where they do no damage. Lastly, the Government has started giving every village a patch of forest to manage as its own responsibility.

This last provision is one of the most effective ways of preserving forests, for if villagers own the trees around them they usually manage them well, so as to preserve their benefits for the future. In Nepal, the hill people managed to conserve their trees and soil, despite rapid population growth, until the late 1950s, by voluntarily restricting felling and grazing in the forests, which had always belonged to them. When the Nepalese Government nationalized the forests, the people no longer felt responsible for them, and the cycle of destruction began.

Now, in an attempt to repair some of the damage, Nepal has been giving forest back to the people. Villagers have been paid to plant trees for fruit and fuel, and grasses to provide fodder for their animals, which were kept in stalls. The land regenerates and their needs are met. Even if land is left alone for only a year or two, the growth of grass and saplings stabilizes the soil, cutting erosion by 90 per cent. When the grass is cut, it feeds four times as many animals as could live by grazing the land. Some villages started doing this even before the Government launched its scheme; now it is hoped their initiative will spread. But worldwide at least 100 million hectares need to be planted with trees to save the world's watersheds.

Temperate Forests under Threat

Before the dawn of agriculture, some 10,000 years ago, trees covered some 6.2 billion hectares, about half of the earth's land surface. About a third of this original forest has been flattened, and most of the rest has been completely changed. Only about a quarter of it, 1.5 billion hectares, remains undisturbed.

Most attention is now focused on the destruction of the tropical rainforests. But many temperate countries felled their own forests centuries ago, and continue to destroy and degrade them. Half the wildwood that once covered 80 per cent of Britain was cleared by the time the Anglo-Saxons took over England, and half of what was left vanished before the Normans came. Only 1.5 per cent of it now remains, and much of that is threatened.

Four fifths of the forests that originally covered Europe had been cut down by the late Middle Ages, and now only fragments are left: Poland's Bialoweiza National Park contains the last major primeval forest in Central Europe. And within a century of independence, three quarters of United States forests had been felled; less than 5 per cent of the original woodlands now remain.

Nowadays, temperate forest is usually replanted, rather than completely cleared. But the new plantations bear little relation to the rich, ecologically-diverse, "old growth" forests that are cut down. They usually consist of regimented stands of just one or two species, planted, in the words of one Canadian forester, as "thick as the bristles on a bear's back."

Temperate rainforests worldwide are far rarer than their more famous tropical counterparts, covering only a fortieth of their area, and yet they are being cut down far faster. The biggest, which once stretched unbroken from Alaska to California, along the Northwest American Pacific coast may be the most productive ecosystem on earth, growing 10 times as much plant matter per hectare as tropical rainforests: it is home to the tallest trees on earth, soaring up to 100 meters high. Eighty per cent of the richest areas of the Tongass forest in the Alaska panhandle, the last remaining large expanse in the US, is scheduled for clear felling; more than 90 per cent of the forest in the lower 48 states has already been cut down. And environmentalists expect that the important parts of the Canadian section of the forest, in the province of British Columbia, will be gone within 15 years, though the Government and industry dispute the estimate. Only eight of Vancouver Island's 91 large watersheds have not yet been logged, and only one of these is protected. But, after a change of provincial government in late 1991, logging in the most controversial areas was suspended for 18 months while a special commission prepared a report.

It is much the same story in Scandinavia, often hailed as a model of modern forestry. Nearly 60 per cent of Sweden is wooded, and the area is actually increasing. But almost all of it is new, relatively sterile, plantations; only about 30,000 square kilometers of the original old growth remains, and national laws force landowners to cut it down. Two thirds of Finland is under forest, but only about 3 per cent of its old growth is left; logging began in one of the last major remnants in early 1992. Virtually no old growth remains in Norway at all. Hundreds of species are in danger including lynx, bear, wolverine and many kinds of birds – and so are the Lapps for their reindeer feed on the lichens in old growth forests, and the new plantations lack them.

Wood production in China doubled between 1976 and 1988, and its remaining harvestable production forests are expected to have been felled entirely within 10 years. Almost all the lowland forests of southern Chile have been cleared, and a Japanese logging firm has started cutting the magnificent southern beech forest of the Magellan Straits. Even the last great expanses of boreal forest, in Canada and Russia, are increasingly under threat. By the end of 1988, one third of the land surface of Alberta – 221,00 square kilometers – had been given over to logging. And new joint ventures threaten greatly to increase the 4 million hectares of Siberian forest felled each year: Japanese, Korean, European and American firms are active from Tomsk in western Siberia to the Chinese border.

Waldsterben – "forest death"

European and North American temperate forests are also under threat from *Waldsterben*, German for "forest death". By 1986, estimates suggested that almost 31 million hectares of coniferous and broadleaved forests were affected in 19 European countries outside the USSR. WWF, using the most recent information, has calculated that 50 million hectares are now affected. Another study estimates that it costs Europe and the European part of Russia at least $29 billion a year in lost timber, tourism, manufactured goods and other social benefits – and such losses threaten to persist for the next 100 years.

Coniferous forests along the spine of the Appalachian mountains, from Georgia to New England, are in decline from a similar dieback process: 60 per cent of the high-elevation red spruce in the states of New York, Vermont and New Hampshire have already perished. Eighty seven per cent of the Ponderosa and Jeffery pines of the San Bernadino Forest in California are sick, and the sugar maples of southeastern Canada are also affected.

There are alarming signs of damage to Third World temperate forests: 90 per cent of one area of China's Sichuan province that was once covered in pines is now bare.

Waldsterben was first spotted in West Germany in the early 1970s. At first, little notice was taken. But in 1982 the Government became alarmed when it found that 8

Temperate forests under threat

Pollution damage in the temperate forests of mainland Europe is well established, and extensive surveys of it have been carried out and published. In North America the situation is different. While it is generally accepted that forests are suffering damage on a scale similar to Europe, surveying is in its infancy, and few definitive data have yet been published.

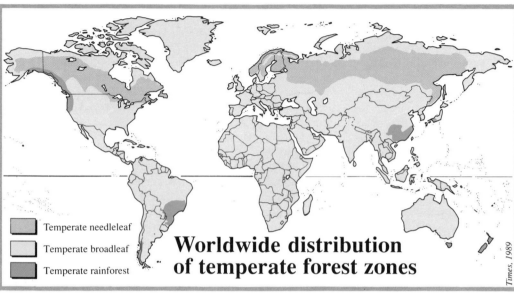

Temperate needleleaf

Temperate broadleaf

Temperate rainforest

Worldwide distribution of temperate forest zones

Times, 1989

Predominantly coniferous forest

Mixed forest

Predominantly hardwood forest

Sclerophyllous woodland

Figures show % of total number of trees in each country suffering from defoliation, 1990
(Numbers suffering from slight to severe defoliation, including dead trees).
* Coniferous trees only

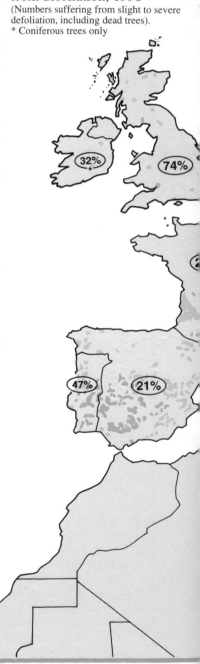

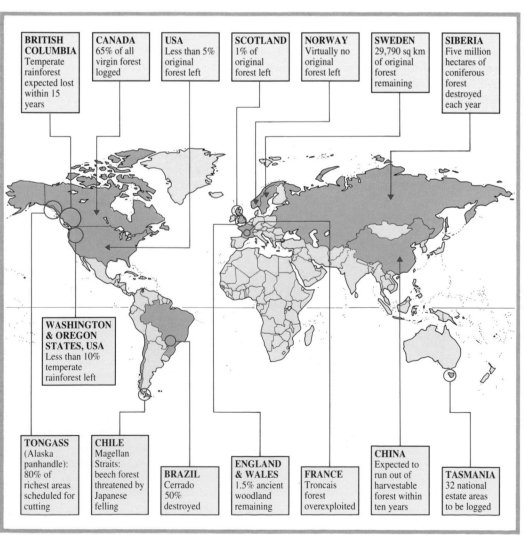

BRITISH COLUMBIA Temperate rainforest expected lost within 15 years

CANADA 65% of all virgin forest logged

USA Less than 5% original forest left

SCOTLAND 1% of original forest left

NORWAY Virtually no original forest left

SWEDEN 29,790 sq km of original forest remaining

SIBERIA Five million hectares of coniferous forest destroyed each year

WASHINGTON & OREGON STATES, USA Less than 10% temperate rainforest left

TONGASS (Alaska panhandle): 80% of richest areas scheduled for cutting

CHILE Magellan Straits: beech forest threatened by Japanese felling

BRAZIL Cerrado 50% destroyed

ENGLAND & WALES 1.5% ancient woodland remaining

FRANCE Troncais forest overexploited

CHINA Expected to run out of harvestable forest within ten years

TASMANIA 32 national estate areas to be logged

82

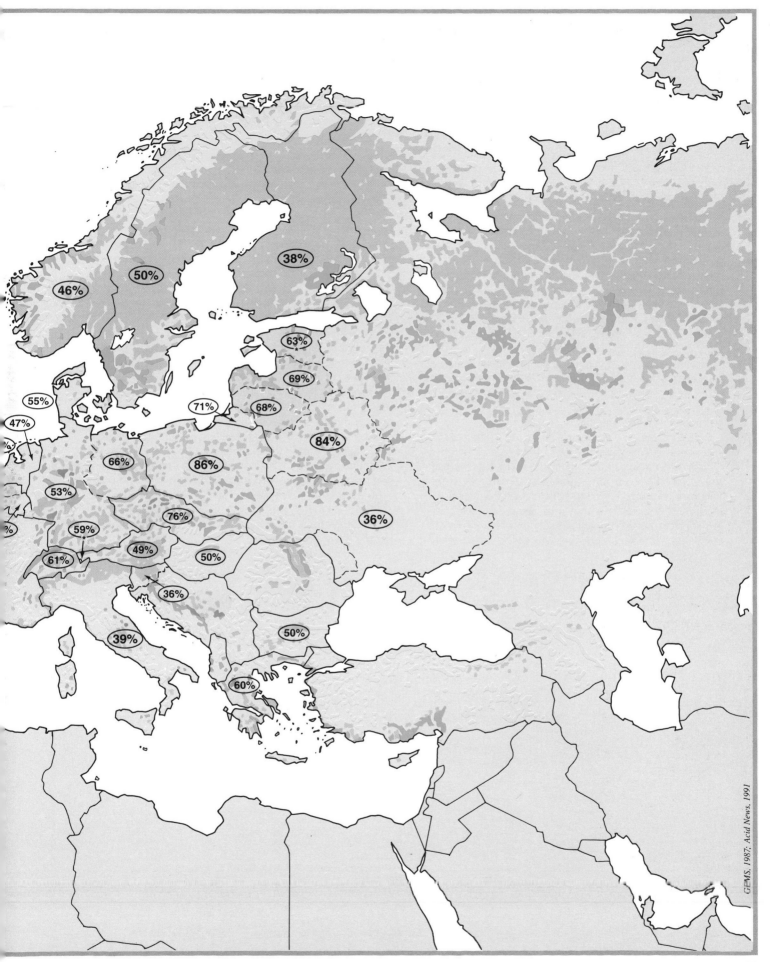

GEMS, 1987; Acid News, 1991

per cent of its forests were damaged. In 1983 that figure had increased to 34 per cent, and in 1984 to 50 per cent. By 1985, 52 per cent of the country's forest stands showed symptoms of decline.

Multiple causes suspected

Researchers in Europe and the US are trying to determine the exact cause of this massive dieback. They say it is one of the most challenging ecological puzzles of the century. *Waldsterben* is often accompanied by damage from such secondary stress factors as insects, fungi, frost, wind and snow. There is as yet no consensus on what triggers it and scientists are looking at a broad range of factors, including a noxious assortment of airborne pollutants in combination with disease and climatic stress. The perplexing array of symptoms and the speed with which entire stands are affected suggest that the causes are multiple – some 180 different causes have been suspected. Emerging evidence points to the combined effects of a number of pollutants working together.

Forest decline in Europe and North America has many similarities. Recent research has revealed that temperate forests (especially coniferous) have been under continuous stress for the past three decades. Increment cores – windows on a tree's growth rate – taken from thousands of conifers over the past few years show that their annual growth rings have become markedly narrower since 1970, compared with the two previous decades.

An ecological puzzle

There are other connections too. In both Europe and North America, the most severely damaged trees are consistently found on higher slopes facing toward the prevailing winds. Those shrouded in cloud or fog much of the time are particularly affected.

"There is an alphabet soup of pollutants up here," says Dr Robert Bruck, a forest pathologist at North Carolina's State University in Raleigh, surveying the crown of Mount Mitchell – the highest peak in eastern North America, now covered with dead and dying trees. "We have found elevated levels of lead, zinc and copper in the forest litter, and seven days out of 10 the mountain is bathed in cloud moisture that is often 100 times more acidic than normal rainwater." In addition, the southern Appalachians are swathed in heavy doses of ozone and other photochemical oxidants, known to damage trees and other vegetation.

Researchers are reasonably certain that this "alphabet soup" of pollutants plays a key role. But which ingredients cause which symptoms? This question has divided the scientists. At least six major schools of thought have emerged to explain the loss of forests, but no "smoking guns" have yet been found linking one pollutant to identifiable symptoms of damage.

Some scientists are skeptical that a clear cause and effect relationship can be established for *Waldsterben*. Dr Bruck is more optimistic. "We will solve this puzzle, but it is going to take time."

Acid Rain

Acid rain, first described in 1872 by an English chemist, remains one of the industrialized world's most intractable problems. What has taken humankind centuries to build, and nature millennia to evolve, is being degraded and destroyed in a matter of a few years.

Acid rain: causes and culprits

Acid deposition is caused mainly by sulfur and nitrogen emissions from the burning of fossil fuels such as coal and oil in power plants, industrial boilers and car engines. When these pollutants combine with water vapor, sunlight and oxygen in the atmosphere, they create a diluted "soup" of sulfuric and nitric acids. In some heavily industrialized regions, hydrogen chloride gases in the atmosphere produce hydrochloric acid, which can also be an ingredient of acid rain. It has been estimated that in northeastern United States 65 per cent of the acid rain is due to sulfuric acid, 30 per cent to nitric acid and 5 per cent to hydrochloric acid.

When this mixture is washed out of the atmosphere by rain, snow crystals, or in the form of dry particles, it increases the acidity of freshwater lakes and streams (and in some cases soils), thereby decreasing their pH values. The pH scale measures acidity and alkalinity. A solution with a pH of 7 is neutral, anything above this is alkaline, anything beneath is acidic. It is a logarithmic scale, so every full point is equivalent to a factor of 10. Thus a pH of 6 is 10 times more acidic than neutral, and a pH of 5 is 100 times more acidic than neutral.

The environmental toll

The magnificent historic buildings of Krakow, Poland, a world heritage site, are being destroyed by acidic smogs; ornate facades are disintegrating, walls and roofs weakening. Acid rain and other airborne pollutants are biting into the marble of Athens' monuments; experts say that more damage has been done to the Parthenon in the last 25 years than in the previous 2,400. The masonry of Cologne cathedral in Germany is being eaten away, many of Europe's 100,000 stained glass windows are fading, and libraries in the United States have had to be fitted with special anti-pollution equipment to preserve precious books.

Nearly a quarter of Sweden's 90,000 lakes are acidified to some extent; 4,000 of them so severely that no fish are thought to survive in them. Some 100,000 kilometers of its rivers and streams are also affected. In the southern half of Norway, four fifths of the lakes and streams are either technically dead or on the critical list; authorities say that fish have been destroyed in lakes covering a total of over 13,000 square kilometers.

Thousands of lakes in the eastern United States – including at least 10 per cent of all those in the Adirondack mountains – are too acid to support fish. More than 300 lakes in Canada's Ontario province are estimated to have pH values lower than 5 (the level at which most fish die); another 48,000 are threatened. Trout and salmon no longer reproduce in nine acidic rivers in Nova Scotia.

Researchers in Germany believe that acid rain is one of the causes of *Waldsterben* (tree death), which afflicts more than half the country's forests. Acid deposition is also thought to be one of the main agents of the decline of Switzerland's forests; 43 per cent of the conifers in its central alpine region are dead or dying.

More than 900,000 square kilometers of European Russia is thought to be affected by acid rain. The former East Germany has the highest per capita sulfur dioxide emissions in the world, and Poland, Czechoslovakia and Romania are among the most polluted countries on earth. China, the world's third largest emitter of sulfur dioxide, has growing problems in its southern provinces, and acid rain damage is spreading in other Third World countries like India, Nigeria, Colombia, Venezuela and Brazil.

A man-made problem

Although acid rain can be caused by natural processes such as volcanic eruptions, nature's own doses of sulfur and nitrogen oxides are dwarfed by industrial pollution. Every year, around 100 million metric tons of sulfur oxides are released across the globe, with Europe and North America accounting for 38 million metric tons. More than 90 per cent of the deposited sulfur is manmade. The countries belonging to the Organization for Economic Cooperation and Development (OECD) generate around 37 million metric tons of nitrogen oxides every year.

Rainfall is naturally acidic, with a pH of around 5.6. But man's pollution routinely increases acidity up to 100 times the natural level in industrialized countries. Data gathered by the European Monitoring and Evaluation Programme show that the average pH values in central Europe are 4.3 or below. According to the OECD, polluted areas in Scandinavia, Japan, central Europe and eastern North America have annual pH values that can fall as low as 3.5.

Every year, Norway experiences some rainfall that is as acidic as lemon juice. In the US, precipitation as acid as vinegar has fallen on Kane, Pennsylvania, and "rain" with a pH value almost equivalent to battery acid once fell on Wheeling, West Virginia.

Acid rain

The pH scale – measuring acidity

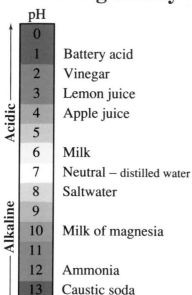

pH	
0	
1	Battery acid
2	Vinegar
3	Lemon juice
4	Apple juice
5	
6	Milk
7	Neutral – distilled water
8	Saltwater
9	
10	Milk of magnesia
11	
12	Ammonia
13	Caustic soda
14	

Acidic (pH 0–6), Alkaline (pH 8–14)

The lower the pH number the **greater** the acidity; the higher the pH number, the **lower** the acidity. Healthy, non-acidified lakes and rivers should have a pH of somewhere between 5.6 and 8.

The pH scale ranges from 0 to 14. Since it is logarithmic, a change of one unit represents a tenfold increase or decrease in pH. Hence, a solution at pH 2 is 10 times as acidic as one at pH 3, 100 times as acidic as pH 4, and 1,000 times as acidic as pH 5. Acid rain has a pH value of 5 or below. Some scientists, however, regard any rainfall below pH 5.6 as acidic. As can be seen from the pH maps, most of Europe and eastern North America are experiencing acid rain throughout the year.

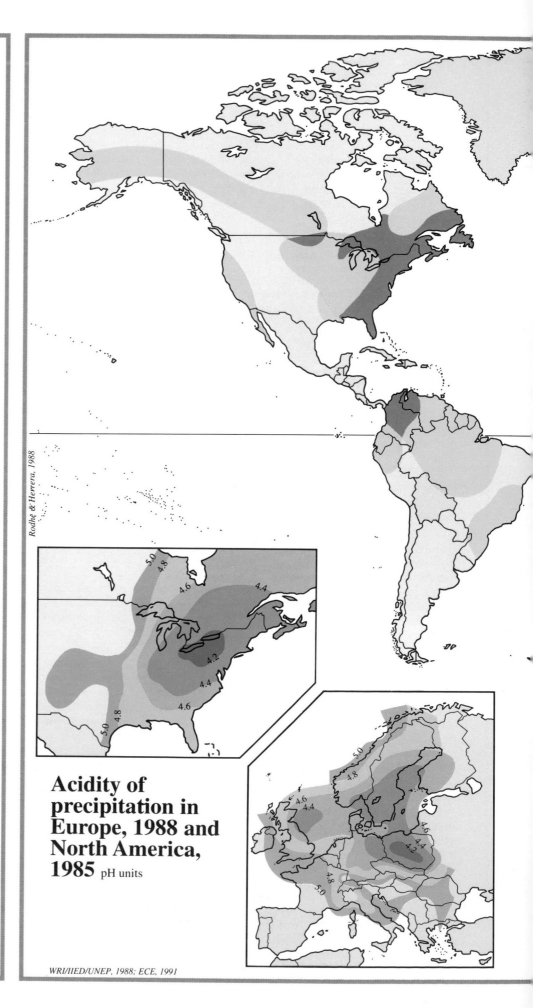

Rodhe & Herrera, 1988

Acidity of precipitation in Europe, 1988 and North America, 1985 pH units

WRI/IIED/UNEP, 1988; ECE, 1991

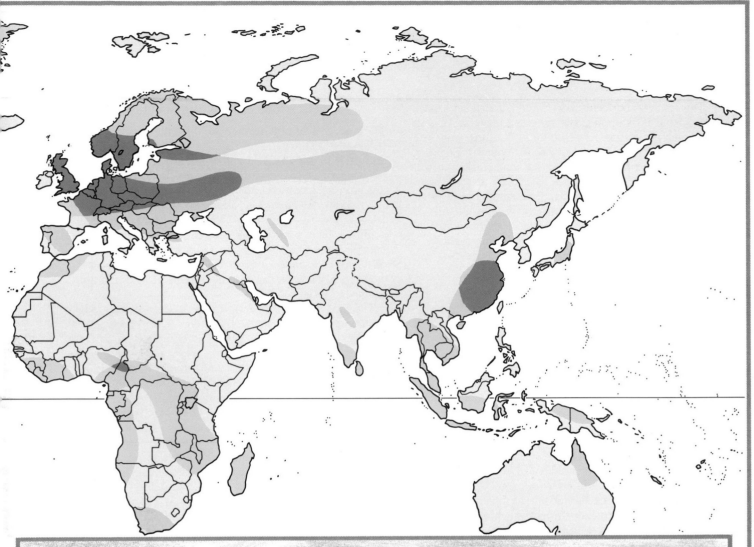

How acid rain is formed

Chemical transformation pan

$NO_2 \rightarrow HNO_3$
$SO_2 \rightarrow H_2SO_4$

Photo-oxidation
Acid pollutants

Hydrocarbons NO_x SO_2
Emissions to the atmosphere

Dry deposition

Particulates   Gases

**Wet deposition:
acid rain and snow**
H_2SO_4
HNO_3

Industry

Transport

Domestic

Electricity generation

The effects of acid rain would be much more widespread if nature had not provided many areas with their own protection. Alkaline soils, like those covering most of Midwest US and much of Southeast England, can resist acid fallout because the calcium in the soil neutralizes – or "buffers" – the acids; so can lakes which have beds of limestone or sandstone. On the other hand, where lakes and earth sit on thin glacial soils or thick slabs of granite – as in most of Scandinavia, Scotland and central Europe – this buffering capacity is greatly reduced. It is these sensitive areas that are hardest hit by acid deposition.

Hidden killers

Acid rain, in itself, kills few fish. Sensitive species like salmon, trout, minnows and arctic char succumb to the lethal water chemistry which acid rain fosters. Acid waters contain increased concentrations of toxic heavy metals such as mercury, aluminum, manganese, lead and zinc. It is aluminum – the most common metal found in soils – leached into lakes and streams that delivers the final blow. Aluminum toxicity peaks at around pH 5. It clogs fishes' gills, causing them to suffocate.

Research shows that acidity of forest soils has increased five- to 10-fold over the past 20 to 50 years across vast areas of Europe and eastern North America. Acid rain leaches crucial nutrients such as potassium, calcium and magnesium from soils, depriving trees and other vegetation of these life-supporting elements. If sufficient stocks of soil nutrients are not available, trees become more susceptible to climatic stresses like frost and winter damage, as well as damage from other pollutants.

Acid rain has been linked to the alarming decline of sugar maples in northeast US and eastern Canada. Surveys carried out in Quebec province showed that nearly 50 per cent of all sugar maple stands were affected by *Waldsterben*-like symptoms of dieback. If not reversed, Canada's maple syrup industry – worth $80 million a year – could collapse.

Conservation efforts: the acid test

The costs of acid rain damage are difficult to assess. But it has been estimated that damage to metals, buildings and paint in the OECD member countries costs around $20 billion a year, and that does not include costs of dead forests, acidified lakes and damaged crops.

By 1990, Sweden had added lime to some 6,000 lakes, covering more than 75 per cent of the total acidified lake area in the country, in an effort to neutralize the acid and bring them back to life – at a cost of nearly $15 million a year. This, however, is at best a stop-gap measure, mitigating the effects of acid rain but failing to tackle the causes.

The Convention on Long-Range Transboundary Air Pollution, promulgated by the UN Economic Commission for Europe, calls for reductions in sulfur dioxide emissions at source. It covers both West and East Europe and was ratified and brought into effect in 1983. At a meeting in Helsinki in 1985, a protocol mandating sulfur dioxide reductions of 30 per cent by 1993 (using 1980 as the base year) was opened for signature and was quickly endorsed by 21 countries. A second protocol freezes nitrogen oxide emissions.

The European Community (EC) has gone further. Large fossil fuel power plants will have to cut emissions of sulfur dioxide by about 40 per cent by 1998 and around 60 per cent by 2003. Austria, Germany and Switzerland have committed themselves to a two-thirds reduction by 1995. Nitrogen oxides will be cut by 30 per cent by 1998 in 12 Western European countries. And by 1993, all new cars sold in EC countries will have to be fitted with catalytic converters that reduce emissions of nitrogen oxides and other pollutants.

Industrialized countries are finally beginning to get to grips with the pollution that causes acid rain; even the US and the UK, which long resisted controls, are taking action to reduce emissions. But environmentalists warn that even greater cuts are needed; sulfur dioxide emissions, for example, need to be cut by 90 per cent. Until this happens, the scourge of acid rain will continue to spread.

Urban Air Pollution

In many of the world's big cities, air pollution is becoming a chronic problem, affecting human health, buildings, monuments and green belts. Increasingly, urban dwellers suffer from pollution-induced ailments: eyes water, sinuses clog up, noses run and throats get sore. Trees die from assaults of chemical pollutants generated by industries and vehicles. Monuments and buildings deteriorate as a virulent combination of pollutants eats away their facades.

The pollution cocktail

Urban air contains a frightening mix of pollutants, including sulfur and nitrogen dioxide (from power plants, industries and incinerators); reactive hydrocarbons (from petrochemical plants and refineries and vehicles); carbon monoxide (mostly from vehicles); heavy metals (from vehicles, industries and metal smelters); and organic compounds (generated mostly from the chemical industry).

These primary pollutants often combine in the atmosphere to form even more deadly secondary contaminants. Oxides of sulfur and nitrogen mix with sunlight, oxygen and moisture to form the diluted sulfuric and nitric acids that fall as acid rain. Ozone, and other so-called photochemical oxidants, are formed when hydrocarbons react with nitrogen oxides and oxygen in the presence of sunlight.

Ozone damages vegetation, crops and human health. And it is one of the major components of urban smog, the cocktail of pollutants – including hydrocarbons, nitrogen oxides and carbon monoxide – that plagues cities in developed and developing countries alike.

The hazards of smog

"Smog alerts" are commonplace in many cities with heavy traffic. In Mexico City, for example, smog levels exceeded World Health Organization (WHO) standards on all but 11 days of 1991, and reached double the WHO level on 192 days out of 365. Breathing its air – contaminated by 2.5 million vehicles and 130,000 factories – is said to be as damaging as smoking two packets of cigarettes a day, and half its children are born with enough lead in their blood to damage their development. Six times as many people die in Greater Athens on heavily polluted days as when the air is relatively clean. Air pollution levels in 102 cities in the former Soviet Union, affecting more than 50 million people, are often 10 times above the country's limits, and some 150 million US citizens breathe air officially designated as unhealthy.

Cities as diverse as Los Angeles, Beijing, London, Milan, Jakarta, Bangkok, Manila, Tehran and Lagos often seem to be locked inside gigantic bags of pollution.

Pollution in Eastern Europe

Eastern Europe suffers the worst air quality in the industrialized world. In cities like Krakow in Poland, Bratislava in Czechoslovakia, Miskolc in Hungary and Tulcea in Romania, residents are exposed to such high levels of sulfur and nitrogen oxides, heavy metals and soot, ash and dust that their health is endangered. Nearly 3 million Poles in Upper Silesia, in the southwest of the country, have to live with up to 1,000 metric tons of dust fallout per square kilometer – four times the maximum permitted level. According to a report prepared by the Institute of Environmental Engineering in Zabrze, residents of Katowice province in Silesia have a 15 per cent higher incidence of circulatory illness, a 47 per cent higher rate of respiratory ailments and 30 per cent more cancers than the rest of the Polish population. In one district, more than a third of the children and adolescents had symptoms of lead poisoning, normally found only in workers in heavily polluted factories.

Dirty air in Bratislava, capital of Slovakia, is blamed for increases in cancer by a third, infant mortality by two thirds, heart complaints by 40 per cent, and miscarriages by half, since 1970.

Between 35 and 40 per cent of Hungary's 10.3 million people live with officially "inadmissible" air and water pollution. Recent studies have shown that every 17th death and every 24th disability is due, directly or indirectly, to air pollution. Scientists at the National Institute of Public Health in Budapest calculate that air pollution will cost Hungary $374 million over the course of the next five years because of illness and premature death.

The Third World problems ignored

Some heavily industrialized cities in the Third World are even more polluted. In China, city dwellers are four to six times more likely than country people to die of lung cancer. Sulfur dioxide concentrations in the center of Beijing are more than four times WHO guidelines. Benxi, near the North Korean border, is China's – and perhaps the world's – most seriously polluted city. Half the city's 420 factories contribute to air pollution, pumping out 213,000 metric tons of smoke and dust and 87 million cubic meters of polluting gases every year. A few years ago, this city of a million people simply vanished from satellite photos, permanently hidden under a pall of pollution.

Few studies have been done of Third World cities, but there are plenty of danger signs. Pollution levels in Kuala Lumpur – mainly from vehicles – are two to three times as bad as in US cities. Bangkok's children lose some four points of IQ by the age of seven because of lead pollution from car exhausts, and air pollution costs

Urban air pollution

Europe
1 Glasgow (UK)
2 Birmingham (UK)
3 London (UK)
4 Dublin (Ireland)
5 Lisbon (Portugal)
6 Madrid (Spain)
7 Paris (France)
8 Brussels (Belgium)

(9-13 Germany):
9 Essen-Düsseldorf-Dortmund
10 Frankfurt
11 Berlin
12 Möbis
13 Leipzig

(14-17 Czechoslovakia):
14 Chomutov
15 Most
16 Prague
17 Bratislava

(18-22 Poland):
18 Warsaw
19 Bogatynia
20 Zabrze
21 Katowice
22 Krakow

23 Milan (Italy)
24 Zagreb (Yugoslavia)
25 Budapest (Hungary)
26 Ozd (Hungary)
27 Miskolc (Hungary)
28 Giurgiu (Romania)
29 Bucharest (Romania)
30 Ruse (Bulgaria)
31 Athens (Greece)

Africa
33 Cairo (Egypt)
34 Accra (Ghana)
35 Lagos (Nigeria)

USSR
36 Arkhangel'sk
37 Cherepovets
38 Moscow
39 Kirishi
40 St Petersburg
41 Kiev
42 Dneprodzerzhinsk, Dnepropetrovsk
43 Krivoy Rog
44 Zaporozh'ye
45 Mariupol
46 Donetsk
47 Volgograd
48 Sterlitamak
49 Salavat
50 Magnitogorsk
51 Nizhniy Tagil
52 Yekaterinburg
53 Chelyabinsk
54 Omsk
55 Novokuznetsk
56 Bratsk
57 Usol'ye-Sibirskoye
58 Angarsk
59 Noril'sk
60 Khabarovsk

Asia
61 Kuwait City (Kuwait)
62 Tehran (Iran)
63 Lahore (Pakistan)
64 Delhi (India)
65 Bombay (India)
66 Calcutta (India)

(67-72 China):
67 Xian
68 Beijing
69 Shenyang
70 Benxi
71 Guangzhou
72 Shanghai

73 Seoul (S. Korea)
74 Osaka (Japan)
75 Tokyo (Japan)
76 Hong Kong
77 Bangkok (Thailand)
78 Kuala Lumpur (Malaysia)
79 Jakarta (Indonesia)
80 Manila (Philippines)
81 Iligan (Philippines)

North & Central America
82 Sudbury (Canada)
83 Hamilton (Canada)

(84-88 United States):
84 Los Angeles
85 Phoenix
86 Denver
87 Chicago
88 New York

89 Mexico City (Mexico)

South America
90 Caracas (Venezuela)
91 Santiago (Chile)
92 São Paulo (Brazil)
93 Rio de Janeiro (Brazil)

• Major centers of urban pollution, 1990

These cities (listed left) all exceed international guidelines in **several** categories of air pollutants. The air in these cities is a clear health hazard, at least for part of the year. Figures are for location purposes only.

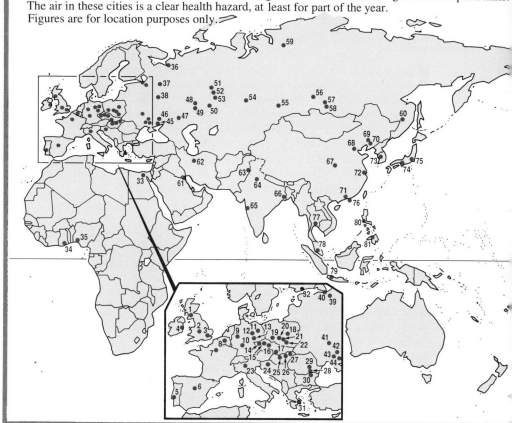

Pollution in selected cities

Annual averages. Micrograms per cubic meter.

- ● 1979 – 82
- ★ 1983 – 86
- ◆ 1987 – 90

Sulfur dioxide

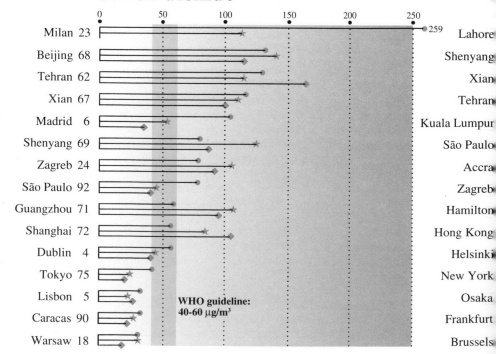

WHO guideline: 40-60 µg/m³

	Milan 23		Lahore
	Beijing 68		Shenyang
	Tehran 62		Xian
	Xian 67		Tehran
	Madrid 6		Kuala Lumpur
	Shenyang 69		São Paulo
	Zagreb 24		Accra
	São Paulo 92		Zagreb
	Guangzhou 71		Hamilton
	Shanghai 72		Hong Kong
	Dublin 4		Helsinki
	Tokyo 75		New York
	Lisbon 5		Osaka
	Caracas 90		Frankfurt
	Warsaw 18		Brussels

259 Lahore

World Bank, 1992

Urban air pollution levels and trends across country income groups

Suspended particulate matter. Micrograms per cubic meter.

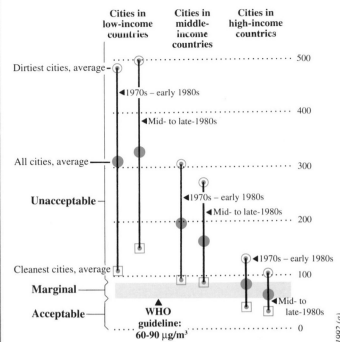

Data are for 20 cities in low-income countries, 15 in middle-income and 30 in high-income countries. "Cleanest" and "dirtiest" cities are the top and bottom 25 per cent of cities when ranked by air quality.

World Bank, 1992 (a)

e graphs show decreasing levels of sulfur dioxide and particulate matter in a number of cities. However, it should be pointed out that levels of other pollutants, as hydrocarbons, heavy metals and nitrogen dioxide, emitted by vehicles and industries, are likely to be high and still rising. Many polluted cities have not yet appraised by the UN-sponsored Global Environmental Monitoring System (GEMS). In many cases, governments have yet to permit international monitoring ghly polluted cities. Figures next to the city names are for location purposes only.

spended particulate matter

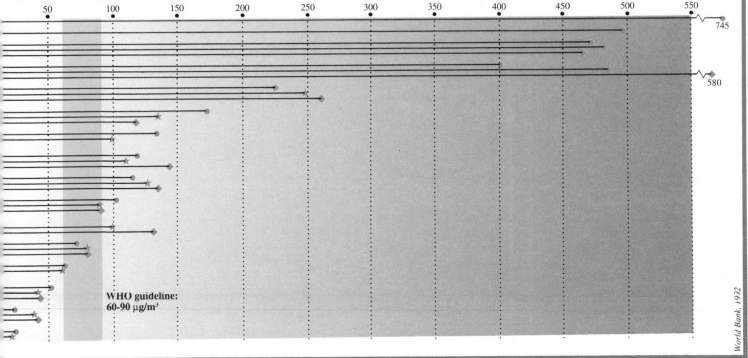

WHO guideline: 60-90 µg/m³

World Bank, 1992

at least $1 billion every year through medical bills, sick workers and damage to buildings. Even when the problems are known, they are often ignored: both Lagos and Ibadan in Nigeria have accumulated pollution data for years, but their air has got worse.

When the WHO and the United Nations Environment Programme evaluated pollution data for cities participating in the Global Environment Monitoring System for the period 1980-84, they discovered that over 600 million people live in urban areas where levels of sulfur dioxide exceed WHO's recommended limits. The same study also revealed that over a billion people are exposed to unhealthy levels of suspended particulate matter (soot, ash, dust, hydrocarbons).

Sulfur dioxide and particulate matter combined to create the great London smogs, one of which killed more than 4,000 people in December 1956. The disaster led to the Clean Air Acts, which reduced Britain's smoke emissions by 85 per cent and increased winter visibility in London from to 2.4 to 6.4 kilometers. But sulfur dioxide still exceeds WHO safety standards in the city.

A breath of fresh air?

Los Angeles introduced pioneering regulations to combat photochemical smog in the 1960s and 1970s which were copied worldwide; its requirement that all cars should be equipped with catalytic converters to reduce pollution is only now being taken up by the European Commission. But its clean-up program stagnated in the 1980s, while the number of vehicles, which cause 70 per cent of the problem, grew. In 1988, Southern Californians breathed air that exceeded health standards on 232 days of the year: on 75 of them the smog was so bad that schoolchildren and people suffering from respiratory and heart trouble had to be advised to stay indoors. Over 100,000 school age children in Los Angeles are routinely exposed to concentrations of ozone three times higher than the federal standard. The health of 98 per cent of the 12 million people of the area is affected in one way or another. Children brought up in the most polluted areas suffer reductions of 10-15 per cent in lung capacity for life.

In 1989, stringent new regulations were brought in to cut air pollution fivefold over the next 20 years. The Air Quality Management Plan introduced 150 regulations that will cost nearly $3 billion a year to implement. They will force firms to cut the numbers of car journeys their employees make to work – on pain of fines of $25,000 a day – by introducing car pools, four-day weeks or more working at home. They will make 40 per cent of Southern California's cars, 70 per cent of its trucks and all its buses abandon petrol for cleaner fuels by the year 2000. They are imposing a clean-up program on the local electricity utility, and they will clamp down on a host of other emissions, including those from lawnmowers, paint, drycleaners, bakeries and barbecues.

Southern California has particularly severe problems, but similar, if less stringent, measures are likely to be adopted in other US cities. There has been progress in the past; the US and Japan have heavily cut sulfur dioxide emissions and eliminated lead, which damages children's brains, from petrol, and Europe has followed. But it is becoming clear that radical action, particularly against vehicle pollution, is going to be needed. Already, driving restrictions are becoming more common. Traffic is banned from the center of Florence in daylight hours, and from central Rome during seven hours of the day. In Santiago a fifth of the cars are kept off the road, by rotation, each day. Strict regulations on industry – including heavy fines and temporary plant closures – are cleaning up Cubatão, near São Paulo in Brazil, which used to be called the "Valley of Death".

So far, such action as has been taken has come late even in developed countries. In most of the Third World, little or nothing has been done, and air pollution is becoming an ever-increasing menace to health, compounding the manifold problems of developing nations' rapidly expanding cities.

The Greenhouse Effect

Scientists are now broadly agreed that the greenhouse effect is bringing about the greatest and most rapid climatic change in the history of civilization. It will have enormous consequences for all life on earth.

Causes and history

Carbon dioxide and other gases in the atmosphere act like glass in a greenhouse, letting the sun's rays through but trapping some of the heat that would otherwise be radiated back into space. Natural levels of carbon dioxide make life possible: without them the average temperature of the planet would be 30°C colder.

In 1896 Svante Arrhenius, the great Swedish chemist, coined the phrase "the greenhouse effect" and predicted that the burning of fossil fuels would increase the amount of carbon dioxide in the atmosphere and lead to a warming of the world's climate.

In 1800 the gas was still at its pre-industrial level, about 280 parts per million (ppm). By the time Arrhenius issued his warning, pollution had already raised this level to about 300 ppm. It has now topped 350 ppm and is growing rapidly. Every year some 24 billion metric tons of carbon dioxide are released, and this is increasing by about 750 million metric tons a year. About four fifths comes from the burning of fossil fuels. The rest is from destroying vegetation, mainly the felling of forests: trees soak up the gas when alive, but release it when they are cut down and burned.

Carbon dioxide accounts for rather more than half of the warming. It is joined by several other "greenhouse gases". Chlorofluorocarbons (CFCs) are responsible for about another quarter and the remainder of the greenhouse effect is caused mainly by two other gases, methane and nitrous oxide. Both are given off by fossil fuels and the burning of vegetation. Nitrous oxide is also emitted by fertilizers and methane by rice paddies and ruminants, like cattle. Concentrations of these gases together, the best scientific estimates suggest, will reach the equivalent of 560 ppm of carbon dioxide, double the natural level, by the year 2030. Ozone near ground level makes an as yet unquantified contribution.

The 1980s were by far the hottest decade ever recorded, and the trend has continued into the 1990s: the 12 years between 1980 and 1991 included the eight warmest since measurements were first taken. Over the last century, the world has warmed by between 0.3°C and 0.6°C. No-one can yet be sure whether this is due to the greenhouse effect, or simply the result of natural variations in the climate, though it would be consistent with the increase in the polluting gases over the period. Scientists are sure, however, that the greenhouse effect will take place. In May 1990, 300 of the world's top experts, reporting for the official United Nations Intergovernmental Panel on Climatic Change (IPCC) said they were "certain that man-made emissions are substantially increasing the atmospheric concentrations of the main greenhouse gases." Their "best estimate" was that by the year 2020 the world will, on average, be 1.3°C warmer than now, rising to 3°C warmer by 2070. After much more research, a further IPCC report in the summer of 1992 confirmed these conclusions.

Effects and consequences

At first sight, this may not seem very much, but apparently small changes have dramatic effects. An increase of 2°C will produce temperatures last seen 125,000 years ago. A rise of 3°C would make the world hotter than it has been for the last 2 million years. Past changes of this size took thousands of years and species could adapt. The greenhouse effect threatens to produce them in decades. In the past, as conditions grew harsher, people could move to more congenial areas. On a crowded planet, divided by national frontiers, it will be hard to find anywhere to go. Sea levels will rise as the world gets warmer because the heat will melt ice and expand the water in the oceans. Over the next century, levels are expected to increase by a meter or more.

Sinking islands and deltas

Almost all the 1,196 islands of the Maldives are less than 3 meters high, and most people there live less than 2 meters above the waves. Six other coral atoll countries – the Cocos Islands, Tuvalu, Tokelau, Kiribati, the Marshall Islands and the Line Islands – face a similar crisis. In all, 300 Pacific atolls are expected to disappear, but will become uninhabitable long before as storms wash over them more frequently and freshwater supplies become salt.

Many more people are at peril from the flooding of deltas and other low-lying coastal areas. Some areas are already subsiding, making them doubly vulnerable to the rising sea. Half of Bangladesh is less than 4.5 meters above sea level. The land is already sinking, partly because some 120,000 wells have been drilled to extract drinking water. Studies suggest that up to 18 per cent of Bangladesh could be under water by the year 2050: by 2100 this could rise to 34 per cent and affect 35 per cent of the population.

The Nile Delta – twice as densely populated as Bangladesh – is sinking rapidly, because the Aswan High dam traps the silt that used to replenish the land. By 2050, up to 19 per cent of Egypt's cultivable land, home to 16 per cent of the population, could disappear.

Global warming

Likely effects of a possible doubling in atmospheric carbon dioxide concentration
(Model for December – January – February)

Temperature increase from average, ˚C:

- 12˚ and over
- 8.0 – 11.9˚
- 6.0 – 7.9˚
- 4.0 – 5.9˚
- 2.0 – 3.9˚
- 0 – 1.9˚

Areas likely to experience a change in soil moisture*:

- ✛ Increase of over 2cm
- + Increase of 1 – 2cm
- − Decrease of 1 – 2cm
- ▬ Decrease of over 2cm

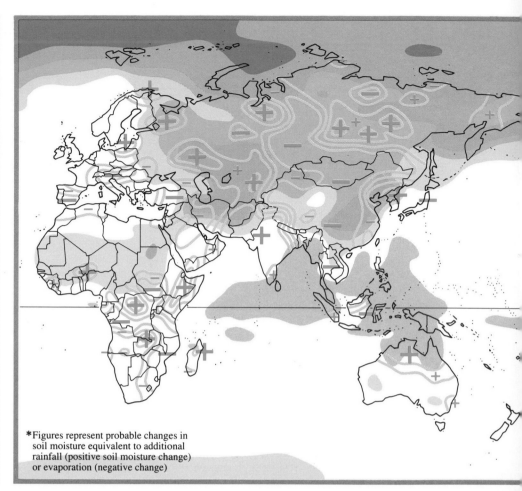

*Figures represent probable changes in soil moisture equivalent to additional rainfall (positive soil moisture change) or evaporation (negative change)

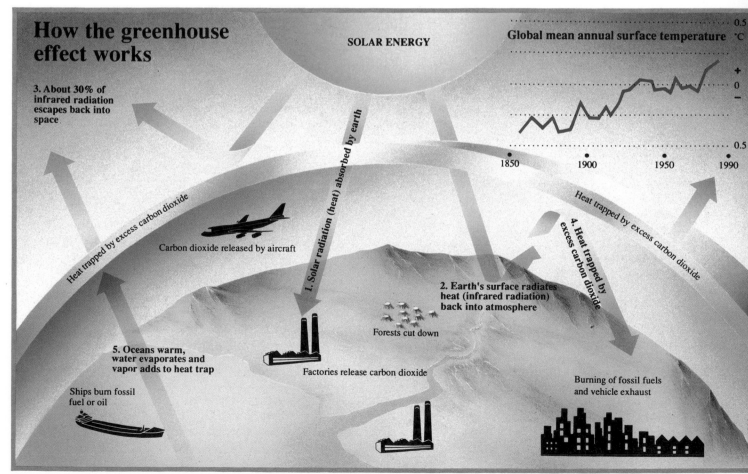

How the greenhouse effect works

SOLAR ENERGY

3. About 30% of infrared radiation escapes back into space

Global mean annual surface temperature ˚C

1850   1900   1950   1990

Heat trapped by excess carbon dioxide

Carbon dioxide released by aircraft

1. Solar radiation (heat) absorbed by earth

2. Earth's surface radiates heat (infrared radiation) back into atmosphere

Forests cut down

4. Heat trapped by excess carbon dioxide

Heat trapped by excess carbon dioxide

5. Oceans warm, water evaporates and vapor adds to heat trap

Ships burn fossil fuel or oil

Factories release carbon dioxide

Burning of fossil fuels and vehicle exhaust

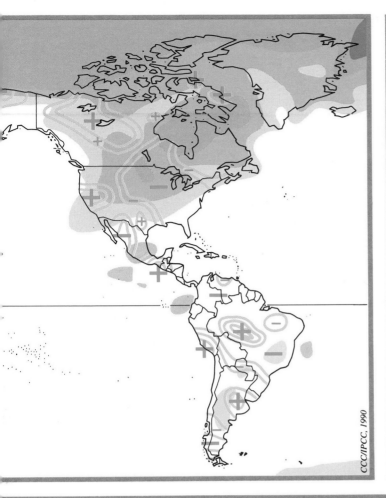

Effects of climate change

Differing predictions

Bolin, 1986

Present extent of
the corn belt ☐

Possible extent based
on 3°C temperature
increase and 8cm
precipitation increase:

Dryland corn ⬚

Irrigated corn ▨

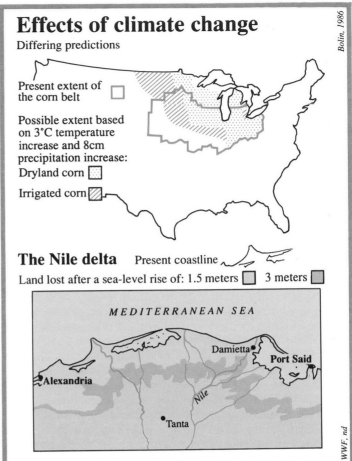

The Nile delta

Present coastline

Land lost after a sea-level rise of: 1.5 meters ☐ 3 meters ▨

MEDITERRANEAN SEA

Damietta

Port Said

Alexandria

Nile

Tanta

WWF, nd

CCC/IPCC, 1990

Contribution of greenhouse and ozone-depleting gases to global warming

1980-90

Other CFCs 7%

Nitrous
oxide 6%

CFCs
11 & 12
17%

Methane 15%

Carbon dioxide 55%

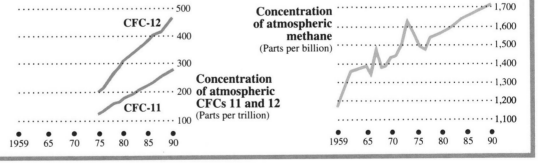

CFC-12 — 500
— 400
— 300
Concentration of atmospheric CFCs 11 and 12
(Parts per trillion)
— 200
CFC-11 — 100

1959 65 70 75 80 85 90

Concentration of atmospheric methane
(Parts per billion)

1,700
1,600
1,500
1,400
1,300
1,200
1,100

1959 65 70 75 80 85 90

IPCC, 1990; WRI/IIED/UNEP, 1992

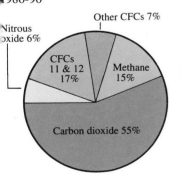

Concentration of atmospheric CO₂
(Parts per million)

— 360
— 350
— 340
— 330
— 320
— 310

1959 65 70 75 80 85 90

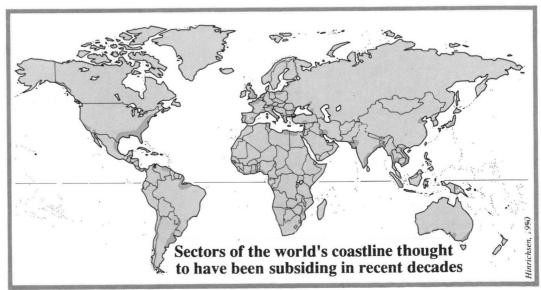

Sectors of the world's coastline thought to have been subsiding in recent decades

Hinrichsen, 1990

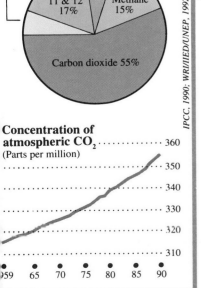

Croplands on the move

It is estimated that a 1-meter rise in sea level could make 200 million people homeless. But even this is likely to be overshadowed by the impact of global warming on harvests. For, as the world heats up, the rain will start to fall at different times and in different places, severely disrupting food production. It is far harder to work out the consequences than to estimate sea level rise, but huge computer programs that try to model the warmer world of around 2030 have come up with some rough predictions.

The American Midwest, which helps to feed 100 nations, may see its harvests cut by about a third. The United States, it is thought, will still be able to feed itself, but exports to the rest of the world could fall by up to 70 per cent. New land will open up in Canada as the weather warms, but the soils are too poor to make up the loss. The Ukraine may lose out to Russia.

Greece and Italy are expected to be badly hit and harvests may decline, if less seriously, in France and Germany. Britain, the Netherlands and Denmark should benefit, at least initially; harvests will increase greatly in Sweden, Norway and Finland, while improved grassland in Iceland may be able to carry two and a half times as many sheep as at present.

Developing countries will be hardest hit. Areas that are already arid – like Tunisia, Algeria, Morocco, Ethiopia, Somalia, Botswana, eastern Brazil and parts of Asia – will probably dry out even further. Some relatively wet regions, including Central America and Southeast Asia, are also likely to suffer. Third World farmers are desperately vulnerable, dependent both on the quality and the timing of the rains. In May 1992, the first ever global study of the impact of the greenhouse effect on harvests concluded that it would cause up to an extra 360 million people to go hungry by the year 2060.

These forecasts are based on snapshot predictions of the world relatively early in global warming. As the world heats up further, even countries which benefited in the early stages may suffer badly.

Wildlife in peril

Wildlife will certainly be seriously affected. With every rise of 1°C, plant and tree species will have to move about 90 kilometers polewards to survive; many will simply not be able to spread fast enough. The strain will be greatest in the higher latitudes because they will heat up fastest; winter temperatures in latitudes between 60° and 90° are expected to warm up more than twice as fast as the global average, and the Arctic tundra may disappear altogether. Changing rainfall patterns will compound the ecological disaster, while rises in sea levels will swamp coastal habitats. As trees and plants die out and habitats disappear, so will the animals that depend on them. And as the world continues to get warmer there will be nowhere for habitats to re-establish themselves.

Facing the climate crisis

Global warming from the greenhouse effect is now inevitable: the build-up of pollution that has already taken place ensures it. The best the world can do is to slow down the rate, in the hope that it will become manageable. Action has to start at once if it is to be effective. Delays will ensure that the greenhouse effect accelerates beyond all control.

Carbon dioxide emissions can best be controlled by conserving energy, halting deforestation and planting more trees. These measures require an unprecedented degree of cooperation, because very few nations, acting alone, can make much of an impact on global emissions. The world will have to cooperate both in cutting pollution in rich countries and in striking a new deal with the Third World to enable it to develop faster, but without relying too much on fossil fuels.

By April 1992, all the major countries of the Organization for Economic Cooperation and Development – except one – were ready to stabilize their emissions of carbon dioxide at 1990 levels by the year 2000. But the United States, which is responsible for a quarter of the world's emissions, refused to agree to the target. As a result, the world agreed to a watered-down treaty, signed at the Earth Summit in Rio de Janeiro in June 1992, which bound rich nations only to "aim" to do this. But the treaty did mark a first step towards tackling global warming, and environmentalists and many nations hope that it will be tightened up in the coming years.

The Ozone Layer

he earth is enveloped in a shroud of poison, on which all living things depend. Ozone, a form of oxygen with three atoms instead of the normal two, is highly toxic: less than one part per million of the blue-tinged gas in air is poisonous to humans. Near ground level, it is a pollutant which helps form photochemical smog and acid rain. But far overhead in the stratosphere, 15-50 kilometers up in the sky, it forms the lifesaving ozone layer.

Ozone is the only gas in the atmosphere which can screen out the lethal ultraviolet rays of the sun. It is spread so finely throughout the stratosphere that if it were collected altogether it would form a ring round the earth as thick as a shoe's sole. Yet if it were not for this fragile filter, ultraviolet radiation would kill all terrestrial life.

Ultraviolet radiation attack

A small amount of harmful radiation gets through doing enough damage to warn against weakening the ozone layer. It is the main cause of skin cancer, a fast increasing disease which already kills some 12,000 people a year in the United States alone. It suppresses the immune system, helping cancers to become established and grow, and increasing susceptibility to herpes and leishmaniasis. It is a major cause of cataracts, which blind at least 12 million people worldwide and dim the sight of at least another 18 million.

Damaging the ozone layer, even slightly, increases this toll on human health. Food supplies will also be hit. More than two thirds of the plant species – mainly crops – tested for their reaction to ultraviolet light have been found to be damaged by it. Fisheries and other marine life may be even more vulnerable. Ultraviolet radiation penetrates underwater, by as much as 20 meters in clear conditions. Plankton, microscopic organisms that drift on the surface of the sea, are particularly vulnerable. They play an all important part in the marine food chain and absorb over half the world's carbon dioxide emissions. So depleting them could both upset the balance of life in the sea and speed up global warming.

Danger from CFCs

The greatest danger to the ozone layer comes from chlorofluorocarbons (CFCs) – outstandingly useful and versatile chemicals that were long seen as miracle substances. Inert and immensely stable, neither flammable nor poisonous, easy to store and cheap to produce, they seemed designed for the modern world, and did much to shape it. They were first developed as coolants, and have since played an essential part in the spread of refrigeration and air conditioning. They were introduced as aerosol propellants during the Second World War to spray pesticides in the fight against malaria, and diversified to power a host of products from medicines to furniture polish to instant freshness and spray-on sex appeal. They were used to blow up foams used in furniture, insulation, carpets, fast-food containers, and in many other applications. And they have aided the computer revolution, because they can clean delicate circuitry without damaging its plastic mountings.

Unfortunately the stability that makes them so useful allows them to attack the ozone layer. They drift upwards, taking about eight years to reach the stratosphere; once there they can survive for about a century. As they break down, under intense ultraviolet radiation, they release chlorine, which reacts with ozone, converting it to ordinary oxygen. The chlorine acts as a catalyst: it does not itself undergo any permanent change. So every CFC molecule lives on to destroy thousands of molecules of ozone.

The "hole" in the ozone layer

CFCs cause the most dramatic example of the depletion of the ozone layer. An ozone "hole" opens over Antarctica every southern spring. As wide as the United States and as deep as Mount Everest is tall, its discovery surprised everyone – not least the scientists who found it.

In October 1982, Dr Joe Farman and a team from the British Antarctic Survey were taking routine measurements at the scientific base at Halley Bay when they found that much of the ozone overhead had apparently vanished. Ozone depletion on such a scale was totally unforeseen. Published ozone measurements from the American weather satellite, Nimbus 7, showed no trace of anything so dramatic and seemed to disprove their findings. Farman thought that his well-worn instruments might be faulty, so he and his colleagues went back with new ones at the same time of year in 1983 and 1984 – to find even less ozone. They reported their findings in the scientific magazine *Nature*, causing a sensation. NASA scientists re-examined the data from Nimbus 7. They found that the satellite had indeed discovered the same massive depletions, but that NASA's computers had been programmed to reject them as spurious, before anyone had a chance to see them. Fortunately the discarded data had been kept and, once retrieved, confirmed the observations made by Dr Farman's old instruments.

Over the next years the hole grew, and so did the scientific effort to discover what was causing it. On October 5, 1987, the total amount of ozone over Halley Bay was less than half of what it had been in the years before the hole opened; at its worst, between 15 and 20 kilometers up, 95 per cent had vanished. At the same

The ozone layer

The two world maps show average trends on ozone decline in the years from 1978 – 1991, versus latitude and longitude. Trends are in % per decade. The **top map** shows trends for December to March, the **bottom map** shows trends for May to August.

The **polar insets** below and right show concentration of ozone at the dates shown.

- 400 Dobson units and over (high concentration)
- 300 – 399
- 200 – 299
- 150 – 199
- Under 150 (low concentration)
- Polar night (Measurements not possible)

Evolution of the Antarctic ozone hole, 1979 – 87 (October)

1979

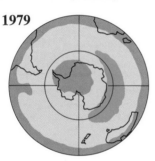

1985

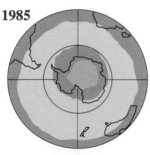

1987

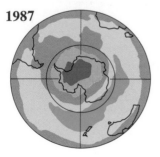

UNEP, 1989 (a); NASA, pers. comm.

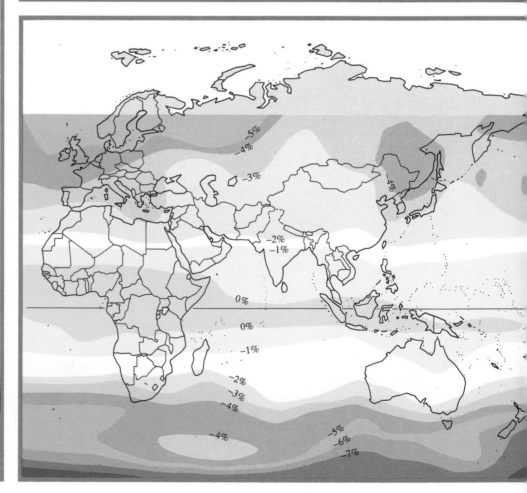

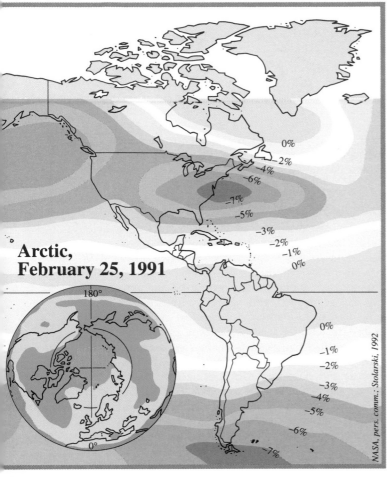

Arctic,
February 25, 1991

0%
−2%
−4%
−6%
−7%
−5%
−3%
−2%
−1%
0%

0%
−1%
−2%
−3%
−4%
−5%
−6%
−7%

NASA, pers. comm.; Stolarski, 1992

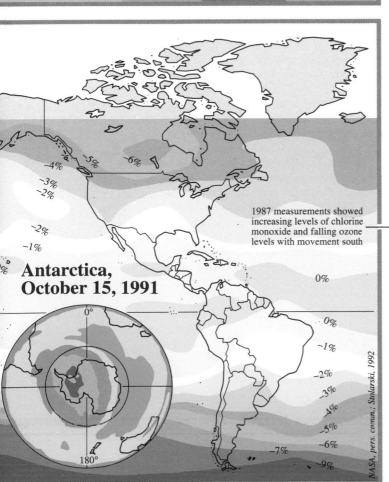

Antarctica,
October 15, 1991

−4%
−5%
−6%
−3%
−2%
−2%
−1%

0%

0%
−1%
−2%
−3%
−4%
−5%
−6%
−7%
−9%

1987 measurements showed increasing levels of chlorine monoxide and falling ozone levels with movement south

NASA, pers. comm.; Stolarski, 1992

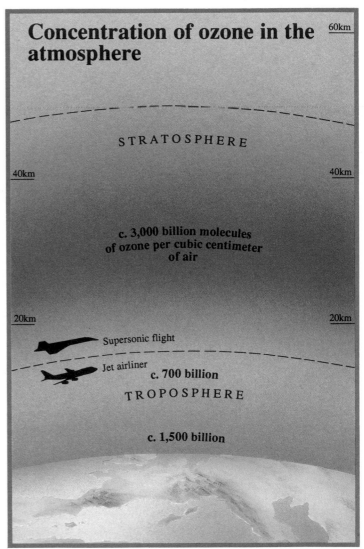

Concentration of ozone in the atmosphere

60km

STRATOSPHERE

40km   40km

c. 3,000 billion molecules
of ozone per cubic centimeter
of air

20km   20km

Supersonic flight

Jet airliner   c. 700 billion

TROPOSPHERE

c. 1,500 billion

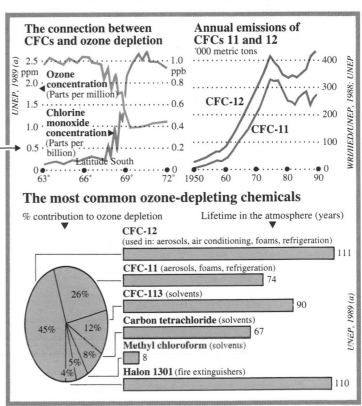

The connection between CFCs and ozone depletion

UNEP, 1989 (a)

2.5 ppm
2.0
1.5
1.0
0.5
0

1.0 ppb
0.8
0.6
0.4
0.2
0

Ozone concentration (Parts per million)

Chlorine monoxide concentration (Parts per billion)

Latitude South
63°   66°   69°   72°

Annual emissions of CFCs 11 and 12
'000 metric tons

400
300
200
100
0

CFC-12

CFC-11

1950 60 70 80 90

WRI/IIED/UNEP, 1988; UNEP

The most common ozone-depleting chemicals

% contribution to ozone depletion ▼   Lifetime in the atmosphere (years) ▼

CFC-12
(used in: aerosols, air conditioning, foams, refrigeration)   111

CFC-11 (aerosols, foams, refrigeration)   74

CFC-113 (solvents)   90

Carbon tetrachloride (solvents)   67

Methyl chloroform (solvents)   8

Halon 1301 (fire extinguishers)   110

45%   26%   12%   8%   5%   4%

UNEP, 1989 (a)

time, 150 scientists from 19 organizations gathered in Puenta Arenas in Chile, the southernmost city of the world, to mount an operation to determine what was causing the phenomenon. They flew two aircraft packed with instruments 175,000 kilometers into the area of the ozone hole, taking measurements which proved beyond reasonable doubt that CFCs were responsible.

In four of the five years to 1991, the ozone hole reached record levels: in 1991, it was 13 times the size it had been 10 years before. By the beginning of 1992, there were ominous reports of blindness in fish, rabbits and sheep in southern Chile – the populated area nearest to Antarctica. Doctors in the region were reporting greatly increased numbers of patients with allergies, skin and eye complaints.

Damage to the ozone layer has also been increasing over the Northern Hemisphere, where most of the world's people live. Between 1979 and 1990, the amount of ozone in springtime above the latitudes between 30 to 50 degrees N – covering the United States, most of Europe, northern China and Japan – had thinned 8 per cent, twice as much as scientists had expected. The United Nations Environment Programme (UNEP) calculates that every 1 per cent loss of ozone results in an extra 50,000 skin cancers and 100,000 cases of blindness from cataracts worldwide.

In February 1992, scientists warned that an ozone hole threatened to open in the Northern Hemisphere because substances released by a volcanic eruption in the Philippines were aggravating the effects of CFCs and other ozone-destroying chemicals. Fortunately, an unusually warm winter helped limit the damage to a 10-15 per cent depletion, but Dr Farman warned that the ozone layer over the most populated areas is likely to shrink by 20-30 per cent on a regular basis by the year 2000.

All this scientific evidence has fuelled a growing international determination to phase out the use of ozone-depleting chemicals. UNEP has been working on the issue since 1975, shortly after two Californian scientists, Professor Sherwood Rowland and Mario Molina, first suggested that CFCs damaged the ozone layer. But international progress was painfully slow.

Eventually, in 1985, UNEP secured agreement to the Vienna Convention for the Protection of the Ozone Layer, but this concentrates mainly on research and exchange of information, and lays down no specific measures to protect ozone.

Action on ozone

Following the discovery of the Antarctic ozone hole, UNEP had more success. After tough, but rapid negotiations, nations agreed the Montreal Protocol in September 1987, arranging to cut the use of damaging CFCs by half by mid-1996. It was a delicate agreement, only just achieved, with special provisions to make things easier for particular groups of countries, most notably the Third World. But it was an important first step. Less than two years later the first meeting of the parties to the Protocol unanimously agreed in principle to phase out damaging CFCs altogether "as soon as possible but not later than the year 2000". In late June 1990, the second meeting of parties, in London, formally agreed to stop using CFCs by the year 2000 and accepted similar phase-out dates for the other main ozone-depleting chemicals: halon and carbon tetrachloride by the year 2000, methyl chloroform by 2005. A special fund was set up to help developing countries, which had 10 years' grace, afford less polluting, but more expensive, technologies – the first of its kind ever established. Between 1988 and 1992, the worldwide consumption of CFCs fell by 40 per cent. And in February 1992 the European Community and the US announced that they would bring forward their deadline for phasing out CFCs and other ozone-depleting substances to the end of 1995.

It will be a long time before the ozone returns. US Environmental Protection Agency analyses suggest that, even if all ozone-depleting chemicals are phased out, it will take a century for conditions in the atmosphere to return even to what they were in 1986. But the world did move fast to agree to phase out the chemicals, once their dangers were demonstrated, and this offers a hopeful precedent for international action on other threats to the global environment.

Toxic and Hazardous Wastes

In Europe, North America and beyond, hazardous waste dumps have displaced entire communities, fouled the air, poisoned surface and ground water, contaminated soils, and adversely affected human health.

The United States generates roughly 240 million metric tons of hazardous and toxic wastes every year – 70 per cent of it from the chemical industry. The worldwide total is hard to ascertain. One estimate by the Organization for Economic Cooperation and Development (OECD) put it at 338 million metric tons; another covering just 19 countries was more than 100 million metric tons higher. The great discrepancies in estimates are due to differences over definitions, the lack of firm medical evidence as to what is toxic, and scientific disagreements over the levels at which various wastes become hazardous to health and the environment.

Lack of information

Many chemical compounds and their products have not been tested for their potential effects on human health or the environment. The task is a daunting one. There are over 70,000 chemicals in trade and each year another 1,000 come into general use. Half are classified as hazardous, or potentially so, but there is little firm information. The US National Academy of Sciences sampled 100 chemicals at random and ran extensive computer and literature surveys to find out if any relevant epidemiological or toxicological data was available. Only 10 per cent of all the pesticides analyzed, 2 per cent of cosmetic ingredients, 18 per cent of drugs and 5 per cent of food additives had enough data on them to enable complete health assessments to be made.

For years most hazardous wastes were dumped with impunity in landfills, pumped into settling ponds, or else discharged into the air, buried in fields, left by roadsides, or pumped untreated into rivers, lakes or the sea. The Minimata Bay disaster in Japan, where over 400 people died from eating fish poisoned with mercury discharged from a nearby factory, was only the first in a long list of horror stories involving improper discharge, storage, transport, or disposal of toxic wastes.

The results of such "wild dumping" have surfaced in practically every industrialized country over the past 20 years. Most legislation governing the generation and disposal of hazardous wastes dates from the mid-1970s.

The cost of clean-up

- In the Netherlands about 4,000 waste sites have been identified; at least 1,000 of them require immediate clean-up as they pose threats to public health and the environment. The Dutch are spending around $80 million a year for the next 15-20 years to try to combat the effects of a long history of illegal and improper disposal. The bill could eventually reach $6 billion.

- West Germany before reunification had the task of cleaning up at least 15,000 of its abandoned waste sites. The total cost could reach $30 billion and take 20 years to complete. And there are thought to be thousands of dangerous, unregulated toxic dumps in the eastern half of the country that will need immediate attention.

- French authorities have identified 450 abandoned waste dumps, 80 of which need immediate clean-up.

- A 1990 study by Friends of the Earth, in cooperation with *The Observer* newspaper, disclosed 4,800 toxic tips in the UK; 1,300 of them were officially identified in the early 1970s as posing a risk to groundwater supplies, but little was done.

- Poland generates roughly 30 million metric tons of hazardous wastes every year, two thirds of it is dumped in unregulated sites.

- The US holds the record for abandoned hazardous waste dumps. The US Office for Technology Assessment estimates that there could be 10,000 sites needing immediate action and that cleaning them up could take 50 years and cost $100 billion. The Environmental Protection Agency's more conservative estimate put the number of dangerous sites at a maximum of 2,500, costing around $23 billion to clean up.

Exporting poisons to the Third World

As regulations tightened in industrialized countries "midnight shipments" increased to some developing countries where legislation governing the disposal of toxic wastes is usually non-existent.

In 1987 and 1988, five boatloads of Italian toxic waste were dumped in Koko, a little Nigerian port. Ten thousand barrels of assorted poisons were piled in the yard of an unsuspecting farmer; half began leaking. Eventually, after Nigeria seized two other Italian ships, the Italian Government took away the wastes. It tried unsuccessfully to export them to the UK, and eventually had to take them back to Italy. But some northern firms still try to dump toxic wastes in the Third World.

Controlling transboundary shipments

By the late 1980s, some 100,000 shipments totalling over 2 million metric tons of hazardous waste were

Toxic wastes

The paucity of data on hazardous waste generation is not surprising. Regulations governing hazardous wastes did not come on the books until the early to late 1970s. Many countries, especially in the developing world, still have little or no information on the amounts of hazardous wastes generated, by whom and where it goes. Industrialized countries are still grappling with various definitions of hazardous waste, not to mention the levels at which chemical and other industrial wastes become hazardous to human health and the environment. So little is known about the health effects of hazardous wastes that setting exposure levels, in efforts to protect human health, often proves ineffective.

Sound management of hazardous wastes is a goal not yet achieved in many countries, even developed ones. And the lack of data on health and environmental effects continues to be a major stumbling block, hampering proper regulation.

Toxic and dangerous substances and materials that require priority consideration

- Arsenic and compounds
- Mercury and compounds
- Cadmium and compounds
- Thallium and compounds
- Beryllium and compounds
- Chromium (VI) compounds
- Lead and compounds
- Antimony and compounds
- Phenolic compounds
- Cyanide compounds
- Isocyanates
- Organohalogenated compounds, excluding inert polymeric materials
- Chlorinated solvents
- Organic solvents
- Biocides and phytopharmaceutical substances
- Tarry materials from refining and tar residues from distilling
- Pharmaceutical compounds
- Peroxides, chlorates, perchlorates and azides
- Ethers
- Chemical laboratory materials
- Asbestos
- Selenium and compounds
- Tellurium and compounds
- Polycyclic aromatic hydrocarbons
- Metal carbonyls
- Soluble copper compounds
- Acids and/or basic substances used in the surface treatment and finishing of metals
- Organochlorines (eg PCBs, DDT)

WHO, 1983

Available data for health hazard assessment

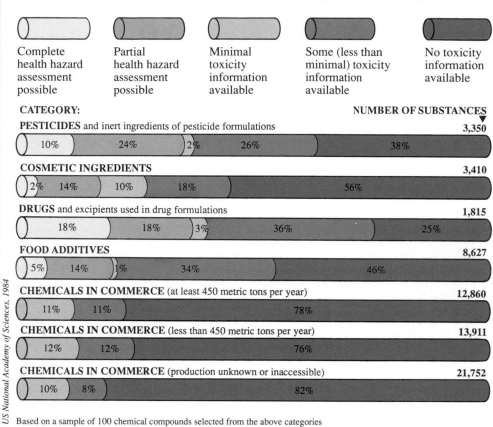

Complete health hazard assessment possible

Partial health hazard assessment possible

Minimal toxicity information available

Some (less than minimal) toxicity information available

No toxicity information available

CATEGORY:				NUMBER OF SUBSTANCES ▼
PESTICIDES and inert ingredients of pesticide formulations				3,350
10%	24%	2%	26%	38%
COSMETIC INGREDIENTS				3,410
2%	14%	10%	18%	56%
DRUGS and excipients used in drug formulations				1,815
18%	18%	3%	36%	25%
FOOD ADDITIVES				8,627
5%	14%	1%	34%	46%
CHEMICALS IN COMMERCE (at least 450 metric tons per year)				12,860
11%	11%			78%
CHEMICALS IN COMMERCE (less than 450 metric tons per year)				13,911
12%	12%			76%
CHEMICALS IN COMMERCE (production unknown or inaccessible)				21,752
10%	8%			82%

US National Academy of Sciences, 1984

Based on a sample of 100 chemical compounds selected from the above categories

Domestic hazardous waste management

Four selected countries, mid-1980s

Landfill | Re-use/recycle | Incineration | Underground | Physical/chemical treatment | Dumping at sea | Discharge to sewers and rivers

REP. OF KOREA
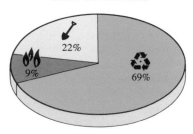
22% 9% 69%

SWEDEN
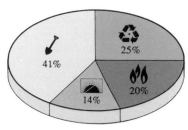
25% 41% 14% 20%

UNITED KINGDOM
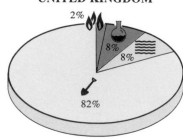
2% 8% 8% 82%

UNITED STATES
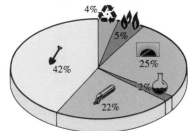
4% 5% 25% 42% 22% 2%

UNEP, 1989; US Congressional Budget Office, 1985

Hazardous waste production

Countries for which reliable statistics are available, late 1980s.
Thousand metric tons per year.

- 10,000 and over
- 1,000 – 9,999
- 100 – 999
- Under 100

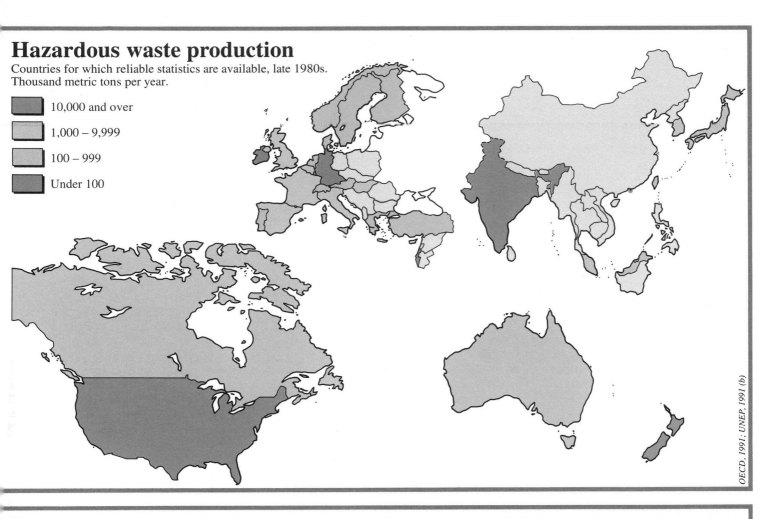

OECD, 1991; UNEP, 1991 (b)

Imports and exports of toxic waste '000 metric tons, late 1980s

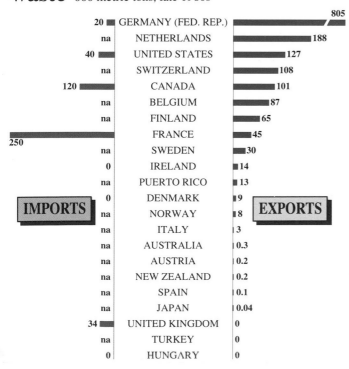

IMPORTS		EXPORTS
20	GERMANY (FED. REP.)	805
na	NETHERLANDS	188
40	UNITED STATES	127
na	SWITZERLAND	108
120	CANADA	101
na	BELGIUM	87
na	FINLAND	65
250	FRANCE	45
na	SWEDEN	30
0	IRELAND	14
na	PUERTO RICO	13
0	DENMARK	9
na	NORWAY	8
na	ITALY	3
na	AUSTRALIA	0.3
na	AUSTRIA	0.2
na	NEW ZEALAND	0.2
na	SPAIN	0.1
na	JAPAN	0.04
34	UNITED KINGDOM	0
na	TURKEY	0
0	HUNGARY	0

The following countries are those with reliable statistics for hazardous waste production but where figures for both exports and imports are not available: Greece, India, Israel, Republic of Korea, Luxembourg, Malaysia, Portugal, Taiwan.

Countries which have banned waste imports November 1990

PROTECTED FROM WASTE IMPORTS UNDER THE LOMÉ IV CONVENTION

Angola	Liberia
Antigua & Barbuda	Madagascar
Bahamas	Malawi
Barbados	Mali
Belize	Mauritania
Benin	Mauritius
Botswana	Mozambique
Burkina Faso	Niger
Burundi	Nigeria
Cameroon	Papua New Guinea
Cape Verde	Rwanda
Central African Republic	St Christopher & Nevis
Chad	St Lucia
Comoros	St Vincent & the Grenadines
Congo	São Tomé & Príncipe
Côte d'Ivoire	Senegal
Djibouti	Seychelles
Dominica	Sierra Leone
Dominican Republic	Solomon Islands
Equatorial Guinea	Somalia
Ethiopia	Sudan
Fiji	Surinam
Gabon	Swaziland
Gambia, The	Tanzania
Ghana	Togo
Grenada	Tonga
Guinea	Trinidad & Tobago
Guinea-Bissau	Tuvalu
Guyana	Uganda
Haiti	Vanuatu
Jamaica	Western Samoa
Kenya	Zaïre
Kiribati	Zambia
Lesotho	Zimbabwe

WASTE IMPORT BAN UNDER NATIONAL POLICY

Antarctica (world ban)
Egypt
Guatemala
Indonesia
Lebanon
Libya
Namibia
Peru
Philippines
Turkey
Venezuela

OECD, 1991; UNEP, 1991 (b); Vallette, 1990

being transported across national frontiers in Western Europe. The amount of waste shipped from Western to Eastern Europe is estimated to be between 200,000 and 300,000 metric tons a year (not including the 600,000 metric tons sent from former West to East Germany). Safety regulations in most of Western Europe and North America are now so stringent that toxic wastes must be followed from point of origin to final disposal. Elaborate tracking systems have been devised, such as West Germany's "trip ticket", a manifest which accompanies the waste to its final destination.

Fears that a significant proportion of waste shipments escape scrutiny by the proper authorities have resulted in action by the OECD, the European Community (EC) and the United Nations Environment Programme. This culminated in the Basel Convention on the Control of Transboundary Movements of Hazardous Wastes and their Disposal signed in Basel, Switzerland, in March 1989 by 34 countries and the EC. By May 1992, when the Convention entered into force, 52 countries had signed it, with 21 ratifications. The treaty requires countries that become party to it to produce as little hazardous waste as possible, to ensure that they have "adequate disposal facilities" of their own, and to cut imports and exports to the "minimum consistent with environmentally sound and efficient management". Parties are not allowed to trade in waste with countries that are not bound by the treaty. They agree not to export hazardous wastes to countries that have banned them, or when they have "reasons to believe that the wastes will not be handled in an environmentally sound manner". Importing countries have to be given full details of the waste, and it can only be exported when they have given their written consent to receiving it. In 1991, African countries adopted a tougher treaty – the Bamako Convention – which bans the import of hazardous wastes into Africa altogether. But they failed to get the backing of the rest of the world at the 1992 Earth Summit.

Improving treatment and disposal

Most countries still put most of their toxic wastes into holes in the ground. The UK, practically alone in the world, relies on "co-disposal", mixing hazardous wastes with domestic rubbish in hundreds of landfill sites, with widely differing standards, all over the country; experts told a parliamentary inquiry in 1989 that "diluting and dispersing" waste in this way was "a time bomb which has yet to explode". Most other countries are increasingly adopting the opposite approach, "concentrating and containing" their most dangerous wastes.

Denmark has had one of the world's most sophisticated systems for more than a decade: 21 special "transfer stations" collect all industrial toxic waste and send it to the Kommunekemi plant at Nyborg in the center of the country, where it is sorted for the most appropriate treatment. Relatively non-toxic substances, about a quarter of the total, are buried and the rest is incinerated or chemically treated to make it harmless. Treated waste helps fuel the incinerators, and the steam from the burners is used to provide half the town with its central heating. Similar systems are in operation in Bavaria, Sweden and Finland, and South Korea is developing one.

The best way to combat the hazards of toxic waste is not to produce it in the first place. Since 1975, the Minnesota Mining and Manufacturing Company has cut its waste production by half, saving almost $300 million. Two pesticide plants – one operated by Duphar in the Netherlands, the other by Du Pont in Colombia – have both cut their waste production by 95 per cent. France and the Netherlands both spend millions of dollars a year on researching ways of reducing waste.

Much toxic waste can be recycled or reused. Japan recycles more than half of all its wastes; another third is incinerated, leaving less than a fifth to be buried in the ground. But in most countries recycling, like all forms of waste minimization, is still far from reaching its potential. Denmark, despite its sophisticated system, recycles only 5 per cent of its industrial waste – less than a quarter of what could be achieved, even using existing technology. In all, the Worldwatch Institute estimates the amount of hazardous waste generated by industry could be cut by a third over a decade; "an ounce of prevention," it says, "is worth a pound of cure."

The Spread of Agrochemicals

Tremendous increases in the use of fertilizers and pesticides have helped bring about great rises in cereal, vegetable and fruit yields over the past four decades. But in many areas, crop yields are now dependent on heavy doses of these agrochemicals.

A growing dependence

World fertilizer use rose from 14 million metric tons in 1950 to 143 million metric tons by 1989. Per capita use quintupled from 5 kilograms to 26 kilograms between 1959 and 1984. Lester Brown, President of the Worldwatch Institute, points out that this made up for the fact that the amount of land used to grow grain actually fell by a third per person over this period. He adds: "As land becomes scarce, farmers rely more on the use of additional fertilizer to expand output, in effect substituting energy in the form of fertilizer for land in the production process."

By the late-1980s, the world used, on average, 91 kilograms of fertilizer on every hectare of cropland, a third more than 20 years before. This concealed great differences, ranging from 5 kilograms, or less, in several African countries to 3,750 kilograms for every hectare of Iceland's tiny quotient of arable land. Fertilizer use rose in Europe and Japan in the 1980s but fell in North America. It increased rapidly in Third World countries. Asia, in particular, has seen a dramatic increase; between 1977 and 1987, fertilizer use more than doubled and now averages 93 kilograms per hectare, the same level as in the United States. Between 1965 and 1976, developing nations, by using much more fertilizer, increased their crop yields by 55 per cent. However, a law of diminishing returns eventually sets in: each metric ton of fertilizer applied to the United States corn belt or Indonesian rice paddies now boosts yields by only half as much as 20 years ago.

Pesticide use multiplied 32-fold between 1950 and 1986, with global sales totalling nearly $16 billion in 1985. Again, use is dominated by developed countries – the US alone applied some 1.2 million metric tons on its crops every year – but the most rapid growth was in developing countries as their agriculture intensified. By 1989, the Third World accounted for one quarter of all pesticide use; while the developed world's share had slipped 17 per cent over the decade.

The environmental cost

If all fertilizer use ceased overnight, it is estimated that the world's harvests would be cut almost in half. But behind the impressive growth lurks mounting environmental damage from misuse and overuse.

Nitrates from fertilizers are seeping into groundwater aquifers. A 1984 survey carried out by the US Environmental Protection Agency showed that, out of 124,000 wells sampled, 24,000 had elevated levels of nitrates and 8,000 were polluted above natural health limits. At least 1.7 million Britons drink water that exceeds European Community legal limits. In Hungary, some 700 small settlements around the country, containing 300,000 people, have water piped in from neighboring areas or brought in by tanker truck, because their own wells are too polluted by nitrates to be drinkable.

Nitrate pollution of groundwater is less of a problem in developing countries, but it is building up in groundwater in South Korea, one of the world's heaviest users of fertilizers, and has been noted in the drinking water supplied to Buenos Aires, Argentina, and São Paulo, Brazil.

It often takes decades for nitrates to work their way down through the earth to contaminate groundwater: so the problem will worsen as pollution from recent heavy applications of fertilizer worldwide reaches water supplies. Meanwhile, fertilizer nutrients run off farmland to pollute rivers, lakes and the sea. They are the main cause of pollution in US lakes and were implicated in blooms of toxic algae in scores of British lakes and reservoirs at the beginning of the 1990s. They were also the main cause of massive algae plagues off Scandinavian coasts in 1988, which killed millions of fish, and in the Adriatic in 1989, which severely damaged the tourist trade.

Heavy pesticide use has an even greater effect, particularly in developing countries where regulations governing their use are weak or non-existent.

After three decades of intensive use, little is known about the effects of pesticides on human health or the environment. Data are crude and few scientific studies have been carried out to see how extensive the problem is. The World Health Organization (WHO) estimates that nearly 3 million people suffer from acute pesticide poisoning ever year: over 200,000 of them die. Most are Third World farmers and agricultural workers, who do not have the training and equipment to handle these toxic substances, and who may well not be able to read instructions and warnings on the labels. One Brazilian study found that almost two thirds of the farmers using pesticides suffered from acute poisoning at some time.

Pesticide residues in food and drinking water may take an even greater toll. The US National Research Council has estimated that up to 20,000 Americans may die of cancer each year from relatively low levels of pesticide in domestically-grown food. Nearly 300 UK water supplies are contaminated with pesticides above the levels allowed under European

Fertilizer use

Kilograms per hectare of cropland, 1987-89

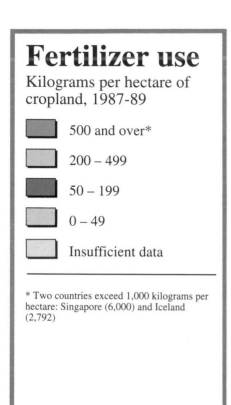

■	500 and over*
▨	200 – 499
▨	50 – 199
▨	0 – 49
▨	Insufficient data

* Two countries exceed 1,000 kilograms per hectare: Singapore (6,000) and Iceland (2,792)

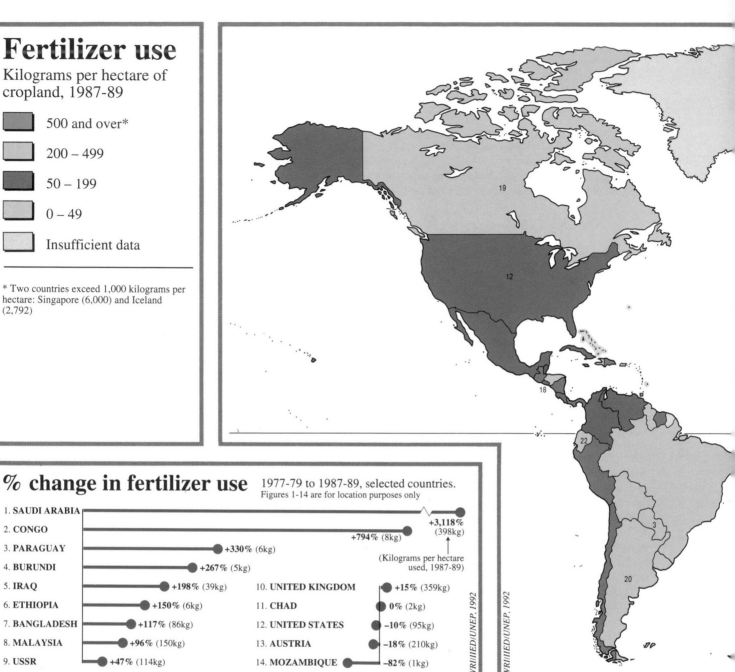

% change in fertilizer use

1977-79 to 1987-89, selected countries.
Figures 1-14 are for location purposes only

1. **SAUDI ARABIA** — **+3,118%** (398kg)
2. **CONGO** — **+794%** (8kg)

(Kilograms per hectare used, 1987-89)

3. **PARAGUAY** — **+330%** (6kg)
4. **BURUNDI** — **+267%** (5kg)
5. **IRAQ** — **+198%** (39kg)
6. **ETHIOPIA** — **+150%** (6kg)
7. **BANGLADESH** — **+117%** (86kg)
8. **MALAYSIA** — **+96%** (150kg)
9. **USSR** — **+47%** (114kg)

10. **UNITED KINGDOM** — **+15%** (359kg)
11. **CHAD** — **0%** (2kg)
12. **UNITED STATES** — **–10%** (95kg)
13. **AUSTRIA** — **–18%** (210kg)
14. **MOZAMBIQUE** — **–82%** (1kg)

WRI/IIED/UNEP, 1992

Growing world market in pesticides

By value (current prices). 1970 = 100

Costa, 1987

Herbicides
Insecticides
Fungicides
Other

1990: % breakdown by value

Herbicides **39%**
Insecticides **33%**
Fungicides **21%**
Other **7%**

% change in pesticide use

1975-77 to 1982-84, selected countries. Figures 4-23 are for location purposes only

15. **YEMEN (N.)** — **+397%** (1,614)

(Metric tons of active ingredient, 1982-84)

16. **MYANMAR** — **+311%** (15,300)
4. **BURUNDI** — **+164%** (59)
17. **REP. OF KOREA** — **+163%** (12,273)
18. **EL SALVADOR** — **+117%** (2,838)
19. **CANADA** — **+103%** (54,767)
20. **ARGENTINA** — **+92%** (14,313)
6. **ETHIOPIA** — **+66%** (993)
9. **USSR** — **+54%** (535,400)
10. **UK** — **+36%** (34,147)
21. **NEW ZEALAND** — **+9%** (1,793)

12. **UNITED STATES** — **–19%** (373,333)
22. **ECUADOR** — **–43%** (3,110)
23. **LIBERIA** — **–75%** (310)

WRI/IIED/UNEP, 1992

Number of species resistant to pesticides

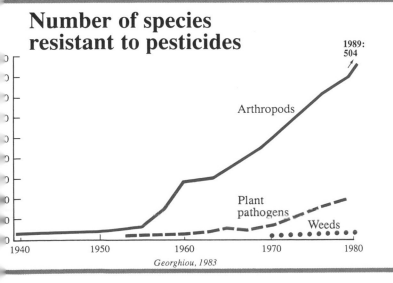

1989: 504

Arthropods

Plant pathogens

Weeds

1940　1950　1960　1970　1980

Georghiou, 1983

Alternatives to pesticides

Selected successful applications of integrated pest management (IPM) and biological control (BC)

● *Integrated pest management* relies on a combination of techniques to control the growth of pests, weeds and pathogens: rotation of crops and intercropping in the same field, careful spraying with pesticides more benign to the environment, and use of biological control.

Country/region	Crop	Effect
Brazil	Soybean	Pesticide use decreased 80-90% over seven years
Jiangsu province, China	Cotton	Pesticide use decreased 90%, pest control costs decreased 84%, yields increased
Orissa, India	Rice	Insecticide use cut by 30-50%

● *Biological control* relies on nature's own set of checks and balances. Natural predators are introduced to keep pests in check, or pest breeding is disrupted by releasing sterilized males.

Country/region	Crop	Effect
Equatorial Africa	Cassava	Parasitic wasp controlling mealy-bug pest on some 65 million hectares
Arkansas, USA	Rice/ Soybean	Commercially marketed, fungus-based "bioherbicide" controlling noxious weed
Costa Rica	Banana	Pesticide use stopped, natural enemies reinvaded to control banana pests

law, and as many as 50 million Americans may be drinking pesticide-polluted water. Meanwhile, improper use of pesticides in the US alone costs nearly $1 billion a year in losses to agriculture, livestock and wildlife.

Industrialized countries have prohibited the use of some pesticides as too dangerous to health or the environment; but they often continue to export them to Third World countries. In the early 1980s, one fifth of all US pesticide exports had been banned at home. Ironically, the banned pesticides then often return to home-country consumers as residues in imported foods.

Immunity and superpests

Pesticides are also becoming increasingly ineffective as more and more species develop immunity to them, particularly as a result of overuse. Fifty years ago, only seven insects were known to be resistant to pesticides, now there are more than 500. Twenty years ago virtually no plant pathogens resisted pesticides, now more than 100 appear to be immune.

Tobacco budworm, which destroys cotton, developed resistance to all known insecticides in northeastern Mexico in the 1960s: the cropped area fell from 280,000 hectares to just 400 hectares as a result. On Long Island, New York, desperate farmers sprayed banned pesticides over potato crops 10 times a season – under a special exemption – in a final effort to control a plague of Colorado beetles, which had grown immune to all legal chemicals. And the worldwide fight against malaria stalled in the late 1960s and early 1970s when overuse of DDT allowed mosquitoes to develop resistance to it. As pests become resistant, more and more chemical is used, increasing the damage to health and nature.

Before the advent of agricultural chemicals, about 30 per cent of the world's harvest was lost to pests and weeds. Now, after the vast expansion in pesticide and herbicide use over the last decades, the proportion of the crop lost is about the same.

A growing solution

Integrated Pest Management (IPM) offers new hope. It uses a variety of means to control pests, by employing natural predators, crop rotation and the use of pest-resistant varieties. Pesticides are used selectively rather than as the first, last and only resort. It aims not to eliminate pests, but to keep their numbers below those that cause serious damage.

IPM can therefore both reduce pesticide use and increase yields. Cotton yields rose in China's Jiangsu province after IPM cut pesticide use by 90 per cent, and cotton growers' profits went up in southern Texas after an 88 per cent fall in insecticide spraying. In 1987, a US Government study concluded that IPM had increased farmers' profits in 15 states by a total of $579 million a year. Apple growers using IPM in New York state were making $528 more per hectare – and Californian almond growers were getting $769 more a hectare – than those who stuck to traditional pesticide use.

Yet IPM is still little used. It is resisted by most of the agrochemical industry; the US Office of Technology Assessment has estimated that pesticide use would fall to one quarter of its present level if it were universally adopted in the United States.

But public opinion is turning against agricultural chemicals, at least in most developed countries. Consumers are increasingly buying food that has been grown organically, without any artificial chemicals at all, and they are often prepared to pay a substantial premium for it. Twenty years ago, the few farmers who tried to grow food without artificial chemicals were derided as cranks. By 1987, there were 400 organic farmers in the UK; by 1990 there were 1,000. Another 9,000 were spread through mainland Europe, mainly in West Germany and France. They remained a small minority, but at a time when European agriculture was in crisis, they – virtually alone – could face the future with confidence.

Natural and Man-Made Disasters

ankind has always been plagued by disasters – volcanic eruptions, hurricanes or typhoons, earthquakes and tidal waves. The explosion of Krakatau in 1883 – the volcano in the Sunda Strait between the Indonesian islands of Sumatra and Java – caused a tidal wave which wiped out 165 coastal villages in Southeast Asia, leaving more than 36,000 people dead. Another tidal wave, triggered by an earthquake, destroyed towns on India's Bengal coast, killing an estimated 200,000 people in 1876. An earthquake which ravaged Shensi province, China, killing 830,000 people in 1556 was perhaps the biggest disaster ever recorded. More recent disasters in China have also been catastrophic: an earthquake in 1976 in Tangshan left 242,000 dead and twice as many homeless. But the last few decades have witnessed the acceleration of human-induced disasters.

The environmental trigger

Environmental degradation plays a critical role in triggering some disasters, while making others worse. Increasingly, countries with severe deforestation, erosion, overcultivation and overgrazing of marginal land, are hit hardest by catastrophe.

In sub-Saharan Africa, for example, severe and prolonged drought affected more than 150 million people in 24 countries in 1984 and 1985. Ten million people had to abandon their land. More than a million died. And in 1992, as famine continued in the Horn of Africa, the south of the continent experienced its worst-ever drought, with crops failing by more than half in eight countries.

Widespread desertification combined to increase the drought. The United Nations Environment Programme (UNEP) reported in 1984 that the highlands of Ethiopia, for example, "have been so overfarmed, overgrazed, and deforested that efforts to scrape a bare living from this land threaten to destroy it permanently. The erosion resulting from overuse causes the Ethiopian highlands to lose a billion tons of topsoil each year." The land had already been pushed to the edge when the drought struck. Drought triggers a crisis, but does not cause it. Overcultivation and overgrazing weaken the land, allowing no margin when drought arrives.

Floods and cyclones

Deforestation of uplands not only contributes to droughts, it also promotes floods. In the Himalayas this process has reached catastrophic proportions. Nepal has lost half its forest since 1953. Every year the rains strip between 35 and 75 metric tons of soil from every hectare of the hillsides. This coats the river beds, raising them by 15 centimeters a year. Then the rains run even more precipitously into shallower watercourses, creating worse floods. In the past, when the Himalayas were covered with trees, overwhelming floods hit Bangladesh only about once every half century; by the 1970s they were happening every four years or so, and their frequency has been increasing since. In India, annual flood losses are more than 14 times what they were in the 1950s. In a typical year, 46,000 Indian villages may be flooded during the monsoons, affecting nearly 9 million hectares of land and inundating crops worth roughly $60 million.

Tropical cyclones increase the danger. In 1970, record floods in the Brahmaputra and Ganges delta met storm waters driven inland by a cyclone. Caught between the rising waters in Bangladesh were several million people. The death toll has never been accurately estimated, but ranges between 150,000 and 300,000. Worldwide, three times as many people were affected by flooding in the 1970s as in the 1960s, largely because of deforestation. In November 1991, 7,000 people were drowned in the Philippines by flooding directly attributed to illegal logging.

Disasters and development

The level of a country's economic development does much to determine suffering from disasters. A study carried out by the Swedish Red Cross, in cooperation with Earthscan, concluded: "When one compares the number of people killed per disaster against the income of the country involved, one finds a steep rise in mortality with decreasing income. There are over 3,000 deaths per disaster in low-income countries, and less than 500 per event in high-income countries."

The differences between rich and poor, in terms of the effects disasters have on populations, is dramatically illustrated by comparing Japan with Peru. Between 1960 and 1981, Japan suffered 43 earthquakes and other disasters, with 2,700 people killed; 63 deaths per disaster. During the same period, Peru suffered 31 disasters with 91,000 dead; 2,900 per disaster. In 1974 hurricanes hit both Honduras and Darwin, Australia: 4,000 died in Honduras, 49 in Darwin.

Within poor countries, it is often the poorest people who are worst affected. The reasons for this vary from country to country but, in general, the poor are forced by circumstances – population pressures, unequal land ownership, the search for cheap housing sites, the necessity to be close to places of potential work – to live on the worst, most hazard-prone land. Many are condemned to live out their lives near garbage dumps and polluting industries, on tidal flats and swamps, on river banks and near sewage outfalls. The poor of Rio de Janeiro, some 3 million of them, live in squalid shacks made from wood, canvas and

Droughts

Number of drought years experienced by each country, 1980 – 1991

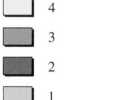

5 and over

4

3

2

1

No reported drought years

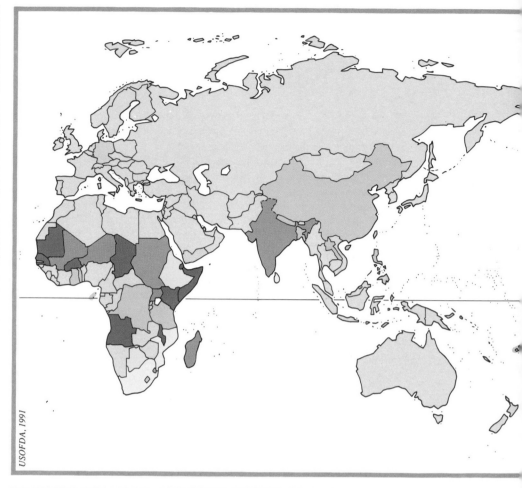

USOFDA, 1991

Floods

Number of people affected by major floods, 1980 – 1991

10 million and over

3.0 – 9.9 million

1.0 – 2.9 million

500,000 – 999,999

Under 500,000

No reported major floods

These figures are cumulative, which means that if two major floods affect the same area, the people affected are counted twice. In Bangladesh, for instance, the number of people affected exceeds the population of the country.

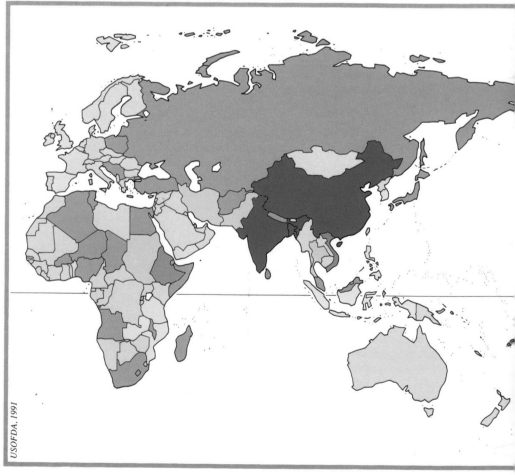

USOFDA, 1991

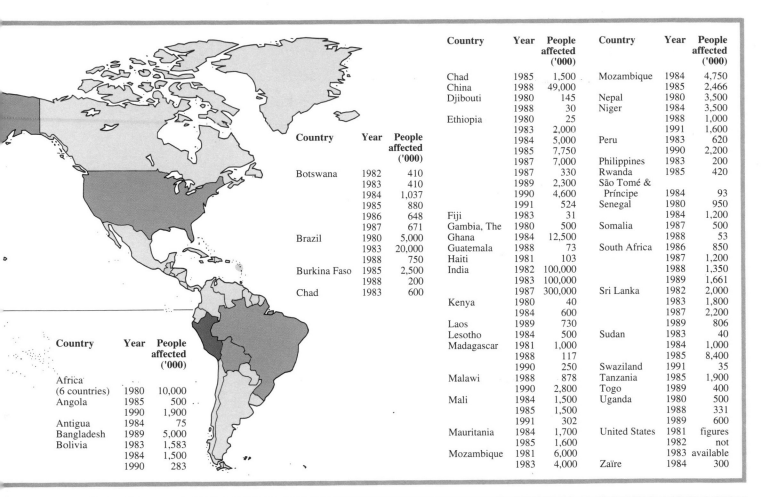

Country	Year	People affected ('000)
Botswana	1982	410
	1983	410
	1984	1,037
	1985	880
	1986	648
	1987	671
Brazil	1980	5,000
	1983	20,000
	1988	750
Burkina Faso	1985	2,500
	1988	200
Chad	1983	600

Country	Year	People affected ('000)
Africa (6 countries)	1980	10,000
Angola	1985	500
	1990	1,900
Antigua	1984	75
Bangladesh	1989	5,000
Bolivia	1983	1,583
	1984	1,500
	1990	283

Country	Year	People affected ('000)	Country	Year	People affected ('000)
Chad	1985	1,500	Mozambique	1984	4,750
China	1988	49,000		1985	2,466
Djibouti	1980	145	Nepal	1980	3,500
	1988	30	Niger	1984	3,500
Ethiopia	1980	25		1988	1,000
	1983	2,000		1991	1,600
	1984	5,000	Peru	1983	620
	1985	7,750		1990	2,200
	1987	7,000	Philippines	1983	200
	1987	330	Rwanda	1985	420
	1989	2,300	São Tomé & Príncipe	1984	93
	1990	4,600	Senegal	1980	950
	1991	524		1984	1,200
Fiji	1983	31	Somalia	1987	500
Gambia, The	1980	500		1988	53
Ghana	1984	12,500	South Africa	1986	850
Guatemala	1988	73		1987	1,200
Haiti	1981	103		1988	1,350
India	1982	100,000		1989	1,661
	1983	100,000	Sri Lanka	1982	2,000
	1987	300,000		1983	1,800
Kenya	1980	40		1987	2,200
	1984	600		1989	806
Laos	1989	730	Sudan	1983	40
Lesotho	1984	500		1984	1,000
Madagascar	1981	1,000		1985	8,400
	1988	117	Swaziland	1991	35
	1990	250	Tanzania	1985	1,900
Malawi	1988	878	Togo	1989	400
	1990	2,800	Uganda	1980	500
Mali	1984	1,500		1988	331
	1985	1,500		1989	600
	1991	302	United States	1981	figures
Mauritania	1984	1,700		1982	not
	1985	1,600		1983	available
Mozambique	1981	6,000	Zaïre	1984	300
	1983	4,000			

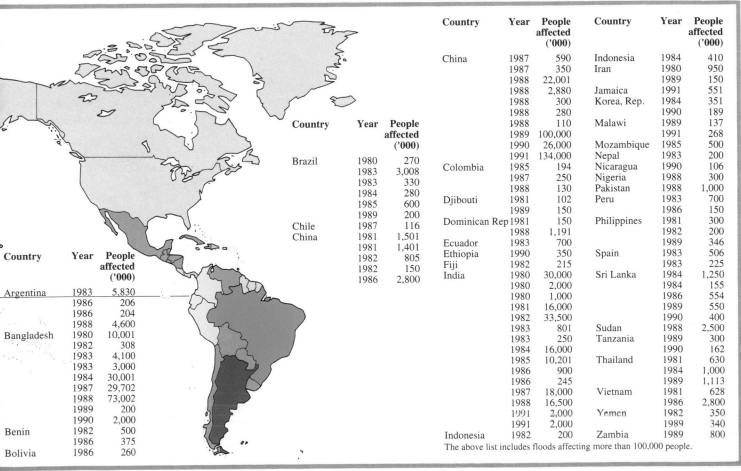

Country	Year	People affected ('000)
Brazil	1980	270
	1983	3,008
	1983	330
	1984	280
	1985	600
	1989	200
Chile	1987	116
China	1981	1,501
	1981	1,401
	1982	805
	1982	150
	1986	2,800

Country	Year	People affected ('000)
Argentina	1983	5,830
	1986	206
	1986	204
	1988	4,600
Bangladesh	1980	10,001
	1982	308
	1983	4,100
	1983	3,000
	1984	30,001
	1987	29,702
	1988	73,002
	1989	200
	1990	2,000
Benin	1982	500
	1986	375
Bolivia	1986	260

Country	Year	People affected ('000)	Country	Year	People affected ('000)
China	1987	590	Indonesia	1984	410
	1987	350	Iran	1980	950
	1988	22,001		1989	150
	1988	2,880	Jamaica	1991	551
	1988	300	Korea, Rep.	1984	351
	1988	280		1990	189
	1988	110	Malawi	1989	137
	1989	100,000		1991	268
	1990	26,000	Mozambique	1985	500
	1991	134,000	Nepal	1983	200
Colombia	1985	194	Nicaragua	1990	106
	1987	250	Nigeria	1988	300
	1988	130	Pakistan	1988	1,000
Djibouti	1981	102	Peru	1983	700
	1989	150		1986	150
Dominican Rep	1981	150	Philippines	1981	300
	1988	1,191		1982	200
Ecuador	1983	700		1989	346
Ethiopia	1990	350	Spain	1983	506
Fiji	1982	215		1983	225
India	1980	30,000	Sri Lanka	1984	1,250
	1980	2,000		1984	155
	1980	1,000		1986	554
	1981	16,000		1989	550
	1982	33,500		1990	400
	1983	801	Sudan	1988	2,500
	1983	250	Tanzania	1989	300
	1984	16,000		1990	162
	1985	10,201	Thailand	1981	630
	1986	900		1984	1,000
	1986	245		1989	1,113
	1987	18,000	Vietnam	1981	628
	1988	16,500		1986	2,800
	1991	2,000	Yemen	1982	350
	1991	2,000		1989	340
Indonesia	1982	200	Zambia	1989	800

The above list includes floods affecting more than 100,000 people.

111

Major disasters

Earthquakes, volcanic eruptions and major accidents, 1980 – 1991

☐	**Areas of highest earthquake risk**	
	Earthquakes (number of people affected):	
●	One million or over	
●	500,000 – 999,999	
●	100,000 – 499,999	
•	Under 100,000	
◆	**Volcanic eruptions**	
★	**Major accidents** (selected)	

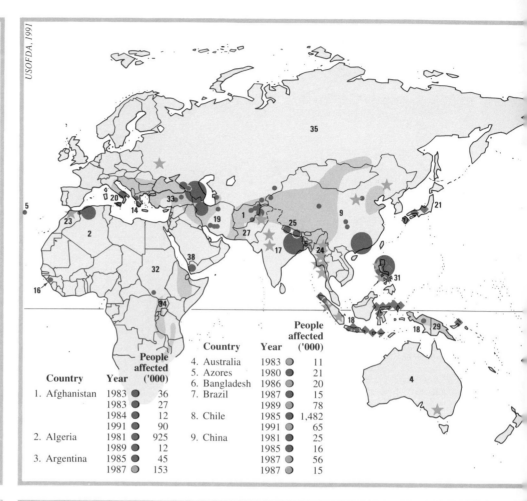

Country	Year	People affected ('000)
1. Afghanistan	1983	36
	1983	27
	1984	12
	1991	90
2. Algeria	1981	925
	1989	12
3. Argentina	1985	45
	1987	153

Country	Year	People affected ('000)
4. Australia	1983	11
5. Azores	1980	21
6. Bangladesh	1986	20
7. Brazil	1987	15
	1989	78
8. Chile	1985	1,482
	1991	65
9. China	1981	25
	1985	16
	1987	56
	1987	15

Storms

Number of people affected by major storms, 1980 – 1991

■	10 million and over
■	1.0 – 9.9 million
■	500,000 – 999,999
☐	250,000 – 499,999
■	Under 250,000
☐	No reported major storms

These figures are cumulative, which means that if two major storms affect the same area, the people affected are counted twice.

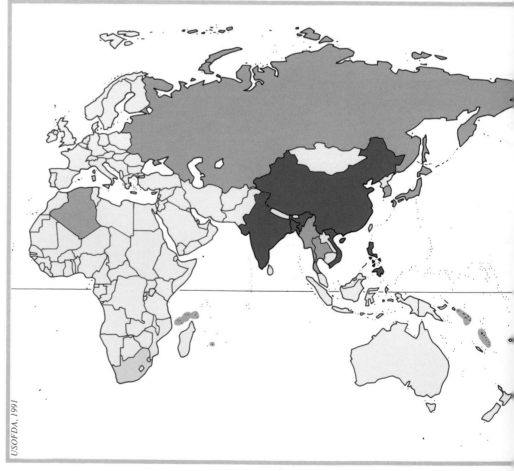

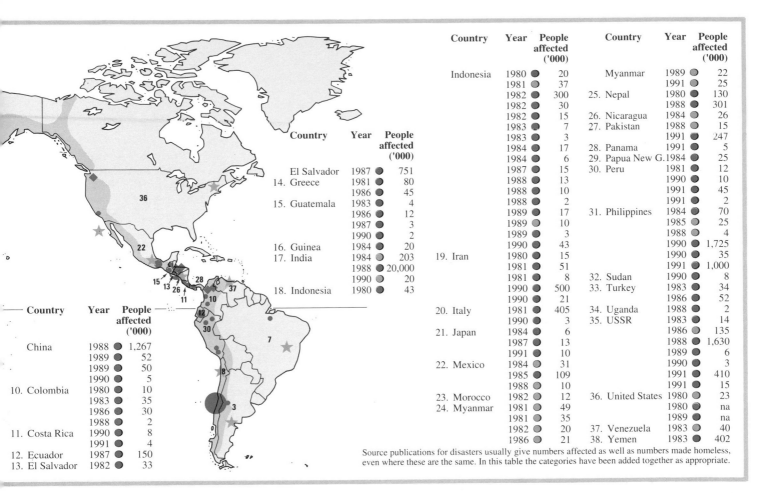

Country	Year	People affected ('000)
China	1988	1,267
	1989	52
	1989	50
	1990	5
10. Colombia	1980	10
	1983	35
	1986	30
	1988	2
11. Costa Rica	1990	8
	1991	4
12. Ecuador	1987	150
13. El Salvador	1982	33

Country	Year	People affected ('000)
El Salvador	1987	751
14. Greece	1981	80
	1986	45
15. Guatemala	1983	4
	1986	12
	1987	3
	1990	2
16. Guinea	1984	20
17. India	1984	203
	1988	20,000
	1990	20
18. Indonesia	1980	43

Country	Year	People affected ('000)	Country	Year	People affected ('000)
Indonesia	1980	20	Myanmar	1989	22
	1981	37		1991	25
	1982	300	25. Nepal	1980	130
	1982	30		1988	301
	1982	15	26. Nicaragua	1984	26
	1983	7	27. Pakistan	1988	15
	1983	3		1991	247
	1984	17	28. Panama	1991	5
	1984	6	29. Papua New G.	1984	25
	1987	15	30. Peru	1981	12
	1988	13		1990	10
	1988	10		1991	45
	1988	2		1991	2
	1989	17	31. Philippines	1984	70
	1989	10		1985	25
	1989	3		1988	4
	1990	43		1990	1,725
19. Iran	1980	15		1990	35
	1981	51		1991	1,000
	1981	8	32. Sudan	1990	8
	1990	500	33. Turkey	1983	34
	1990	21		1986	52
20. Italy	1981	405	34. Uganda	1988	2
	1990	3	35. USSR	1983	14
21. Japan	1984	6		1986	135
	1987	13		1988	1,630
	1991	10		1989	6
22. Mexico	1984	31		1990	3
	1985	109		1991	410
	1988	10		1991	15
23. Morocco	1982	12	36. United States	1980	23
24. Myanmar	1981	49		1980	na
	1981	35		1989	na
	1982	20	37. Venezuela	1983	40
	1986	21	38. Yemen	1983	402

Source publications for disasters usually give numbers affected as well as numbers made homeless, even where these are the same. In this table the categories have been added together as appropriate.

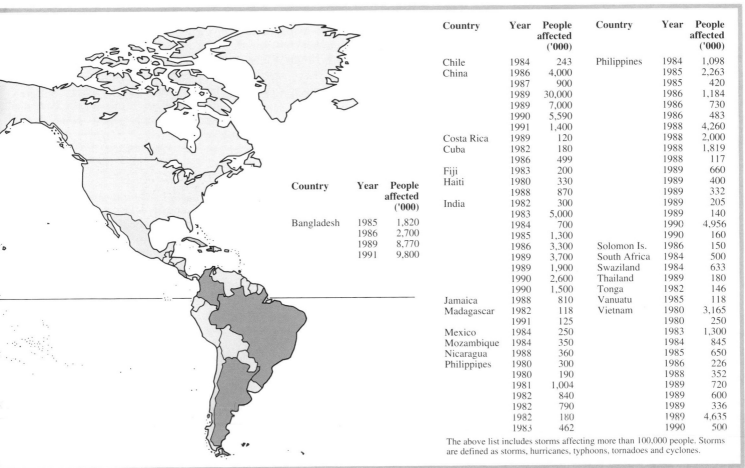

Country	Year	People affected ('000)
Bangladesh	1985	1,820
	1986	2,700
	1989	8,770
	1991	9,800

Country	Year	People affected ('000)	Country	Year	People affected ('000)
Chile	1984	243	Philippines	1984	1,098
China	1986	4,000		1985	2,263
	1987	900		1985	420
	1989	30,000		1986	1,184
	1989	7,000		1986	730
	1990	5,590		1986	483
	1991	1,400		1988	4,260
Costa Rica	1989	120		1988	2,000
Cuba	1982	180		1988	1,819
	1986	499		1988	117
Fiji	1983	200		1989	660
Haiti	1980	330		1989	400
	1988	870		1989	332
India	1982	300		1989	205
	1983	5,000		1989	140
	1984	700		1990	4,956
	1985	1,300		1990	160
	1986	3,300	Solomon Is.	1986	150
	1989	3,700	South Africa	1984	500
	1989	1,900	Swaziland	1984	633
	1990	2,600	Thailand	1989	180
	1990	1,500	Tonga	1982	146
Jamaica	1988	810	Vanuatu	1985	118
Madagascar	1982	118	Vietnam	1980	3,165
	1991	125		1980	250
Mexico	1984	250		1983	1,300
Mozambique	1984	350		1984	845
Nicaragua	1988	360		1985	650
Philippines	1980	300		1986	226
	1980	190		1988	352
	1981	1,004		1989	720
	1982	840		1989	600
	1982	790		1989	336
	1982	180		1989	4,635
	1983	462		1990	500

The above list includes storms affecting more than 100,000 people. Storms are defined as storms, hurricanes, typhoons, tornadoes and cyclones.

cardboard perched precariously on steep hillsides. When heavy rains come, entire communities can be buried alive in mud avalanches, and thousands are left homeless.

In Bangladesh, millions of the poor cling to tiny islands, called "chars", formed by the sediment washed out of the Himalayas and deposited in the vast delta of the Ganges and Brahmaputra rivers. When the flood waters churn down from the mountains, or storm-surges pound in from the sea, tens of thousands can be swept to their deaths. Often, the islands they lived on are completely washed away.

Industrial accidents and the poor

Industrial accidents also, as a rule, take a greater toll of life in slums and squatter settlements than in the community at large. When a liquefied petroleum plant blew up in Mexico City in 1984, it devastated the poor neighborhood of San Juanico: 452 people were killed, over 4,000 injured and more than 31,000 left homeless.

Shortly after midnight on December 2, 1984, a cloud of poisonous gas – methyl isocyanate – leaked from a pesticide factory in the city of Bhopal, India, operated by a subsidiary of the giant multinational, Union Carbide. The gas, heavier than air, hugged the ground, spreading into nearby shanty towns and on into the city. More than 2,500 people died. Years later, another 25,000 were still ill.

The casualties were so high partly because poor people had been allowed to settle near the factory. The shanty town of Jayaprakash Nagar, home to 4,000 people, began only five meters across a road from the plant. Since the 1960s, as migrants fled to the city, the slums grew northwards towards the site of the factory. They encroached illegally on unauthorized land, but the state government allowed the migrants to stay, provided electricity and water, and finally gave them title to their land. Legalizing settlements like this is enlightened policy, the best way to give squatters security and to improve the slums, but in Bhopal it legitimized a death trap. The situation is common throughout the Third World; everywhere slums press in around potentially dangerous factories. It is only a matter of time before another accident takes an equally horrendous toll.

The problems of prevention

When disasters happen, they receive worldwide attention and ignite global concern. Funds are set up and people give generously to help the victims. But there is far less emphasis on preventing them happening in the first place, or taking precautionary action to minimize their effects.

Prevention is far harder than assistance after the event, because the factors that turn such floods, hurricanes, droughts or accidents into disasters are often complex. People live in vulnerable areas for good reasons; it is often the only land they can find or the only place where they can get the facilities they need. Moving squatter settlements away from dangerous factories or land liable to flood, to safer areas with higher rents and beyond walking distance to work, would condemn their poor inhabitants to a permanent economic disaster – in order to protect them from a hazard that may never materialize.

Unsafe buildings or inadequate defenses may help turn earthquakes or floods into catastrophes, but these may just be symptoms of what is wrong. The cause may lie in such social factors as neglect by absentee landowners or corruption. The environmental factors which contribute to so many disasters usually have even more complicated roots. Attempts to ward off disaster must tackle these underlying causes of catastrophe, if they are to have much effect. And they must involve the vulnerable people, rather than be imposed on them.

Studies show that attempts to prevent disaster only work when local people are involved in planning what should be done; when they are mobilized in this way, local people are able to take cooperative action to mitigate catastrophe if it does strike.

"Most disaster problems in the Third World are unsolved development problems," state Anders Wijkman and Lloyd Timberlake in *Natural Disasters: Acts of God or Acts of Man?* "Disaster prevention and mitigation is thus primarily an aspect of development. In some regions, disaster prevention should be the major goal of development assistance from northern countries." By husbanding natural resources more effectively and managing the environment in sustainable ways, the effects of natural disasters can be lessened considerably.

Fossil Fuel Resources and Energy Conservation

Fossil fuels – oil, gas and coal – have made the modern world. They fired the industrial revolution, fuelled the boom after the Second World War, and continue to power most industry and transport.

Patterns of use

Oil, gas and coal account for 90 per cent of the world's commercial energy production; the rest is provided by nuclear power, hydro-electricity and other renewable resources. By 1990, oil provided 38.6 per cent of it, down from 47.4 per cent in 1973, before the first oil price shock. Coal accounted for 27.2 per cent and natural gas for 21.6 per cent, both increasing slightly since 1973.

The countries of the Organization for Economic Cooperation and Development (OECD) use rather less coal and more hydro-electric and nuclear power than the world average. The former Soviet Union consumes 20.48 per cent coal, 29.92 per cent oil and 42.21 per cent natural gas. Coal dominated China's commercial energy: it has the world's largest coal reserves, has become its biggest producer of the fuel, and plans to double production in the next decade. Other developing countries particularly depend on oil: more than half depend on it to meet over three quarters of all their demand for commercial energy.

Energy giants and pygmies

The industrialized countries, with less than a quarter of the world's people, burn about 70 per cent of its fossil fuel. The United States alone consumes about a quarter of the world's commercial energy; the former Soviet Union about a fifth. Canadians burn the most – in 1987 they used 9.15 metric tons of oil equivalent (MTOE) per capita a year followed by Norway with 8.9 MTOE and the US with 7.3 MTOE. Russians and Australians consume about 4.8 MTOE apiece, while the British, French and Danes all consume about 3.8 MTOE.

By contrast, developing countries as a whole use only about 0.5 MTOE of commercial energy a head. This global figure conceals vast disparities. Trinidad and Tobago used 5.2 MTOE per head in 1987, more than most industrial countries, while on average each Venezuelan consumed 2.4 MTOE. China, Zimbabwe, Tunisia and Peru used about the Third World average. But each Ethiopian, Nepali, and Burundian only consumed 0.2 MTOE, and Lesotho's per capita consumption was only 0.1 MTOE – less than one ninehundredth of the Canadian level. Most Third World countries get most of their energy from non-commercial sources like wood, charcoal and dung.

They are the main source of energy for 2.7 billion people, half the population of the world.

Diminishing reserves

Fossil fuels were created over geological time out of the remains of prehistoric life. Resources are for all practical purposes finite; they cannot be recreated in any shorter period. Nature took about a million years to produce the amount of fossil fuel that humanity now burns every 12 months.

Despite prophecies that fossil fuels would soon run out, new reserves were discovered faster during the 1970s and 1980s than the old ones were consumed. By 1989, there were enough known oil reserves to last for 41 years at current consumption rates (compared to 31 years in 1970) and enough natural gas for 60 years (compared to 38). There are hundreds of years' worth of coal.

Further reserves will be discovered and it will also become economic to extract less accessible deposits as scarcity drives prices up. But almost all the easily available reserves are being exploited. Oil, the most versatile of the fossil fuels, is the most unevenly distributed and in the shortest supply. Twenty countries control 95 per cent of proven reserves; more than half are in the Middle East. The US, which has burned four fifths of all the oil it has so far discovered, spent $250 billion looking for new domestic deposits between 1980 and 1984 – and found almost nothing; proven reserves dropped 5 per cent while it searched. The Exxon Corporation believes that this will become a worldwide phenomenon and that production will probably level off early next century, and then fall.

New discoveries of fossil fuels often present major problems. The USSR, for example, announced in the mid-1980s that it had discovered natural gas reserves big enough "to see us through the 21st century". Russia has vast coal resources, but most of the known, and likely, deposits are in Siberia; it will be difficult to exploit them and transport them to where they are needed. And, around the world, most of the discovered but unexploited oil and gas reserves contain relatively low quality fuel which gives off more pollution when burned.

Pollution and global warming

Fossil fuels are already by far the greatest source of industrial pollution. They were responsible for the great 1956 London smog, which killed thousands of people, and are the main cause of modern photochemical smogs that plague cities around the world. Pollution from burning oil in cars alone is thought to cause up to 30,000 deaths in the US each year.

Burning fossil fuels is responsible for acid rain and

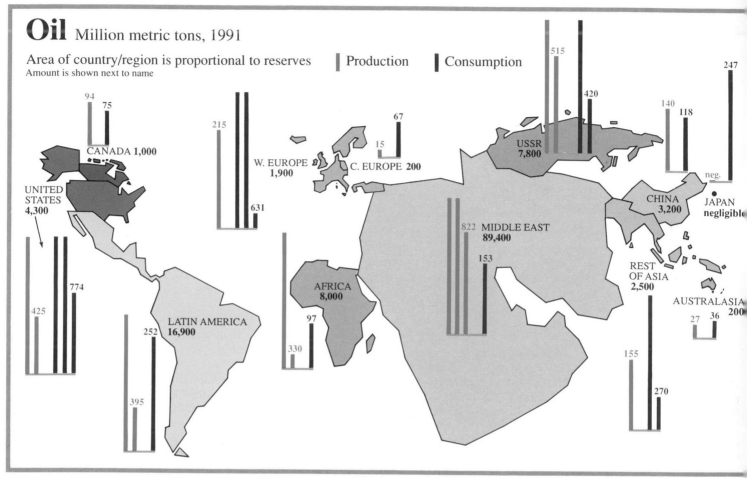

Oil Million metric tons, 1991

Area of country/region is proportional to reserves
Amount is shown next to name

▌Production ▐ Consumption

CANADA 1,000 — 94, 75

UNITED STATES 4,300 — 425, 774

LATIN AMERICA 16,900 — 395, 252

W. EUROPE 1,900 — 215, 631

C. EUROPE 200 — 15, 67

USSR 7,800 — 515, 420

CHINA 3,200 — 140, 118

JAPAN negligible — neg.

AFRICA 8,000 — 330, 97

MIDDLE EAST 89,400 — 822, 153

REST OF ASIA 2,500 — 155, 270

AUSTRALASIA 200 — 27, 36

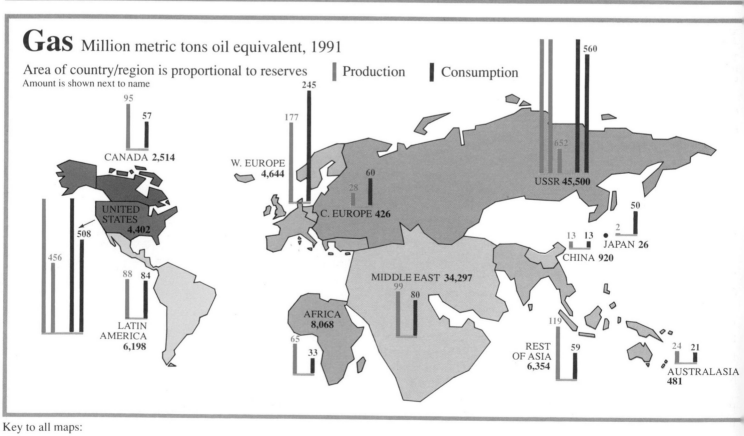

Gas Million metric tons oil equivalent, 1991

Area of country/region is proportional to reserves
Amount is shown next to name

▌Production ▐ Consumption

CANADA 2,514 — 95, 57

UNITED STATES 4,402 — 456, 508

LATIN AMERICA 6,198 — 88, 84

W. EUROPE 4,644 — 177, 245

C. EUROPE 426 — 28, 60

USSR 45,500 — 652, 560

CHINA 920 — 13, 13

JAPAN 26 — 2, 50

MIDDLE EAST 34,297 — 99, 80

AFRICA 8,068 — 65, 33

REST OF ASIA 6,354 — 119, 59

AUSTRALASIA 481 — 24, 21

Key to all maps:

▢ CANADA ▢ UNITED STATES ▢ LATIN AMERICA ▢ WESTERN EUROPE ▢ AFRICA ▢ MIDDLE EAST
▢ CENTRAL EUROPE ▢ USSR ▢ CHINA ▢ JAPAN ▢ REST OF ASIA ▢ AUSTRALASIA

Coal 1991

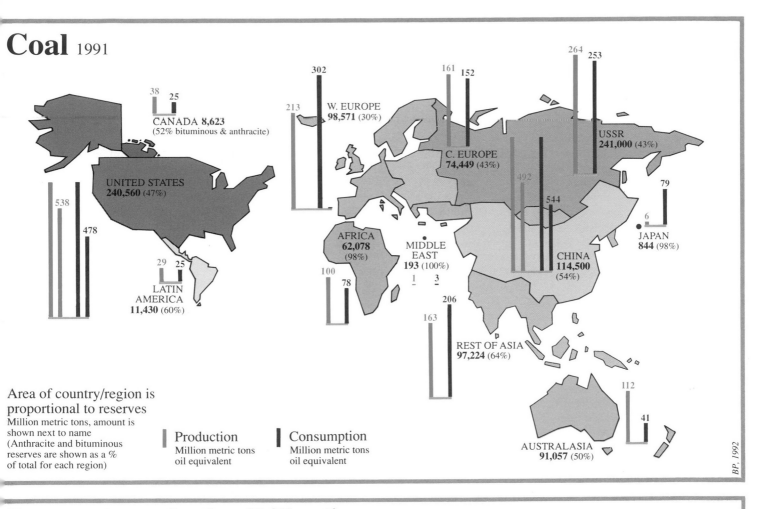

38 25
CANADA **8,623**
(52% bituminous & anthracite)

213 302 W. EUROPE
98,571 (30%)

161 152

264 253

USSR
241,000 (43%)

538

478

UNITED STATES
240,560 (47%)

C. EUROPE
74,449 (43%)

492

544

79

6
JAPAN
844 (98%)

29 25
LATIN
AMERICA
11,430 (60%)

AFRICA
62,078
(98%)

100 78

MIDDLE
EAST
193 (100%)
1   3
—   —

CHINA
114,500
(54%)

206

163

REST OF ASIA
97,224 (64%)

Area of country/region is
proportional to reserves
Million metric tons, amount is
shown next to name
(Anthracite and bituminous
reserves are shown as a %
of total for each region)

Production	Consumption
Million metric tons	Million metric tons
oil equivalent	oil equivalent

112

41

AUSTRALASIA
91,057 (50%)

BP, 1992

Reserves-to-production (R/P) ratios End-1991, world

The reserves-to-production ratio gives the amount of time reserves will last at current rates of production. This ratio is not definitive. Despite the fact that reserves are finite, new discoveries are made. Furthermore, changing technological and economic factors mean that reserves previously considered exhausted or too expensive to work can become viable. Alternative energy sources may also decrease the dependence on fossil fuels, thereby 'increasing' the life of existing reserves.

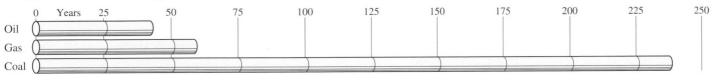

0   Years   25   50   75   100   125   150   175   200   225   250

Oil

Gas

Coal

Energy consumption per capita Metric tons oil equivalent

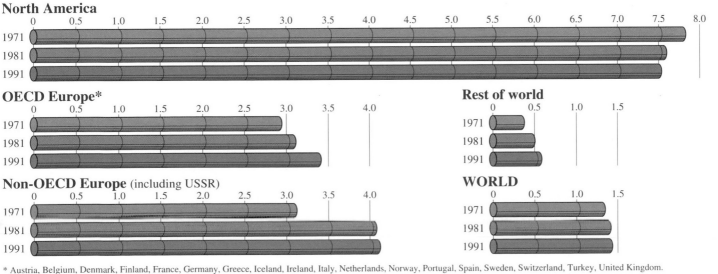

North America

0   0.5   1.0   1.5   2.0   2.5   3.0   3.5   4.0   4.5   5.0   5.5   6.0   6.5   7.0   7.5   8.0

1971

1981

1991

OECD Europe*

0   0.5   1.0   1.5   2.0   2.5   3.0   3.5   4.0

1971

1981

1991

Rest of world

0   0.5   1.0   1.5

1971

1981

1991

Non-OECD Europe (including USSR)

0   0.5   1.0   1.5   2.0   2.5   3.0   3.5   4.0

1971

1981

1991

WORLD

0   0.5   1.0   1.5

1971

1981

1991

* Austria, Belgium, Denmark, Finland, France, Germany, Greece, Iceland, Ireland, Italy, Netherlands, Norway, Portugal, Spain, Sweden, Switzerland, Turkey, United Kingdom.

BP, 1992

the cocktail of pollutants that is killing forests all over Europe and in North America. New technologies are being developed to burn coal more cleanly, but these are slow to catch on and even they can do nothing to reduce emissions of carbon dioxide, the main cause of the greenhouse effect. Fossil fuels account for four fifths of the 24 billion metric tons of carbon dioxide emitted into the atmosphere each year.

Alternative sources of energy have only a limited part to play in replacing fossil fuels in the foreseeable future. Nuclear power avoids the regular, devastating pollution caused by burning coal, oil and gas, but only at the cost of risking catastrophic accidents; it is also in deep decline, and is unlikely to recover fast enough to have much effect in replacing fossil fuels. Renewable sources of energy, such as solar and wind power, are environmentally benign but are far too little developed to make much difference. There is some hope for substitute fuels for cars, at present 99 per cent dependent on oil. Southern California, home to one in every 20 US citizen, aims to replace gasoline with cleaner fuels in 40 per cent of its cars, 70 per cent of its trucks, and all of its buses by the year 2000. But much the best potential for cutting back pollution and saving scarce resources of fossil fuels lies in greater efficiency in the use of energy.

Energy efficiency

For decades it was thought that increasing energy use and economic growth were indivisible, and many industries and some governments still seem to believe that economies cannot expand without burning more fuel. But experience has shown this to be false. Between 1973 and 1985, in the decade after the first oil price shock, per capita energy use in the OECD countries fell by 5 per cent, while per capita GDP (Gross Domestic Product) grew by a third.

This has been achieved by saving energy and increasing the efficiency with which it is used. Buildings in OECD countries as a whole now use a quarter less energy per person than they did before the first oil shock, while the energy efficiency of industry has improved by about a third. Worldwide, cars now get 25 per cent more miles to the gallon than they did in 1973, though increased traffic – more than 30 million new cars come off the world's assembly lines each year – has caused the total amount of fuel burned by automobiles to rise. In all, increased efficiency since 1973 has already saved the industrialized countries $250 billion.

Far more could be achieved, also without impairing economic growth, by a vigorous energy conservation program. The United Nations Economic Commission

for Europe estimates that industrialized countries could save a fifth of their energy by the year 2000, using presently available techniques. But this is a conservative estimate.

Buildings, industry and transport could all increase their efficiency by at least 50 per cent, using existing techniques. New houses in the US, designed to save energy, can use two thirds less fuel than existing ones to keep warm, without sacrificing any comfort, and homes being designed in Sweden can cut energy use by 90 per cent. Architects and engineers surveyed in one US study estimated that energy used in shops, offices and other commercial buildings could be cut in half by the year 2000. New light bulbs have been developed which use only a quarter as much electricity as existing ones, and fridges and other electrical appliances are on the market that use a half to two thirds less power than older models. One experiment in the Netherlands, using existing technologies, cut the electricity consumption of an office building by up to 84 per cent.

Savings in transport and industry

Cars that achieve more than 50 miles per gallon (mpg) are already on the road – and prototypes that get about 100 mpg have been tested. Encouraging public transport would save more energy; buses and trains use about three quarters less fuel per passenger mile than cars; trains and ships use two thirds less energy than trucks to transport the same amount of freight. Swedish steel mills have cut their energy use by more than a quarter, Japanese ones by up to a half – and both were already efficient by world standards.

The technology for massive savings in energy – and pollution – is available. It is much cheaper, as well as much more responsible, to save energy than to exploit new sources. But the will to use the technologies varies from country to country. The former Soviet Union uses energy twice as wastefully as the average OECD country. Japan, which has particularly concentrated on saving energy, uses less than half as much of its national product to pay for fuel as the US and gets a 5 per cent cost advantage on its manufactured goods as a result.

Structural changes, such as the growth of services and the decline of energy-intensive industry, will also lead to greater efficiency. The World Resources Institute has shown that these changes and a drive to save energy could cut the industrialized countries consumption of energy to nearly half its present level by the year 2000. Developing countries cannot afford to cut their energy consumption and will need to use more to develop. But they could achieve a Western European standard of energy services with only a 20 per cent increase in fuel consumption, if they used it efficiently.

Nuclear Energy

The nuclear age dawned at 2.20 pm on December 2, 1942, when scientists activated the first nuclear reactor in a disused squash court at the University of Chicago. This produced the first self-sustaining nuclear reaction, showing that humanity would be able to harness atomic energy. Scientists and politicians were soon urging that this apparently unlimited power should be used for peaceful purposes. The potential seemed enormous; just one kilogram of uranium could produce as much energy as 3,000 metric tons of coal, and enthusiasts predicted that atomic power would generate electricity "too cheap to meter".

In 1956 the Calder Hall pow ~ station opened in Cumbria, in the northwest of gland. It produced plutonium for bombs, but it also was the first full-scale plant to generate electricity for commercia e. By the end of 1990, there were 420 commercia nuclear reactors in 25 countries, producing 17 per cent of the world's electricity. In a few nations they dominated power production; France got 73 per cent of its electricity from the atom, Belgium 59 per cent, Hungary, Sweden and South Korea about half. So great an expansion of such a complex technology in little over 30 years was an extraordinary achievement and, for the first half of this period, nuclear power enjoyed virtually unanimous support. But in the 1970s, opponents emerged and by the late 1980s they commanded majority support in almost every Western country.

The problems of radiation

Most of the earliest opposition centered on low-level releases of radioactivity from nuclear installations. This was easy to dismiss: even today the average person receives over 2,000 times more radiation from inescapable natural sources (from the earth and outer space) than from nuclear power. But there are localized problems. Sellafield, on the UK's Cumbrian coast, which "reprocesses" used nuclear fuel to recover plutonium and uranium for bombs and power plants, has discharged vast amounts of radioactive waste into the Irish Sea. Some of the waste is returning to land in sea spray and sediments. Local children are 10 times more likely to get leukemia than normal, and similar "leukemia clusters" have been found around other nuclear installations. Research suggested in 1990 that some leukemias occur in the children of Sellafield workers whose genes may have been damaged by radiation.

However, the normal operation of nuclear power stations causes little environmental damage. Their tiny radioactive emissions contrast sharply with massive pollution from fossil fuel power stations which contributes greatly to acid rain and the threat of global warming. But accidents at nuclear power stations pose infinitely greater dangers than those at conventional ones.

Accidents: the threat of catastrophe

By the mid-1970s, opponents of nuclear power were fiercely debating the threat of accidents. Everyone agreed that a major accident could kill thousands of people; the main argument was over how likely it was to happen. One major US Government study, endorsed by the world's nuclear industry, concluded that a major accident would occur only once in every million years of reactor operation.

Even before the 1986 Chernobyl disaster, two serious nuclear accidents had already taken place within just two weeks of each other in 1957; a fire at Sellafield (then called Windscale) which sent a radioactive cloud over northwest Britain, and an accident at Kyshtym in the USSR which contaminated thousands of square kilometers of the Urals. There were many near misses; one US Government study showed that there had been 169 incidents that could have led to a catastrophic accident in the US alone between 1969 and 1979. They culminated in the accident at Three Mile Island, Pennsylvania, in March 1979; an official commission concluded that a catastrophe had only been avoided by sheer luck. After Chernobyl some scientists estimated that a major nuclear accident might in fact happen every decade.

Managing nuclear waste

Nuclear power produces intensely radioactive waste, some of which remains lethal for thousands of years. In 1976 the UK's Royal Commission on Environmental Pollution said that it would be "irresponsible and morally wrong" to proceed with a major expansion of nuclear power until a way had been found of dealing with this.

No country in the world has yet worked out how to dispose of the most dangerous waste, 10 per cent by volume but containing 99 per cent of the radioactivity. At present it is being stored and cooled while industries work out what to do with it. France, Germany and the US – the most optimistic countries – hope to begin disposal programs around the year 2010, others are waiting even longer. Meanwhile many countries find it hard to deal with even the less virulent waste: the UK has had to abandon many proposals and has still not evolved a publicly acceptable strategy for its disposal.

Costs: concealed and revealed

For decades the nuclear industry won official backing on the grounds that it produced cheap electricity. Critics challenged this, but government-owned nuclear monopolies could hide the full costs of atomic energy, or pass it on to consumers.

In the US, however, the nuclear industry is dominated by private electrical utilities, which picked up the true

Nuclear electricity supply and types of reactor

(Figures refer to top map). na: not available

Nuclear electricity as % of total electricity supplied, 1991

Types of reactor in operation, 1991
( ): under construction

	%	PW	BW	GC	PHW	LWG	FB	Other
1. FRANCE	73%	52 (5)		2			2	
2. BELGIUM	59%	7						
3. SWEDEN	52%	3	9					
4. HUNGARY	48%	4						
5. REP. OF KOREA	48%	8 (2)			1 (1)			
6. SWITZERLAND	40%	3	2					
7. TAIWAN	38%	2	4					
8. SPAIN	36%	7	2					
9. BULGARIA	34%	6						
10. FINLAND	33%	2	2					
11. CZECH.	29%	8 (6)						
12. GERMANY	28%	14 (1)	7					
13. JAPAN	24%	19 (3)	21 (6)	1		(1)		1
14. USA*	22%	74 (3)	37					
15. UK	21%	(1)		36			1	
16. ARGENTINA	19%				2 (1)			
17. CANADA	16%				20 (2)			
18. USSR*	13%	24 (21)				20 (4)	1	
19. SOUTH AFRICA	6%	2						
20. YUGOSLAVIA	6%	1						
21. NETHERLANDS	5%	1	1					
22. MEXICO	4%		1 (1)					
23. INDIA	2%		2		5 (7)			
24. PAKISTAN	1%				1			
25. BRAZIL	1%	1 (1)						
26. CHINA	0%	1 (2)						
27. CUBA	0%	(2)						
28. IRAN	0%	(2)						
29. ROMANIA	0%				(5)			

PW: Pressurized water reactor   BW: Boiling water reactor   GC: Gas cooled reactor
PHW: Pressurized heavy water reactor   LWG: Light water graphite moderated reactor
(ie Chernobyl type)   FB: Fast breeder reactor   Other: Type unknown

* The number in USA and USSR is variable due to changing classifications, eg conversion to military use etc.

IAEA, 1991; IAEA, 1992

Nuclear energy production

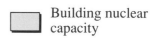 Nuclear power on stre

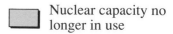 Building nuclear capacity

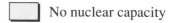 Nuclear capacity no longer in use

No nuclear capacity

Former USSR republics with nuclear reactors connected to the electricity grid:
Armenia*, Lithuania, Russia, Ukraine

* Reactor shut down in 1989, reopened in 199

Former Yugoslav republic with nuclear reactor connected to the electricity grid:
Slovenia

Chernobyl: first-year radiation

Dosage above natural background radiation in the first year after the Chernobyl accident, microsieverts (µsv)

- 1,000 – 2,000
- 500 – 999
- 200 – 499
- 100 – 199
- 10 – 99
- Under 10

The world average annual dose from natural background radiation is 2,400µsv. The map shows, by country and region, the average additional doses in the first year after the accident.

CANADA

UNITED STATES

JAPAN

CHINA

INDIA

●Chernobyl

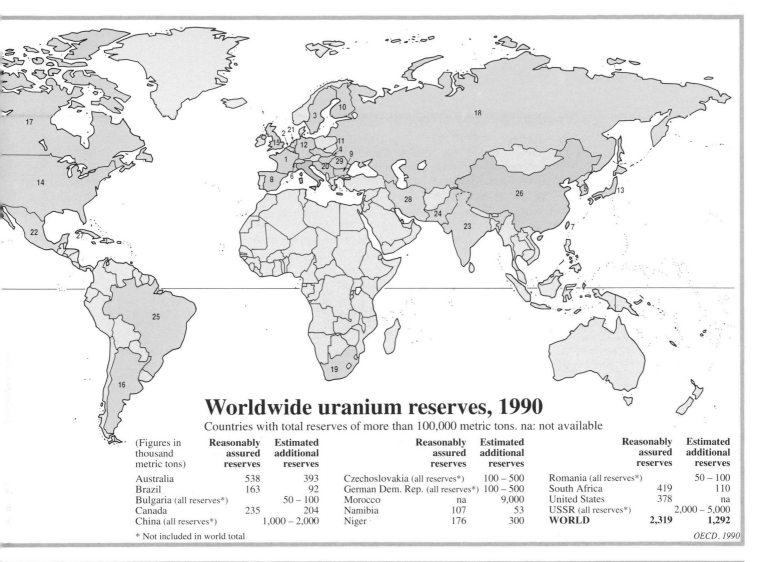

Worldwide uranium reserves, 1990

Countries with total reserves of more than 100,000 metric tons. na: not available

(Figures in thousand metric tons)	Reasonably assured reserves	Estimated additional reserves		Reasonably assured reserves	Estimated additional reserves		Reasonably assured reserves	Estimated additional reserves
Australia	538	393	Czechoslovakia (all reserves*)		100 – 500	Romania (all reserves*)		50 – 100
Brazil	163	92	German Dem. Rep. (all reserves*)		100 – 500	South Africa	419	110
Bulgaria (all reserves*)		50 – 100	Morocco	na	9,000	United States	378	na
Canada	235	204	Namibia	107	53	USSR (all reserves*)		2,000 – 5,000
China (all reserves*)		1,000 – 2,000	Niger	176	300	**WORLD**	**2,319**	**1,292**

* Not included in world total

OECD. 1990

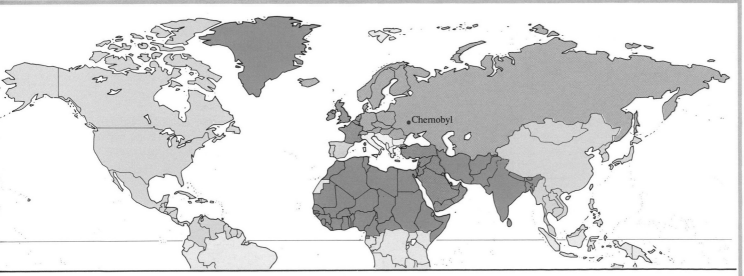

Chernobyl: total radiation received

Total dosage above natural radiation per capita, microsieverts (μsv)

☐ 1,000 – 2,000	▨ 100 – 499	☐ Under 10
▨ 500 – 999	▨ 10 – 99	

The map shows the **total** average additional doses, by area, likely to be received by members of the public as a result of the Chernobyl accident. These include the first and subsequent years.

UNSCEAR, 1988

cost of generating nuclear power much more quickly. They have not ordered a single new nuclear reactor since 1978, and all the orders made in the four preceding years have been cancelled – mainly for economic reasons. It now costs twice as much to generate electricity from new nuclear power stations in the US as from new coal-fired ones.

In 1987 a four-year official UK inquiry into building a nuclear power station at Sizewell on the Suffolk coast concluded that it was more than 97 per cent certain to provide cheaper electricity than a coal-fired plant. Then the Government decided to privatize the electricity industry, exposing its finances to independent scrutiny. This revealed that nuclear power was, in fact, three times as expensive as electricity from coal.

Fading prospects

By the mid-1980s the nuclear industry was already fading. The International Atomic Energy Agency (IAEA) had cut its estimates of future capacity ninefold. In 1974 it estimated that by the year 2000 there would be 4,450 thousand megawatts of nuclear capacity around the world: in 1986 it was expecting merely 505 thousand megawatts, and even this will prove a great overestimate.

National nuclear programs were stagnating even outside the US. Only one West German power station had been ordered since 1975. Sweden, which generates half its electricity from the atom – one of the highest proportions in the world – decided in 1980 to phase out nuclear power altogether. Japan cut its program by a third in 1984, and Denmark decided in 1985 never to build any reactors.

Chernobyl

On April 26, 1986, operators at the Chernobyl plant, thought to be one of the best-run in the USSR, made a mistake when carrying out a test on one of its reactors. Attempting to correct it, they deliberately overrode a series of safety systems designed to prevent an accident. After the last was switched off, the reactor's power surged several hundredfold in a second. Two huge explosions blasted the 1,000 metric ton lid clean off the reactor and lit a fire that blazed for 10 days, releasing several quadrillions of becquerels of radioactive isotopes.

By 1992, as a result, seven times as many children in the Ukraine and Byelorussia were contracting thyroid cancer as before the accident and the incidence of leukemia was also rising. At the Earth Summit in Rio de Janeiro that June, Dr Yuri Sherback, Ukraine's Environment Minister, estimated that more than 6,000 people may have already died as a result of the accident

in his country, and that the death toll in the Ukraine alone would eventually reach 40,000. Tens of thousands more will die of cancer further afield in the former Soviet Union and in Europe.

Restrictions on foodstuffs were imposed in almost every Western European country and imports from Eastern Europe were banned. Four years after the accident, hundreds of sheepfarms in Cumbria and Wales were still so radioactive that they could not sell meat for consumption from sheep that grazed on the pastures, and the Lapps of Scandinavia faced the destruction of their culture because their reindeer were contaminated by eating lichen which accumulate radioactive materials.

The end of the nuclear dream

After Chernobyl, public opposition to building new nuclear plants rose from 33 to 64 per cent in Finland, from 44 to 82 per cent in West Germany, from 65 to 83 per cent in the UK. Opposition in France, previously tiny, soared to 59 per cent. Austria announced that its only nuclear plant, completed in 1978 but never operated, would be taken to pieces. The Philippines scrapped its only reactor and Italy, Spain and Yugoslavia halted any further expansion.

In 1989 the UK Government, one of the most pro-nuclear in the world, cancelled all plans for new atomic power stations for at least five years. France, once even more convinced about the atom, has only ordered one new plant since 1987. Its nuclear expansion ran up multimillion dollar debts and a secret Government report concluded in 1989 that it had caused massive overcapacity and should have been slowed down 10 years before. The USSR and Eastern Europe were long committed to nuclear expansion, but the Chernobyl accident and the rise of democracy changed this, and plans for reactors are being abandoned.

In 1992, the IAEA called four Russian and Eastern European plants "very dangerous" and called for one – Kozloduy in Bulgaria – to be closed immediately. Only Japan still plans a major expansion of nuclear energy and opposition is rising there too.

The amount of electricity generated by nuclear power is still rising because plants ordered long ago are still coming onstream. But the rate of increase is rapidly slowing. During the 1990s nuclear power will almost certainly decline as old plants are closed and not replaced, but problems in dealing with the radioactive parts of closed power stations will arise. No large power plant has yet been dismantled, and decommissioning will include disposing of over 100,000 metric tons of high-level waste, which is expected to cost up to $270 billion. The rise and fall of nuclear power has been encompassed in a single generation.

Renewable Energy Resources

Solar energy powers the world, causing living things to grow, driving the wind and waves, and drawing water into the sky to fall as rain and run to the sea. The sun has supplied almost all of humanity's energy: the wood for the first fire, the wind and water that powered mills, and the fossil fuels – the remains of long-dead plants – that have fired modern economic development. As pollution increases and the most accessible coal, gas and oil go up in smoke, renewable energy looks increasingly attractive. It does not cause acid rain, add to the greenhouse effect, or share the problems of nuclear power.

Every year the equivalent of about 60,000 billion metric tons of oil lands on the earth's surface as solar energy: if only 1 per cent could be tapped at just 5 per cent efficiency, the whole of the world's population could achieve the same level of energy consumption as the United States. The difficulty is in exploiting it. Solar energy is spread so widely that it is hard to collect and concentrate in large amounts. Yet its diffuseness can be an advantage; the energy is already distributed free and can be tapped by small-scale installations near where it is to be used.

Hydro-power: a mixed blessing

Much of the world's energy already comes from exploiting the power of water tumbling to the sea. Hydro-electric power provides more than a fifth of all the world's electricity; its output, the World Energy Conference has predicted, could rise sixfold by the year 2020.

Most potential lies in developing countries. North America, which accounts for about a third of global production, has already used 59 per cent of its potential for large dams; but the Third World has harnessed only 7 per cent of its resources. Developing countries have indeed been building big dams while construction has been coming to a halt in the United States. Venezuela opened the Guri Dam, the biggest in the world, in 1986. Brazil, which tripled its hydro-electric capacity between 1973 and 1983, is planning an even bigger one. And China, which already gets a third of its electricity from hydro-power, is planning to triple its capacity by the year 2000, including the Three Gorges project, the biggest dam of all.

Such giant dams flood good land, or – as in Brazil – priceless rainforest. In hot climates, dams aid the spread of waterborne diseases like malaria, river blindness and schistosomiasis. They also trap silt that would otherwise fertilize land downstream; this builds up in the reservoirs, often cutting their useful lives by half. The water pressure generated by huge dams may cause earthquakes, and dams may dry up water supplies in nations downstream, causing conflict. They also drive people from their homes: China's Three Gorges Dam alone would displace several million. On the other hand, hydro-electric plants already spare the atmosphere over 2 billion metric tons of carbon dioxide a year – the amount that would be produced if they were replaced by coal-powered electricity.

Small-scale hydro-power plants avoid the disadvantages. They also provide electricity where it is needed and on an accessible scale; large projects often produce more electricity than can be readily absorbed by Third World economies. China now has 90,000 small hydro-power turbines providing two fifths of its country towns and a third of its cities with most of their power. The former West Germany has some 3,000, and the United States and France have more than 1,000 each.

Harnessing the waves

A commercial tide-power plant has operated successfully at the mouth of the river Rance in Brittany since 1967. More than 30 good sites for generating energy have been identified around the world. Canada's Bay of Fundy is considered the world's best site – one small plant is already working there. The second most promising resource is the Severn Estuary, which could provide 5 per cent of the UK's entire electricity demand. Big schemes, however, can damage wildlife habitats and alter sea levels down the coast.

The UK Department of Energy estimated in the late 1970s that about half the country's electricity supplies could be met from waves off northwest Scotland. British scientists pioneered research, but the Department inexplicably ended funding in the late 1980s. Norway brought the world's first wave-power plants into operation near Bergen, and the UK is now planning to build one based on Norwegian experience. Indonesia, Portugal, Puerto Rico and Western Australia also plan plants, and Japan, Sweden, the US and former Soviet Union are all developing the technology.

Reaping the wind

Since the 1973 oil crisis, more than 50,000 electricity-generating wind turbines have been installed worldwide. Ninety five countries now get electricity from the wind, and the US and the Netherlands have windfarms with at least 10 turbines – and some with more than 100. California generates 90 per cent of the world total, mostly from windfarms on three mountain passes – Altamont, San Gorgiono and Tehachapi. The California Energy Commission expects the wind to supply at least 8 per cent of the state's power by the year 2000.

Steady wind speeds of more than 25 kilometers per hour are needed to produce electricity: they are commonest on mountains and coasts. The flat lands of

Electricity generation from renewable sources

WRI/IIED/UNEP, 1992; WEC, 1989

Gigawatt hours, 1989

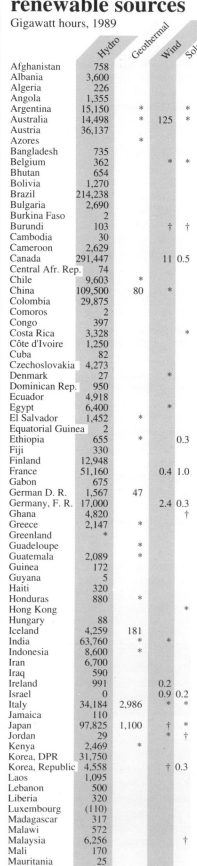

	Hydro	Geothermal	Wind	Solar‡
Afghanistan	758			
Albania	3,600			
Algeria	226			
Angola	1,355			
Argentina	15,150	*		*
Australia	14,498	*	125	*
Austria	36,137	*		
Azores		*		
Bangladesh	735			
Belgium	362		*	*
Bhutan	654			
Bolivia	1,270			
Brazil	214,238			
Bulgaria	2,690			
Burkina Faso	2			
Burundi	103		†	†
Cambodia	30			
Cameroon	2,629			
Canada	291,447		11	0.5
Central Afr. Rep.	74			
Chile	9,603	*		
China	109,500	80	*	
Colombia	29,875			
Comoros	2			
Congo	397			
Costa Rica	3,328			*
Côte d'Ivoire	1,250			
Cuba	82			
Czechoslovakia	4,273			
Denmark	27		*	
Dominican Rep.	950			
Ecuador	4,918			
Egypt	6,400		*	
El Salvador	1,452	*		
Equatorial Guinea	2			
Ethiopia	655	*		0.3
Fiji	330			
Finland	12,948			
France	51,160		0.4	1.0
Gabon	675			
German D. R.	1,567	47		
Germany, F. R.	17,000		2.4	0.3
Ghana	4,820			†
Greece	2,147	*		
Greenland	*			
Guadeloupe		*		
Guatemala	2,089	*		
Guinea	172			
Guyana	5			
Haiti	320			
Honduras	880	*		
Hong Kong				*
Hungary	88			
Iceland	4,259	181		
India	63,760	*	*	
Indonesia	8,600	*		
Iran	6,700			
Iraq	590			
Ireland	991		0.2	
Israel	0		0.9	0.2
Italy	34,184	2,986	*	*
Jamaica	110			
Japan	97,825	1,100	†	*
Jordan	29		*	†
Kenya	2,469	*		
Korea, DPR	31,750			
Korea, Republic	4,558		†	0.3
Laos	1,095			
Lebanon	500			
Liberia	320			
Luxembourg	(110)			
Madagascar	317			
Malawi	572			
Malaysia	6,256			†
Mali	170			
Mauritania	25			
Mauritius	102			

Renewable energy resources

Energy generated from renewable sources, gigawatt hours, 1989:

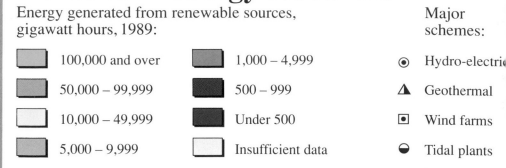

100,000 and over	1,000 – 4,999
50,000 – 99,999	500 – 999
10,000 – 49,999	Under 500
5,000 – 9,999	Insufficient data

Major schemes:

- ⊙ Hydro-electric
- ▲ Geothermal
- ⊡ Wind farms
- ◓ Tidal plants

All figures refer to commercially traded energy resources only, and exclude biomass, which is the dominant energy resource in the developing world. In some nations, such as Tanzania, it represents over 90% of all energy resources. In many cases, but by no means all, biomass means fuelwood (see pages 73 – 76). Other main types of biomass include crop wastes, animal dung, sugar cane, and sewage. Biomass fuels are usually burnt directly, fermented for alcohol, or subjected to anaerobic digestion. Country by country production figures are practically non-existent. The World Energy Conference has estimated that the total world production of biomass (*including* fuelwood) for 1987 was 1,257 million metric tons of oil equivalent (MTOE), of which 169 MTOE were produced by the developed world and 1,088 MTOE by the developing world.

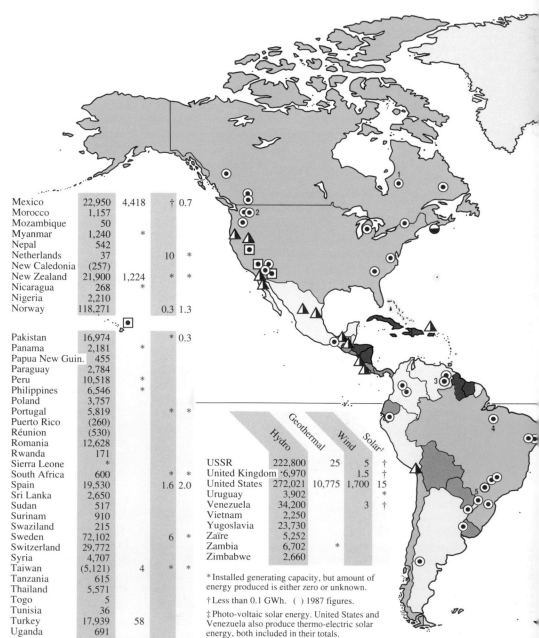

	Hydro	Geothermal	Wind	Solar‡
Mexico	22,950	4,418	†	0.7
Morocco	1,157			
Mozambique	50			
Myanmar	1,240	*		
Nepal	542			
Netherlands	37		10	*
New Caledonia	(257)			
New Zealand	21,900	1,224	*	*
Nicaragua	268	*		
Nigeria	2,210			
Norway	118,271		0.3	1.3
Pakistan	16,974		*	0.3
Panama	2,181	*		
Papua New Guin.	455			
Paraguay	2,784			
Peru	10,518	*		
Philippines	6,546	*		
Poland	3,757			
Portugal	5,819		*	*
Puerto Rico	(260)			
Réunion	(530)			
Romania	12,628			
Rwanda	171			
Sierra Leone	*			
South Africa	600		*	*
Spain	19,530		1.6	2.0
Sri Lanka	2,650			
Sudan	517			
Surinam	910			
Swaziland	215			
Sweden	72,102		6	*
Switzerland	29,772			
Syria	4,707			
Taiwan	(5,121)	4	*	*
Tanzania	615			
Thailand	5,571			
Togo	5			
Tunisia	36			
Turkey	17,939	58		
Uganda	691			

	Hydro	Geothermal	Wind	Solar‡
USSR	222,800	25	5	†
United Kingdom	6,970		1.5	†
United States	272,021	10,775	1,700	15
Uruguay	3,902			*
Venezuela	34,200		3	†
Vietnam	2,250			
Yugoslavia	23,730			
Zaïre	5,252			
Zambia	6,702		*	
Zimbabwe	2,660			

* Installed generating capacity, but amount of energy produced is either zero or unknown.

† Less than 0.1 GWh.

‡ Photo-voltaic solar energy. United States and Venezuela also produce thermo-electric solar energy, both included in their totals.

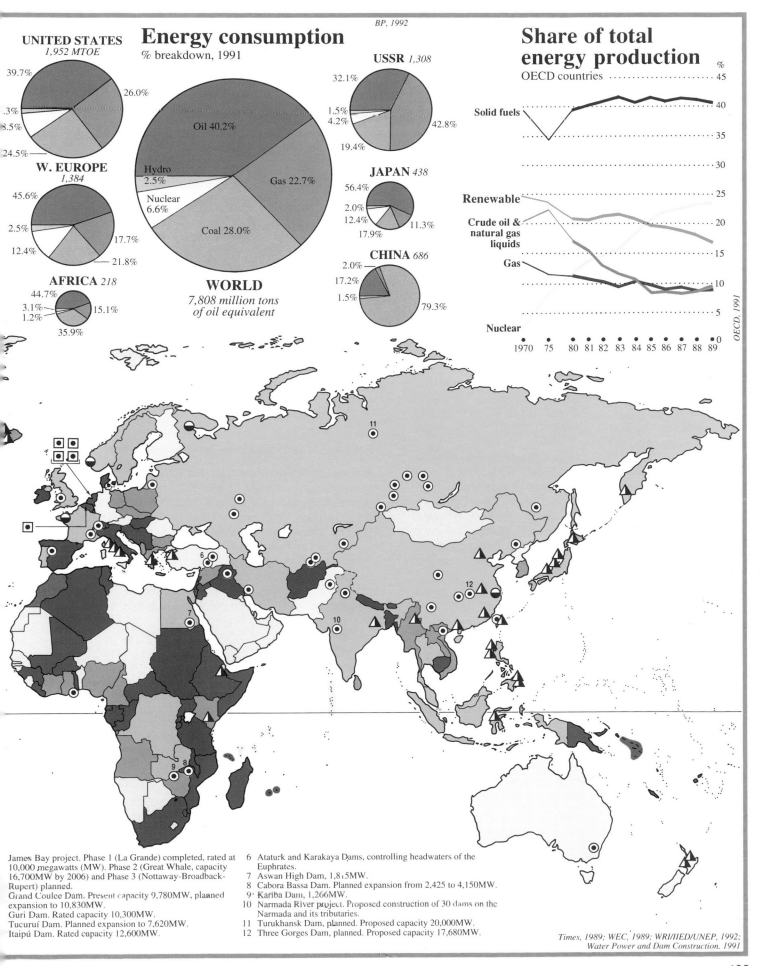

Energy consumption
% breakdown, 1991

BP, 1992

UNITED STATES
1,952 MTOE

39.7%
26.0%
.3%
8.5%
24.5%

W. EUROPE
1,384

45.6%
2.5%
12.4%
17.7%
21.8%

AFRICA *218*

44.7%
3.1%
1.2%
15.1%
35.9%

WORLD
*7,808 million tons
of oil equivalent*

Oil 40.2%

Hydro
2.5%

Nuclear
6.6%

Coal 28.0%

Gas 22.7%

USSR *1,308*

32.1%
1.5%
4.2%
42.8%
19.4%

JAPAN *438*

56.4%
2.0%
12.4%
17.9%
11.3%

CHINA *686*

2.0%
17.2%
1.5%
79.3%

Share of total energy production
OECD countries .... %

45

Solid fuels

40

35

30

25

Renewable

Crude oil &
natural gas
liquids

20

Gas

15

10

Nuclear

5

1970   75   80 81 82 83 84 85 86 87 88 89   0

OECD, 1991

James Bay project. Phase 1 (La Grande) completed, rated at 10,000 megawatts (MW). Phase 2 (Great Whale, capacity 16,700MW by 2006) and Phase 3 (Nottaway-Broadback-Rupert) planned.
Grand Coulee Dam. Present capacity 9,780MW, planned expansion to 10,830MW.
Guri Dam. Rated capacity 10,300MW.
Tucuruí Dam. Planned expansion to 7,620MW.
Itaipú Dam. Rated capacity 12,600MW.

6 Ataturk and Karakaya Dams, controlling headwaters of the Euphrates.
7 Aswan High Dam, 1,815MW.
8 Cabora Bassa Dam. Planned expansion from 2,425 to 4,150MW.
9 Kariba Dam, 1,266MW.
10 Narmada River project. Proposed construction of 30 dams on the Narmada and its tributaries.
11 Turukhansk Dam, planned. Proposed capacity 20,000MW.
12 Three Gorges Dam, planned. Proposed capacity 17,680MW.

*Times, 1989; WEC, 1989; WRI/IIED/UNEP, 1992;
Water Power and Dam Construction, 1991*

Denmark are particularly windy: the longest period of calm ever recorded is seven days. Denmark now produces 6 per cent of the world's wind-generated electricity – more than half the power produced anywhere outside California.

Giant windmills have been built in Sweden and West Germany, as well as in Denmark and the US, and are planned in Canada, the UK and the Netherlands. Some have technical problems: lots of small turbines may do the same job better. China has installed thousands of them. India plans to produce enough electricity from the wind by the year 2000 to meet the needs of 15 million of its people, and the UK Department of Energy has estimated that windpower could provide a fifth of the country's electricity by 2020.

Power from the sun

The simplest way of using the sun's energy is to heat water. Ninety per cent of Cyprus' houses have solar panels. Two thirds of all Israel's domestic water heating is provided by 700,000 panels and 4 million are in use in Japan.

The sun's rays have to be concentrated to produce electricity. Israel completed the world's first solar power station in 1979, using special ponds which absorb the sun's energy in saltwater. Nearly 20 solar power stations using reflectors have been built with government help, half of them in California. The most promising technique uses U-shaped mirrors placed in troughs to concentrate the sun's energy onto pipes filled with oil or water.

Solar cells are likely to prove particularly promising: some 15,000 homes worldwide are powered by them. At present they are too expensive to use except in remote locations, away from national grids. But costs fell 50-fold between 1970 and 1990, and are continuing to fall. The World Bank estimated in 1992 that both solar cells and solar power stations were on the verge of becoming competitive with nuclear power.

Plants and trees store the sun's energy. Power stations can be run on wood grown on special plantations or on crop wastes: a third of Hawaii's electricity comes from sugar residues, while India, Malaysia, Thailand, the Philippines, Surinam and the US all run power plants that burn rice husks. Brazil used ethanol, produced from sugar cane, for half the fuel used in its vehicles in 1986 and, although its program has hit problems, such alcohol fuels are becoming increasingly attractive, particularly in the US, for combating pollution.

Geothermal power

The top 5 kilometers of the earth's crust contain 40 million times as much energy as its oil and gas reserves. But geothermal energy, too, is widely dispersed and is only used when concentrated by hot water, trapped in rock. Most of Iceland's homes are heated by such reservoirs.

If the water is hot enough, it can generate electricity and has done throughout this century at Larderello, Italy. The world's geothermal electrical capacity almost tripled in the 1980s and there are now plants in 18 countries. It provides more than a sixth of the power consumed in Central America. Japan plans to increase its use 10-fold in the 1990s. The total potential is enormous and new technologies to exploit hot dry rocks promise to expand it even further.

Renewable potential: the benign contribution

Together, renewable sources could make a great, and overwhelmingly benign, contribution to energy supplies. The Philippines, Brazil and Norway already get at least half their energy from them. By the year 2000, small-scale renewable technologies will overtake nuclear power in India. Israel, Japan, Sweden, Denmark, Greece and Germany all have major programs.

But government and energy utilities in most countries still resist using renewable sources, particularly small-scale applications. Although quick to install and largely avoiding environmental drawbacks, they are far less easy to control than a few big power stations.

Renewables will only prosper when they are given encouragement. A first step would be to reduce subsidies for fossil fuels; in China these amount to almost $20 billion, 7 per cent of its GNP. Tax incentives for renewables have proved their worth in California and in Denmark. With a fraction of the support that has been poured into nuclear power – which still gets some 60 per cent of the money for energy research in western countries, compared to 6 per cent for renewables – the world would be on its way to getting much of its energy from the sun.

Biological Diversity and Genetic Resources

The next decade will see a wildlife holocaust. By the turn of the century, a million kinds of animals, plants and insects are expected to be driven to extinction; by the year 2050, half of all the species alive today could be lost forever.

The disaster threatens to rival the mass extinction of 65 million years ago. Pollution, overhunting and overfishing, and the trade in wildlife all play a part. But by far the greatest cause of the extinctions is the destruction of wild habitats for farming, fuel, industry and a host of other uses.

The world's tropical rainforests, which contain at least half the world's species, are falling fast: only half of their original expanse still remains, and an area over twice the size of Austria is cleared each year. Half the world's wetlands – other abundant habitats – have been drained or developed and species-rich coral reefs are being destroyed throughout the earth's warmer seas.

Species are now becoming extinct at 25,000 times the natural rate; by the year 2000, several are likely to be vanishing every hour and the tempo of destruction will be increasing. The loss of one plant can cause the loss of as many as 30 kinds of animals and insects which depend upon it.

Genetic storehouses threatened

Even the loss of a single species is a tragedy, because each form of life is a storehouse of irreplaceable substances. Every civilization has been rooted in the wealth of nature. The domestication of wild species made the first farming possible; selective breeding made them more productive; and natural resources enabled the agricultural and industrial revolutions to take off. Genetic resources, taken from the wild, still sustain modern societies, providing medicines, food and raw materials for industry.

Worldwide, medicines from wild products are worth some $40 billion a year. Foxgloves have saved the lives of millions of sufferers from heart disease, by providing digitoxin and digitalis. Bee venom is used to treat arthritis. Codeine and morphine come from poppies; an Amazonian tree and liana respectively provide quinine to fight malaria and curare, used to relax muscles for surgery and to treat multiple sclerosis and Parkinson's disease. And an alkaloid from Australia's Moreton Bay chestnut shows promise in combating the HIV virus.

The rosy periwinkle, a pretty but unremarkable plant discovered in a Madagascan forest, has transformed the prospects of children with leukemia. Before its discovery less than a fifth of them survived; now two drugs taken from the plant, together with other treatments, have increased the remission rate to over 80 per cent. The Pacific yew, which was burned as a weed by foresters cutting down the temperate rainforest of the Northwest United States, was found in 1991 to contain the most important anti-cancer drug found in the last 15 years. In all, 1,400 tropical forest plants and 500 marine organisms yield chemicals with the potential to fight cancer; but many may be driven to extinction before their promise can be assessed or tapped.

The health of harvests

The health of the world's harvests is even more dependent on genetic resources. Just three species – wheat, rice and maize – provide half the world's food: another four – potato, barley, sweet potato and cassava – bring the total to three quarters. Such overwhelming dependence on a few crops is dangerous; disease can spread rapidly through monocultures – as it did through the Irish potato harvest in the 1840s, causing a fifth of the country's people to die.

Crops need to be given new protection every five to 15 years, because pests and disease develop ways around their existing defenses. The only effective way to confer it is to interbreed them with other strains, often wild ones.

In the 1960s an epidemic of the wheat disease, stripe rust, struck the US; the state of Montana repeatedly lost a third of its harvest. Genes from a wild wheat from Turkey saved the situation, providing resistance to this and 50 other diseases. In 1970 an even more virulent plague spread by up to 150 kilometers a day across the American corn belt. It devastated a sixth of the maize crop, wiping out half the harvest of some Southern states, at a cost of some $2 billion.

Maize has been particularly vulnerable to such disasters, as inbreeding has given it an almost uniform genetic pattern – and new genes from wild varieties have been urgently needed. Two ancestors of the plant were found in Mexico in the late 1970s; they can confer resistance to seven of the domestic crop's major diseases, and can turn it into a perennial crop, allowing it to spring up every year like grass, without resowing. These maizes have been called the botanical find of the century; ominously, just a few stalks of them were discovered in a tiny area now threatened with destruction.

Most of Brazil's coffee traces its ancestry back to a single tree, imported from East Africa via the Caribbean. In 1970 disease struck its crop and spread through Latin America, threatening several national economies with disaster: a wild coffee from Ethiopia's fast-disappearing forests was used to prevent a recurrence. The story has been repeated many times over.

Interbreeding with wild varieties increases yields and extends the area under crops. Nearly 10 million square kilometers of land around the world, for example, is too

Biological diversity and genetic resources

- Unesco biosphere reserves

- ▲ IUCN/WWF centers of plant diversity

Around 250 sites and regions of high plant diversity, ranging in size from just 2 square kilometers to over 500,000, have been identified in the IUCN/WWF Centers of Plant Diversity project. Conservation of these sites would offer protection to up to 90% of the world's plant species.

Biosphere reserves

- ▨ Core zone representing one of earth's major ecosystems
- ▨ Multi-use buffer zone
- ▢ Transition area
- ▨ Experimental research area
- ◗ Research, training or educational facilities
- • Human settlements

Simple biosphere reserve

Cluster biosphere reserve

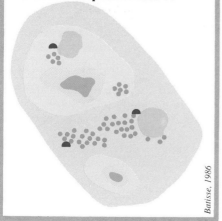

Batisse, 1986

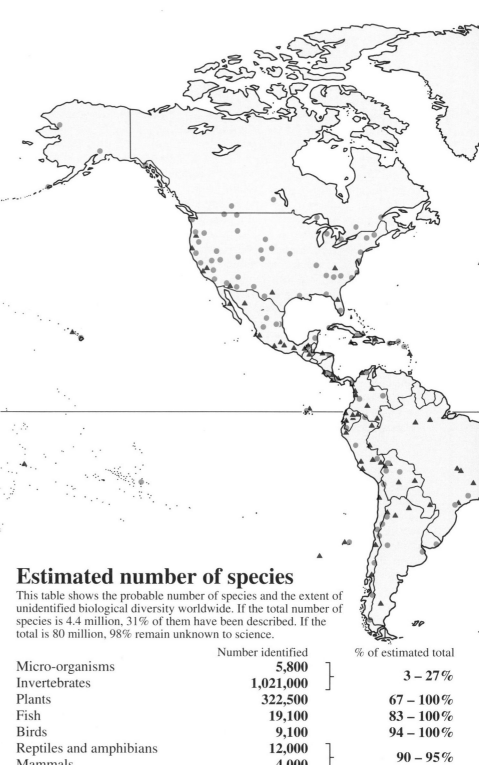

Estimated number of species

This table shows the probable number of species and the extent of unidentified biological diversity worldwide. If the total number of species is 4.4 million, 31% of them have been described. If the total is 80 million, 98% remain unknown to science.

	Number identified	% of estimated total
Micro-organisms	**5,800**	
Invertebrates	**1,021,000**	**3 – 27%**
Plants	**322,500**	**67 – 100%**
Fish	**19,100**	**83 – 100%**
Birds	**9,100**	**94 – 100%**
Reptiles and amphibians	**12,000**	**90 – 95%**
Mammals	**4,000**	
TOTAL	**1,393,500**	

	Number of species	% identified
Low estimate of all species	**4.4 million**	**31%**
High estimate of all species	**80 million**	**2%**

Wilson, 1988, updated by editors

Estimates of the number of species vary widely. This is because many species-rich ecosystems like tropical forests contain unexplored and unstudied habitats, and because the likely number of habitats may have been underestimated. In a study in Panama, 80% of 1,200 different species of beetle found in a group of 19 trees were new to science.

Marine systems are also revealing surprising amounts of li with new communities of organisms being discovered containing hundreds of new species.
The figures given here are the most recent and widely accepted. The percentage values are based on the averages a number of estimates and are intended as a guide only.

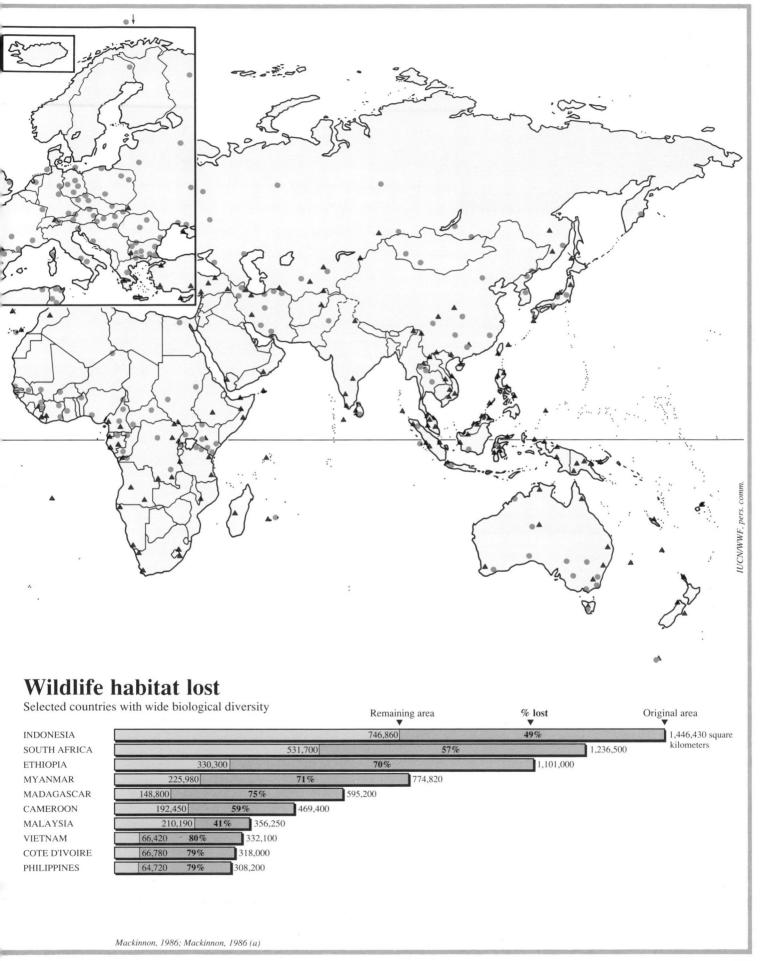

Wildlife habitat lost

Selected countries with wide biological diversity

	Remaining area ▼	% lost ▼	Original area ▼
INDONESIA	746,860	49%	1,446,430 square kilometers
SOUTH AFRICA	531,700	57%	1,236,500
ETHIOPIA	330,300	70%	1,101,000
MYANMAR	225,980	71%	774,820
MADAGASCAR	148,800	75%	595,200
CAMEROON	192,450	59%	469,400
MALAYSIA	210,190	41%	356,250
VIETNAM	66,420	80%	332,100
COTE D'IVOIRE	66,780	79%	318,000
PHILIPPINES	64,720	79%	308,200

Mackinnon, 1986; Mackinnon, 1986 (a)

IUCN/WWF, pers. comm.

129

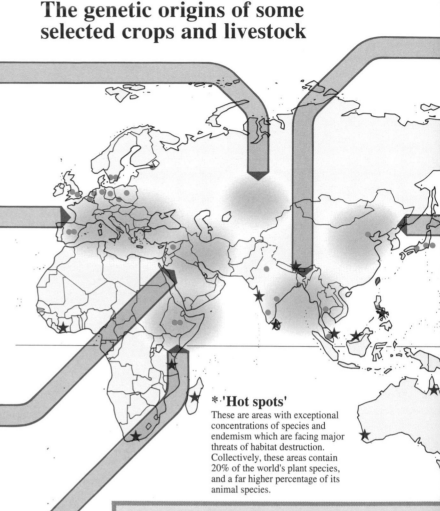

CENTRAL ASIA

Almond	Cucumber	Opium poppy
Apple	Flax/linseed	Rhubarb
Carrot	Garlic	Spinach
Common grape	Onion	

EUROPEAN – SIBERIAN REGION
Cattle	Kale	Watercress
Chicory	Lettuce	
Gooseberry	Licorice	

MEDITERRANEAN

Asparagus	Common grape	Oat
Broad bean	Globe	Opium poppy
Cabbage	artichoke	Parsnip
Cauliflower	Lavender	Rape
Celery	Mint	Sugar beet

BELGIUM
Brussel sprout

NORTH AFRICA
Cattle	Marjoram

NEAR EAST

Barley	Leek	Shallot
Cabbage	Lentil	Sheep
Einkorn wheat	Pea	Spelt wheat
Fig	Pig	Sugar beet
Goat	Plum	Sweet cherry
Hazelnut	Rye	

HORN OF AFRICA

Black-eye pea	Date palm	Short staple
Bread wheat	Egyptian cotton	cotton
Castor bean	Finger millet	Sorghum
Coffee	Mustard	Yam
Cowpea	Pearl millet	

The genetic origins of some selected crops and livestock

* 'Hot spots'
These are areas with exceptional concentrations of species and endemism which are facing major threats of habitat destruction. Collectively, these areas contain 20% of the world's plant species, and a far higher percentage of its animal species.

Major drugs derived from plants

Plant	Drug	Use
Amazonian liana	Curare	Muscle relaxant
Annual mugwort	Artemisinin	Antimalarial
Autumn crocus	Colchicine	Antitumor agent
Belladonna	Atropine	Anticholinergic
Coca	Cocaine	Local anesthetic
Common thyme	Thymol	Antifungal
Ergot fungus	Ergotamine	Analgesic
Foxglove	Digitoxin, digitalis	Cardiotonic
Indian snakeroot	Reserpine	Antihypertensive
Meadowsweet	Salicylic acid*	Analgesic
Mexican yam	Diosgenin	Birth-control pill
Nux vomica	Strychnine	CNS stimulant
Opium poppy	Codeine†, morphine	Analgesic (†& antitussive)
Pacific yew	Taxol	Antitumor agent
Recured thornapple	Scopolamine	Sedative
Rosy periwinkle	Vincristine, vinblastine	Antileukemia
Velvet bean	L-Dopa	Antiparkinsonian
White willow	Salicin*	Analgesic
Yellow cinchona	Quinine	Antimalarial, antipyretic

* Compound formed from salicylic acid and acetic acid is called acetylsalicylic acid: better known as aspirin.

Improving the world's crops
Origins of the wild germplasm used to improve 15 of the world's major crops

Crop	Sources of wild germplasm
Apple	Japan (*Malus floribunda*); USSR (*M. baccata*)
Barley	Turkey (*Hordeum spontaneum*)
Cassava	Brazil (*Manihot glaziovii*)
Cotton	Central Africa (*Gossypium anomalum*); Mexico (*G. hirsutum mexicanum*); USA (*G. tomentosum*)
Grape	USA (*Vitis aestivalis, V. berlandieri, V. lincecumii, V. riparia, V. rupestris, V. labrusca*); USSR (*V. amurensis*)
Maize	Mexico (*Zea diploperennis*); USA, Venezuela (*Tripsacum dactyloides*)
Oat	Algeria, Israel, Portugal, Tunisia (*Avena sterilis*)
Oil palm	Côte d'Ivoire, Nigeria, Zaïre (*Elaeis guineensis*)
Pea	Israel, Jordan, Lebanon, Syria, Turkey (*Pisum fulvum*
Potato	Argentina (*Solanum spegazzinii, S. vernei*); Argentina, Bolivia, Peru (*S. acaule*); Mexico (*S. demissum, S. stoloniferum*)
Rice	India (*Oryza nivara*)
Rubber	Brazil (*Hevea brasiliensis*)
Sunflower	USA (*Helianthus annuus, H. petiolaris*)
Tomato	Ecuador (*Lycopersicon cheesmanii*); Ecuador, Peru (*L. esculentum cerasiforme, L. pimpinellifolium, L. hirsutum*); Peru (*L. chmieleswkii*); Peru, Chile (*L. peruvianum*)
Wheat	Israel, Turkey (*Triticum turgidum dicoccoides*); Italy, Spain (*Aegilops ventricosa*); Turkey (*Ae. umbellulata*

Prescott-Allen, 1983

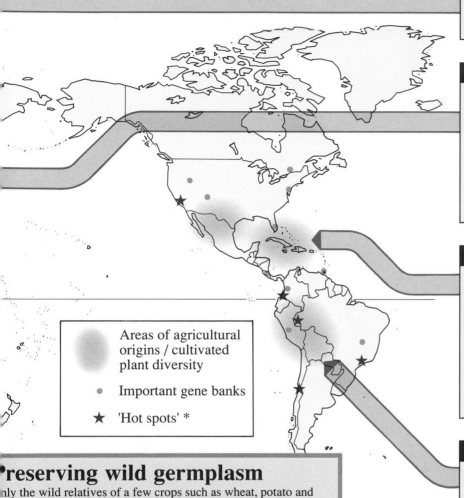

INDIA / INDO-MALAYA

Bermuda grass	Dwarf wheat	Rice
Black pepper	Lime	Tree cotton
Cardamon	Mango	
Chicken	Moth bean	

CHINA

Camphor tree	Litchi	Sweet orange
Chive	Peach	Tea
Foxtail millet	Radish	Turnip
Ginseng	Soybean	Water chestnut

SOUTHEAST ASIA

Apricot	Clove	Lemon
Banana	Coconut palm	Mung bean
Cinnamon &	Eggplant	Sugar cane
cassia	Indian almond	Tangerine

MEXICO / CENTRAL AMERICA

Avocado	Papaya	Tabasco
Common bean	Pecan	pepper
Hemp/sisal	Sweet potato	Vanilla
Maize/corn	Tomato	

NORTH AMERICA

Cranberry	Jerusalem	Muscadine
Sunflower	artichoke	grape
		Turkey

CARIBBEAN
Grapefruit

ANDES / SOUTH AMERICA

Cashew	Lima bean	Quinine
Cayenne	Pepper	Rubber
Cocoa	Pineapple	Upland cotton
Groundnut/	Potato	
peanut	Pumpkin	

Legend:
- Areas of agricultural origins / cultivated plant diversity
- Important gene banks
- ★ 'Hot spots' *

Hoyt, 1988; Myers, 1987; UNFPA, 1991; IBPGR, 1989, updated by pers. comm.; WWF, pers. comm.

Preserving wild germplasm

Only the wild relatives of a few crops such as wheat, potato and tomato have been widely collected and preserved in seed banks. In most cases, wild germplasm represents less than 2% of the seed bank holdings and most wild relatives of crops still thrive only in the wild.

Crop	% of holdings in seed banks that are wild species	% of wild species still to be collected (estimated)
CEREALS		
Barley	5%	0-10%
Maize	5%	50%
Minor millets	0.5%	90%
Pearl millet	10%	50%
Rice	2%	70%
Sorghum	0.5%	9%
Wheat	10%	20-25%
ROOT CROPS		
Cassava	2%	80%
Potato	40%	30%
Sweet potato	10%	40%
LEGUMES		
Beans	1.2%	50%
Chickpea	0.1%	50%
Cowpea	0.5%	70%
Groundnut	6%	30%
Pigeonpea	0.5%	40%

Hoyt, 1988

Foods of the future?

FRUITS AND VEGETABLES		From:
Buffalo gourd	*Cucurbita foetidissima*	Western North America
Cape gooseberry	*Physalis peruviana*	South America
Durian	*Durio zibethinus*	Southeast Asia
Eelgrass	*Zostera marina*	Mexico
Mangosteen	*Garcinia mangostana*	South America
Naranjilla	*Solanum quitoense*	South America
Narrowleaf lupin	*Lupinus angustifolius*	Australia
Oca	*Oxalis tuberosa*	South America
Pummelo	*Citrus grandis*	Southeast Asia
Rambutan	*Nephelium lappaceum*	Southeast Asia
Tree tomato	*Cyphomandra betacea*	South America
Wax gourd	*Benincasa hispida*	Asia
Winged bean	*Psophocarpus tetragonolobus*	New Guinea
Yeheb nut	*Cordeauxia edulis*	Somalia

ANIMALS		
Green iguana	*Iguana iguana*	South America
Kouprey (cattle)	*Bos sauveli*	Thailand
Pigmy hog	*Sus salvanus*	India
Tilapia (freshwater fish)		Asia/Africa

saline for agriculture, much of it ruined by faulty irrigation schemes. Pressing it into service could increase the world's cropland by two thirds. Some strains of wild wheat, rice, barley, millet and sorghum grow well on salty land; they could be used to create new crops for the ruined fields.

Resources for the future

New foods, too, are to be found in the wilderness. Some 3,000 plant species have been used as food at some time, and another 75,000 – more than a quarter of all known species – are edible. A grass called Job's Tears is an extremely nutritious, but undeveloped cereal. A Paraguayan plant produces calorie-free substances 300 times sweeter than sugar, and a coffee entirely free from caffeine has been discovered on the Comoros Islands near Madagascar. Some species could provide much needed food in arid areas. One Australian grass can yield good crops even if it is only watered once. Such genetic resources will be needed more and more as human numbers grow and marginal land is increasingly pressed into service – and the greenhouse effect alters the world's climate and rainfall patterns.

Industry, too, relies on wild species. Wood and rubber play an enormous part in daily life; rubber was originally used by South American Indians to make toys for their children. Gums, like gum arabic or gum tragacanth are used in inks and cosmetics, sweets and pharmaceuticals, liqueurs and dyes. Frankincense and myrrh are still used in incense and perfumes. Palm oil is used in a hundred products from lipstick to tinplate, ice cream to jet engines; but hardly any of the world's 28,000 species of palm have been investigated by scientists.

Squandering and saving the unknown

Only a fraction of 1 per cent of the world's species has been properly studied for its potential value to humanity in medicines, food or industry. So far scientists have managed to name about 1.4 million of them, but most remain anonymous and unknown. There may be a total of 10 million species on the planet or even, according to the latest estimates, more than 80 million. Remedies for incurable diseases, miracle crops and industrial products wait to be discovered and are destroyed without even being named.

A few reserves have been set aside to protect wild relatives of crop species, and national gene banks have been set up in some 60 countries, usually keeping seeds to save space. Half of the world's strains of rice are stored at the International Rice Research Institute in the Philippines; 12,000 types of wheat and maize from 47 countries are kept at the International Maize and Wheat Improvement Center in Mexico. But stores are costly and can never replicate the wild. Seeds cannot be stored forever without deteriorating, and are vulnerable to disease.

Now that the enormous value of genetic resources is being realized, countries and companies are fighting over who owns them. Some developed and developing countries regard their genetic resources as their property and stop them being exported. Some, including the US, claim proprietorial rights over all the genetic material they keep from anywhere in the world. Private companies are buying up seed firms; 10 of them control a third of all the cereal crop species listed by the Organization for Economic Cooperation and Development. They may jettison the less profitable species, even though these may have huge genetic potential and are particularly useful on specialized terrain. As seed and chemical companies combine, there are fears that they may design crops that require their pesticides – and only theirs – to fight off disease.

In a more hopeful development, Merck, a US drug company, has struck a deal with Costa Rica's National Biodiversity Institute: the company funds the Institute to search for promising genes in the country's rich forests and promises it more than half the royalties on any medicines developed, while retaining the rights to their patents.

In May 1992, after almost four years of negotiations, rich and poor nations agreed on a convention to protect biological diversity. But the treaty was disappointing, dealing only in broad principles. Poor countries, which contain most of the world's wild species, resisted calls from the rich to agree a list of especially valuable areas for protection as "eco-imperialism", while the rich were reluctant to share technologies for exploiting wild genes. Dr Mostafa Tolba, Executive Director of the United Nations Environment Programme, which conducted the negotiations, called the treaty "the minimum on which the international community can agree." Yet the United States refused to sign and campaigned at the Earth Summit in June 1992 for the rest of the developed world to follow suit. In the event, it was isolated with 153 countries signing the convention by the time the Summit ended.

Meanwhile the wildlife massacre continues, interrupting 4 billion years of evolution. In 1986 a group of nine leading American biologists warned that it was a threat to civilization "second only to thermonuclear war" in its severity. The consequences are quite literally incalculable. Life on earth will, at best, take millions of years to recover.

Areas of Endemism

Almost every corner of the earth supports some form of life but some appear to have it in superabundance. Ecuador has many more plant species than the whole of Europe, which is more than 30 times as big. Madagascar has five times as many kinds of trees as the whole of temperate North America. The United States contains fewer woody species of plant than a single volcano, Mount Makiliang in the Philippines – and the entire 20 million square kilometers of the North American continent contain fewer bird species than a 2,000-square kilometer national park in Costa Rica.

Endemism: the uniqueness of species

Some areas have many endemic species – ones that occur nowhere else in the world. Indonesia has one sixth of the world's bird species, and nearly a quarter of them are endemic. Half of Papua New Guinea's birds, half of the Philippines' mammals, and about 80 per cent of Madagascar's plants are unique to them.

Tropical rainforests contain the greatest diversity of species; the US National Academy of Sciences reports that a typical patch, just 10 kilometers square, contains as many as 1,500 species of flowering plant, up to 750 species of tree, 400 different types of bird, 150 butterfly species, 100 kinds of reptile, and 60 species of amphibian. Insects are so abundant that no-one has yet been able to count them, but the Academy estimates that there may be as many as 42,000 in a single hectare. The giant Amazonian rainforest helps to make South America the richest continent for wildlife. It covers only an eighth of the world's land surface, but is home to a third of the world's birds, at least a third of its plants, and probably the same proportion of other species.

Coral reefs are the rainforests of the oceans; the Great Barrier Reef, for example, contains around 350 species of hard corals, 1,500 species of fish and 4,000 different kinds of mollusks. The rainforests' nearest rivals on land are areas with a Mediterranean climate – such as coastal California, the southern part of Western Australia, and the Mediterranean basin itself. These lack the rainforests' diversity of large animals, but have a huge number of endemic plant species.

Species and the changing world

Climate and geography play a part in determining why some areas have so many more species than others. For example, areas with high temperatures and rainfall and little seasonal variation – like tropical rainforests and coral reefs – can support many more species than cold, dry places with distinctly different seasons.

History is important too in determining why similar habitats in different parts of the world have different species. Once areas became isolated from each other – as a result of continental drift, mountain formation and so on – their animal and plant life evolved in different ways. Generally, the longer an area has been isolated, the more distinct and different its inhabitants will be. The best examples are islands and super-islands, such as Madagascar and Australia, with highly distinct fauna and flora.

The theory of continental drift suggests that all the major land areas were once joined in a single "supercontinent". Between 200 and 80 million years ago this broke up, first into two land masses – now called Laurasia and Gondwanaland – and then into areas recognizable as the modern continents. Australia and Antarctica broke off relatively early, while Madagascar probably did not separate entirely until about 60 million years ago. As these land masses separated they carried plants and animals with them.

From this point evolutionary paths began to diverge, and different species formed to fill the available ecological niches. For example, Madagascar and Gondwanaland probably had very similar kinds of primitive primates when they separated. However, over vast periods of time, Madagascar's primates (sheltered from the fierce competition that species still faced on the mainland) developed into lemurs, lower primates found nowhere else. Gondwanaland's primates, subjected to greater pressures, evolved into higher forms – including modern monkeys, apes and, ultimately, man.

Madagascar has more than 6,000 unique flowering plants and half the world's species of chameleons are endemic to the island. Australia's unique array of species evolved similarly in isolation from the rest of the world. Antarctica probably set out on a similar path, but became ever less able to support a wide variety of life as it drifted to its present, icy location.

Volcanic islands and colonization

Some islands, however, have never been attached to the continents. Usually volcanic, they started out as sterile outcrops of rocks. Their animal and plant life consists entirely of species which have colonized them from outside. Birds, bats and winged insects are obvious examples, as are plants with seeds able to resist the ravages of the sea, or those carried in the digestive tracts of birds or bats. The majority of accidental landfalls perish, but a few survive and reproduce. Over time, they evolve into species unique to their islands.

Almost 900 species of bird – 10 per cent of the world's total – have a range of only one island. Old, isolated islands such as the Hawaiian and the Galapagos Islands in the Pacific, and the Mascarene Islands in the Indian Ocean, will have relatively low species diversity compared with continents. Some groups of species will

Endemism & species wealth

Percentage of plant species endemic to country/region:

- 100%
- 80%
- 60%
- 40%
- 20%

(No bar: insufficient data)

- 0%

Number of plant species per 10,000 square kilometers:

- 6,000 and over
- 4,000 – 5,999
- 2,000 – 3,999
- Under 2,000
- Insufficient data

Note: Although no % figure for endemic species is available for Brazil, it is likely to be very high, and the number of plant species per 10,000 square kilometers to be greatly in excess of 6,000.

* Island regarded as a single entity (eg island of Hispaniola instead of Haiti and Dominican Republic).

Cuba
c.6-7,000 plant species, 3-4,000 of these are endemic

Hispaniola
c.5,000 plant species, 1,800 of these are endemic

Azores
56 endemic plant species, 57% are under threat

Canary Island
612 endemic plan species, 67% are under threat

California
c.5,000 plant species, 30% of which are endemic

Galapagos Is.
222 endemic plant species, 59% are under threat

Juan Fernandez
119 endemic plant species, 81% are under threat

Ascension
11 endemic plant species, 10 species are under threat

The drift of the continents

Sedimentary troughs

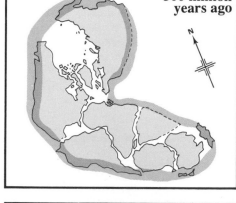

300 million years ago

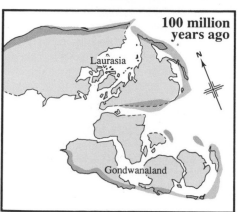

100 million years ago

Laurasia

Gondwanaland

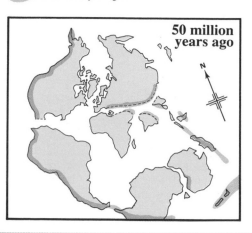

50 million years ago

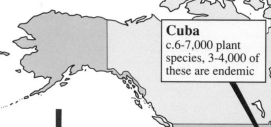

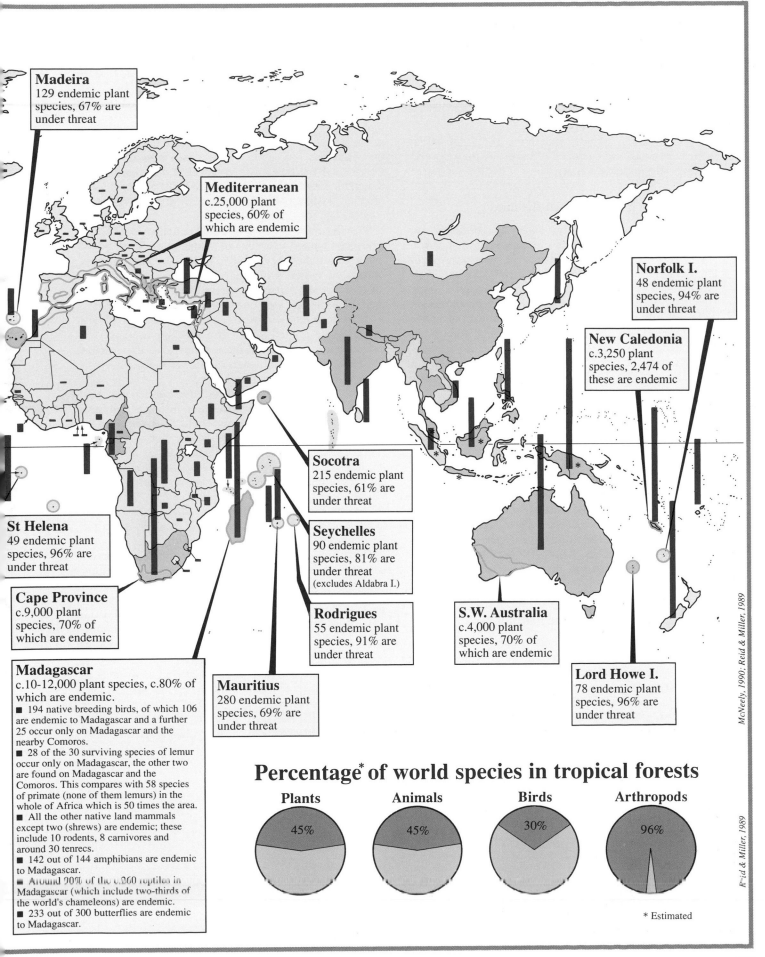

Madeira
129 endemic plant species, 67% are under threat

Mediterranean
c.25,000 plant species, 60% of which are endemic

Norfolk I.
48 endemic plant species, 94% are under threat

New Caledonia
c.3,250 plant species, 2,474 of these are endemic

Socotra
215 endemic plant species, 61% are under threat

Seychelles
90 endemic plant species, 81% are under threat
(excludes Aldabra I.)

St Helena
49 endemic plant species, 96% are under threat

Cape Province
c.9,000 plant species, 70% of which are endemic

Rodrigues
55 endemic plant species, 91% are under threat

S.W. Australia
c.4,000 plant species, 70% of which are endemic

Madagascar
c.10-12,000 plant species, c.80% of which are endemic.
■ 194 native breeding birds, of which 106 are endemic to Madagascar and a further 25 occur only on Madagascar and the nearby Comoros.
■ 28 of the 30 surviving species of lemur occur only on Madagascar, the other two are found on Madagascar and the Comoros. This compares with 58 species of primate (none of them lemurs) in the whole of Africa which is 50 times the area.
■ All the other native land mammals except two (shrews) are endemic; these include 10 rodents, 8 carnivores and around 30 tenrecs.
■ 142 out of 144 amphibians are endemic to Madagascar.
■ Around 90% of the c.260 reptiles in Madagascar (which include two-thirds of the world's chameleons) are endemic.
■ 233 out of 300 butterflies are endemic to Madagascar.

Mauritius
280 endemic plant species, 69% are under threat

Lord Howe I.
78 endemic plant species, 96% are under threat

McNeely, 1990; Reid & Miller, 1989

Reid & Miller, 1989

Percentage* of world species in tropical forests

Plants
45%

Animals
45%

Birds
30%

Arthropods
96%

* Estimated

be entirely missing, but of those present many will be endemic.

Young oceanic islands, and those like coral atolls, which are small and frequently flooded by the sea, tend to have very limited diversity.

The islands on the land

Any ecosystem or habitat surrounded by a different one is, in a sense, an island for the species which live there. And very similar mechanisms of evolution and immigration apply. The flower-rich areas with Mediterranean climates are ecological islands, since they have been separated from each other by enormous areas with quite different habitats for millions of years. They support very diverse flora, with a high percentage of endemic species. The same applies to isolated mountainous regions in the tropics, such as the highlands of Ethiopia, Cameroon and the eastern side of the rift valley in central Africa, which between them support a high proportion of the rare species of Africa. Similarly, the rift valley lakes in Africa are isolated from each other and each has evolved its own highly diverse kinds of fish.

Recent studies have revealed 18 "hot spots" of biological diversity around the world endowed with a superabundance of unique species of plants and animals. Fourteen are in tropical forests, the other four in Mediterranean-type zones. They are thought to contain nearly 20 per cent of the world's plant species on a mere half a per cent of its land. All are under intense development pressures and should be a top priority for preservation.

Extinction and the hand of man

Natural extinctions have always occurred; indeed the number of extinct species far outweighs those alive today. Humankind has altered its environment for millennia, causing some species to prosper and others to suffer. In some ways this has caused incalculable benefits to humanity, such as in the spread of crops and domesticated animals around the world. But it has also led to extinction. Many large species of mammals and birds died out about 10,000 years ago, and some scientists believe that this was caused, at least partly, by the activities of early man. The arrival of the Polynesians on the Hawaiian Islands 1,500 years ago is thought to have consigned as many as 40 bird species to oblivion, nearly three times as many as have perished over the last four centuries.

Most recently some species, like Steller's sea cow, have been hunted to extinction. Others have been wiped out because they have fallen victim to other species, introduced either deliberately or accidentally to their habitats. Others have disappeared because the habitats themselves have been destroyed, and it is this that is fuelling most extinctions today. Human activity is now changing the natural world at a rate unprecedented in evolutionary history, and mass extinctions are a virtual certainty. By some estimates, a million species may die out by the end of the century.

Most concern is focused on the continental land masses, and particularly on tropical rainforests, but most of the extinctions so far documented have occurred on islands. About 108 bird species are known to have become extinct in the last 400 years; 97 of them were island-dwellers. Similarly, islands account for about 75 per cent of all mammals driven to extinction. More than 90 per cent of the endemic plants on St Helena, Ascension Island and Lord Howe Island face extinction, along with more than 80 per cent of those in the entire Seychelles island group.

There are many possible reasons for this apparent paradox. It is hard to be accurate about extinctions, since most are unrecorded; the vast majority of the world's species have neither been named nor classified. But island species are particularly vulnerable if they are endemic. They have nowhere else to go when they come under attack. And they may have few defenses against predators, like cats, rats or pigs, that are introduced onto the islands, because they have evolved in environments where they have no natural enemies. One such species was the proverbial dodo which, during its long evolution on the island of Mauritius, lost both its power of flight and its sense of fear when neither was needed.

Protected Areas and National Parks

About 5 per cent of the world's land surface is now covered by nature reserves, national parks, wildlife sanctuaries and protected landscapes designed to safeguard the rich diversity of plant and animal life on earth.

An old concept

Thousands of years ago, hunter-gatherer societies in India gave special protection to patches of forests. Hundreds of these "sacred groves" still survive and offer vital protection to wildlife and water sources. Kings and princes later protected areas as royal hunting grounds. William the Conqueror set up one of the earliest – the New Forest in southern England, in 1079 AD. Over the last 100 years or more, forest reserves have been set up to provide timber and safeguard soil and water resources.

Yellowstone, the world's first national park, was created in 1872 as a pleasure ground for the American people, the first of many established primarily for recreation. Some of the earliest Canadian national parks were set up by railway companies around scenic parts of their lines to lure tourists onto their trains. But often the protected areas were established on land unfavorable for farming or settlement. The national parks of England and Wales, for example, were set up in beautiful, but poor, high ground. In Zambia most protected areas are sited in tsetse fly-infested areas where big game is still abundant.

The modern networks

Over the last few decades, protection has been more systematic. Countries try to cover the full range of their natural treasures. Canada and New Zealand, which started by protecting scenic and mountainous areas, have since concentrated on expanding their networks to cover the Canadian prairies and the New Zealand lowland. Many countries set out to protect species threatened with extinction or those that are endemic – confined to particular areas. Brazil has put its highest priority on protecting 30 areas to cover endemic species of birds, plants and lizards.

IUCN–The World Conservation Union classifies the world's protected areas into eight categories – from large undisturbed reserves, usually closed to the public, to areas that provide for sustainable production of timber, game and other natural resources. Areas in the first five categories go onto the United Nations List of National Parks and Protected Areas. By the beginning of 1992, there were over 8,000 of them, covering 8.5 million square kilometers. They range in size from small coral islands of 1,200 hectares to the vast Greenland National Park, which covers 70 million hectares. But the numbers still fall far short of what is needed to safeguard what remains of the world's genetic wealth. The Fourth World Congress on National Parks and Protected Areas, held in Caracas, Venezuela, in February 1992, called for the world's protected areas to be doubled in a decade.

Some 100 particularly valuable areas are internationally recognized as World Heritage Sites because of their natural, or combined natural and cultural, importance. Contracting states have a legal duty to protect them; if they become threatened they can be put on a special danger list, which provides for emergency measures.

Nature reserves have to be big enough; preserving small patches of original life as isolated islands in a sea of destruction is ultimately self-defeating, because species need to interreact with the surrounding area. The larger the amount preserved, the smaller and more gradual the loss of species will be. Experiments in the Brazilian rainforest show that isolated patches of forest of just a single hectare deteriorate within months. Even reserves which protect as much as 10 per cent of an original area eventually lose about half their species.

Twelve nature reserves have been set up in the mountains of western China, on the eastern edge of the Tibetan Plateau, to try to protect the Giant panda. More than half of the surviving world population of the species (estimated at about 1,100 animals in the mid-1970s) live in the reserves. But these are fragmented and surrounded by cultivated land. The pandas will almost certainly not be able to survive in the long term unless the reserves are enlarged and linked by special "corridors" of suitable habitat that will allow the animals to move between them to mate or to seek food when supplies are short.

Parks on paper and under threat

Some parks only exist on paper, designated in areas already set aside for intensive agriculture, forestry or other projects. Many have been severely damaged by human activities, or are under threat. By 1989 there were 89 entries on an official list of Threatened Protected Areas of the World, identified by the IUCN Commission on National Parks and Protected Areas and the World Conservation Monitoring Centre.

In 1989, 13 areas were newly identified as at risk. The 1992 Parks Congress decided not to draw up a similar list because the danger to protected areas is so widespread. "Nearly every protected area and park in the world is threatened to some extent, either from development pressures, bad management or a number of other potential dangers," points out the Secretary-General of the Congress, Jeffrey McNeely, Chief Conservation Officer at IUCN.

Protected areas

% of country area classified as protected, 1990

▨	20.0% and over
▨	10.0 – 19.9%
▨	5.0 – 9.9%
▨	1.0 – 4.9%
▨	Under 1.0%
▨	Insufficient data

Protected areas under 1,000 hectares are not included. Protected sites owned privately and those administered locally are also excluded.

World Heritage Sites

●   Natural sites

◆   Combined natural and cultural sites

A World Heritage natural site may exemplify a stage of the world's evolutionary processes, or be representative of biological evolution, or contain the natural habitats of endangered animals. It may be a scene of exceptional beauty, a spectacular view or a reserve for large numbers of wild animals. In addition to the natural and combined sites shown on the map, there are over 200 sites of a purely cultural nature, ranging from archaeological remains (eg Petra in Jordan) to modern cities (eg Brasilia).

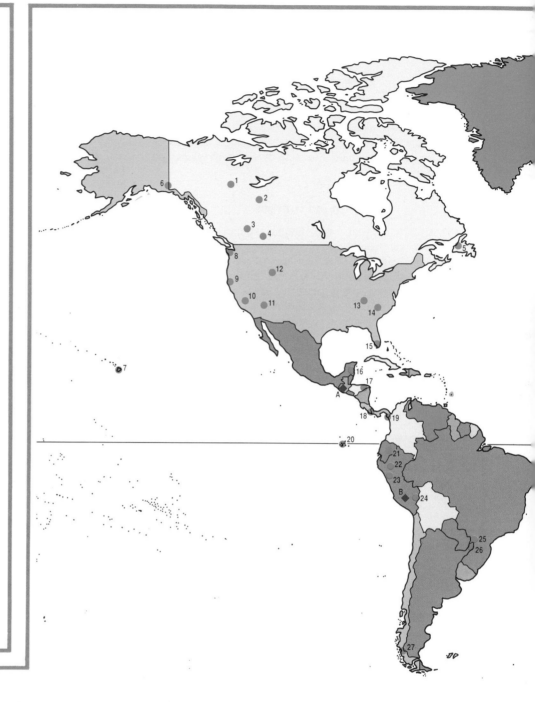

World Heritage Sites: natural ●

1 Nahanni National Park (Canada)
2 Wood Buffalo National Park (Canada)
3 Canadian Rocky Mountains Parks (Canada)
4 Dinosaur Provincial Park (Canada)
5 Gros Morne National Park (Canada)
6 Kluane and Wrangell-St Elias Parks (Canada/United States)
7 Hawaii Volcanoes National Park (United States)
8 Olympic National Park (United States)
9 Redwood National Park (United States)
10 Yosemite National Park (United States)
11 Grand Canyon National Park (United States)
12 Yellowstone National Park (United States)
13 Mammoth Cave National Park (United States)
14 Great Smoky Mountains National Park (United States)
15 Everglades National Park (United States)
16 Sian Ka'an Biosphere Reserve (Mexico)
17 Río Platano Biosphere Reserve (Honduras)
18 Talamanca Range and La Amistad Reserve (Costa Rica)
19 Darien National Park (Panama)
20 Galapagos Islands National Park (Ecuador)
21 Sangay National Park (Ecuador)
22 Rio Abiseo National Park (Peru)
23 Huascaran National Park (Peru)
24 Manu National Park (Peru)
25 Iguaçu National Park (Brazil)
26 Iguazu National Park (Argentina)
27 Los Glaciares National Park (Argentina)
28 St Kilda Island (UK)
29 The Giant's Causeway and its coast (UK)
30 Bialowieza National Park (Poland)
31 Skocjan Caves (Yugoslavia: Slovenia)
32 Plitvice Lakes National Park (Yugoslavia: Croatia)
33 Durmitor National Park (Yugoslavia: Montenegro)
34 Pirin National Park (Bulgaria)
35 Srebarna Nature Reserve (Bulgaria)
36 Danube Delta (Romania)
37 Mount Athos (Greece)
38 Girolata and Porto Gulfs and Scandola Reserve, Corsica (France)
39 Ichkeul National Park (Tunisia)
40 Garajonay National Park, Canary Islands (Spain)
41 Banc d'Arguin National Park (Mauritania)
42 Djoudj Bird Sanctuary (Senegal)
43 Niokolo-Koba National Park (Senegal)
44 Mount Nimba Reserves (Guinea/Côte d'Ivoire)
45 Taï National Park (Côte d'Ivoire)
46 Comoë National Park (Côte d'Ivoire)
47 Aïr Ténéré Reserve (Niger)
48 Dja Faunal Reserve (Cameroon)
49 Parc National du Manovo-Gounda St Floris (Central African Republic)

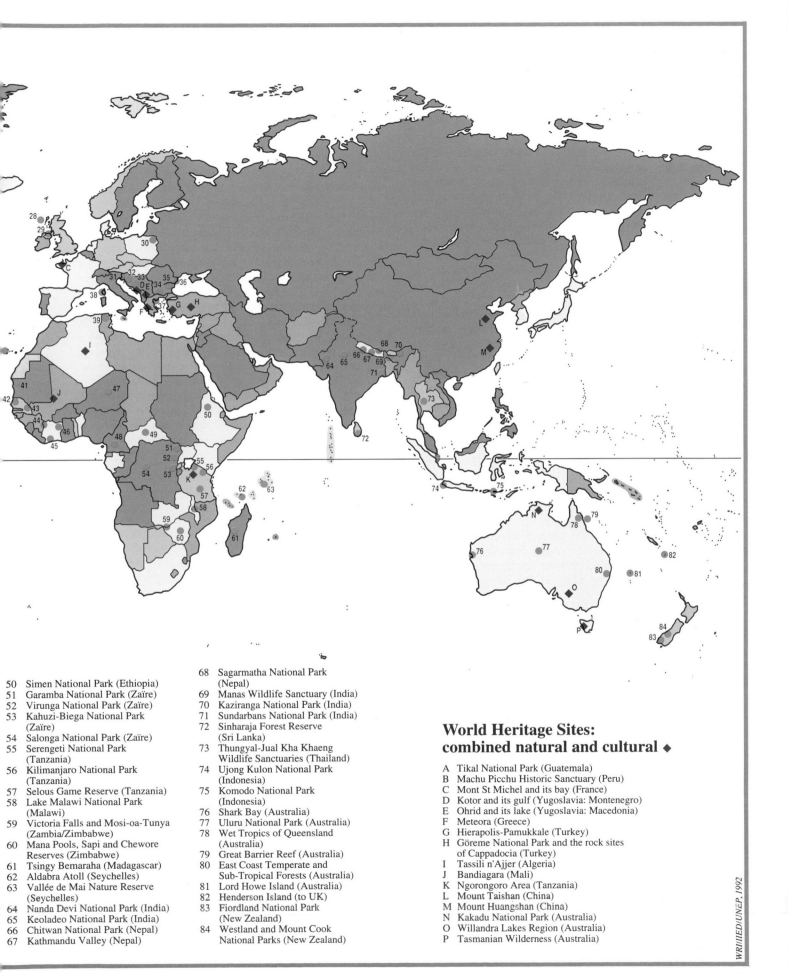

50	Simen National Park (Ethiopia)	
51	Garamba National Park (Zaïre)	
52	Virunga National Park (Zaïre)	
53	Kahuzi-Biega National Park (Zaïre)	
54	Salonga National Park (Zaïre)	
55	Serengeti National Park (Tanzania)	
56	Kilimanjaro National Park (Tanzania)	
57	Selous Game Reserve (Tanzania)	
58	Lake Malawi National Park (Malawi)	
59	Victoria Falls and Mosi-oa-Tunya (Zambia/Zimbabwe)	
60	Mana Pools, Sapi and Chewore Reserves (Zimbabwe)	
61	Tsingy Bemaraha (Madagascar)	
62	Aldabra Atoll (Seychelles)	
63	Vallée de Mai Nature Reserve (Seychelles)	
64	Nanda Devi National Park (India)	
65	Keoladeo National Park (India)	
66	Chitwan National Park (Nepal)	
67	Kathmandu Valley (Nepal)	

68	Sagarmatha National Park (Nepal)
69	Manas Wildlife Sanctuary (India)
70	Kaziranga National Park (India)
71	Sundarbans National Park (India)
72	Sinharaja Forest Reserve (Sri Lanka)
73	Thungyal-Jual Kha Khaeng Wildlife Sanctuaries (Thailand)
74	Ujong Kulon National Park (Indonesia)
75	Komodo National Park (Indonesia)
76	Shark Bay (Australia)
77	Uluru National Park (Australia)
78	Wet Tropics of Queensland (Australia)
79	Great Barrier Reef (Australia)
80	East Coast Temperate and Sub-Tropical Forests (Australia)
81	Lord Howe Island (Australia)
82	Henderson Island (to UK)
83	Fiordland National Park (New Zealand)
84	Westland and Mount Cook National Parks (New Zealand)

World Heritage Sites: combined natural and cultural ◆

A	Tikal National Park (Guatemala)
B	Machu Picchu Historic Sanctuary (Peru)
C	Mont St Michel and its bay (France)
D	Kotor and its gulf (Yugoslavia: Montenegro)
E	Ohrid and its lake (Yugoslavia: Macedonia)
F	Meteora (Greece)
G	Hierapolis-Pamukkale (Turkey)
H	Göreme National Park and the rock sites of Cappadocia (Turkey)
I	Tassili n'Ajjer (Algeria)
J	Bandiagara (Mali)
K	Ngorongoro Area (Tanzania)
L	Mount Taishan (China)
M	Mount Huangshan (China)
N	Kakadu National Park (Australia)
O	Willandra Lakes Region (Australia)
P	Tasmanian Wilderness (Australia)

WRI/IIED/UNEP, 1992

Benefits and costs

Protected areas have many benefits besides simply conserving nature. They protect wild species which could provide future crops, medicines or industrial raw materials. They can generate tourism and safeguard resources like rubber and brazil nuts harvested from the wild. They prevent erosion of coasts and watersheds, conserve water resources and stabilize local climates. They provide shelter for birds that control pests in surrounding farmland, and contribute to the natural balance of whole regions. And they provide facilities for scientific research and education.

These benefits, however, have to be balanced against costs, which fall particularly heavily on local people. They may be prevented from exploiting the areas as freely as they would like. And they may blame protected areas for harboring "pests" – such as the big cats which may kill their cattle, or elephants that may damage their crops.

The Sundarbans, the world's biggest mangrove forest, stretching for nearly 6,000 square kilometers across the delta of the Ganges, Brahmaputra and Meghna rivers, was established as a reserve for the Royal Bengal tiger. It also protects the coasts of India and Bangladesh from the fierce cyclones that sweep up the Bay of Bengal, and provides vital spawning grounds for fish and shellfish. But local people are not allowed to collect honey and wood from the core of the reserve, as they used to, and when they go to get them from peripheral areas they are at great risk from the tigers; about 45 of them are killed every year, as are large numbers of livestock.

A question of balance

No reserve can survive long unless it is at least tolerated by the local people. Protected areas are now increasingly being designed to win local support, by providing benefits for local people as well as for wildlife – and by involving them in their management. In New Guinea, for example, protected areas are often established on tribal lands and run by local management committees. Local people are given incentives to develop profitable ways of exploiting the forest without destroying it. Butterfly farming has proved a great success in Papua New Guinea and is now being spread to the Arfak Mountain Reserve in Irian Jaya, the Indonesian half of the island.

Zambia's Kafue Flats wetlands have been threatened by the poaching of a local antelope, overgrazing by cattle, and hydro-electric dams. The Government's Department of National Parks and Wildlife has worked with WWF to involve local people: annual quotas are set for the antelope and the people then decide how many they want to kill themselves and how many to reserve for trophy-hunting tourists at a charge of $500 apiece. The numbers of antelope have grown rapidly. Similar local cooperation has reduced the numbers of cattle to combat overgrazing. There are plans to mobilize this local support to persuade the dam operators to minimize damage to the wetlands.

Mainland Africa's richest rainforest, Korup in Cameroon, now has a national park surrounded by a "buffer zone". Small-scale hunting, fish farming, forestry and other ways for local people to make an income are encouraged in this zone, protecting the integrity of the park itself. Similar concentric zones are a feature of the global network of "biosphere reserves" (see map on pages 128-129), which aim to strike a balance between strict protection and human use of natural resources.

The fragile future

Ultimately, however, no system of reserves will be able to protect the richness and diversity of the world's wild places unless governments act to reduce poverty, control population growth and cut pollution. The pressure of poor people seeking land on which to grow crops to feed their families is increasing. If, as now, they are excluded from the best land, they will inevitably destroy most of the remaining rainforests. Similarly, governments shackled with debt, and faced with falling prices for their exports, will plunder natural resources to stay afloat economically.

Global warming will also affect wildlife habitats in protected areas. Plant species will have to move about 90 kilometers towards the poles for every 1°C rise in temperature, if they are to survive. If the warming takes place rapidly, they will not be able to move fast enough and habitats will disappear. Wetlands will be inundated as sea levels rise. The world will only be able to prevent a wildlife disaster if it can slow down the rate of warming and allow natural habitats time to spread. If they fail to do so then most of the work undertaken to establish protected areas will have been in vain.

Major Conservation Efforts

Conservation issues have become major popular and political concerns. National and international conservation organizations are increasingly able to enrol the support of governments, international development agencies, corporations and commercial organizations and – most important of all – the general public. Awareness has never been higher: the question is whether it can be translated into action in time to avert disaster.

Saving the species

Campaigns by conservation organizations to save whales and elephants have achieved considerable success. Decisions by the International Whaling Commission have effectively halted commercial whaling – though a few countries continue large-scale whaling for what they describe as scientific purposes.

In 1989, the nations belonging to the Convention on International Trade in Endangered Species of Wild Fauna and Flora (CITES) approved a worldwide ban on the ivory trade after hearing evidence that, if it continued, the African elephant faced extinction within 20 years. Although seven countries formally exempted themselves from the ban, a report in June 1990 showed that the trade had effectively collapsed, and the ban was renewed in 1992.

Four major organizations have joined forces to establish an international migratory birds conservation program. Other similar efforts are underway or in the pipeline.

Saving the habitats

Conservationists are moving into areas relatively undisturbed by human activities, in order to protect them from the degradation experienced elsewhere and show how they can be managed sustainably.

WWF is working with the Government of Bhutan, a tiny and mainly undamaged kingdom in the Himalayas, to plan conservation ahead of development. Around 60 per cent of the country is still covered with forest. The Government has developed a National Forest Policy and a Forest and Nature Conservation Act, to ensure that development will not degrade the environment and that virtually all wildlife is protected. It has also established Bhutan's first national park – Royal Manus Park.

Irian Jaya, the western (Indonesian) portion of New Guinea has some of the world's best, most extensive and least disturbed wetlands – covering 12.5 million hectares, about 30 per cent of the land – and the second largest number of endemic species in the Pacific region. WWF, IUCN–The World Conservation Union and other agencies have cooperated with the Government to produce a development program to benefit both wildlife and the local people who depend on the wetlands. It includes setting up protected areas and developing management plans for them. Conservationists and local people have together come up with a system of "use zones", ranging from areas of strict nature protection to buffer zones in which traditional activities are pursued, under community participation management schemes.

In Zambia, the Kafue Flats and Bangweulu Basin, one of Africa's largest and most important wetland areas, is under threat from overhunting and hydro construction projects. A comprehensive management plan is being prepared to protect the area both for wildlife and local communities, with the active participation of local government and people.

Saving the forests

The International Tropical Timber Agreement (ITTA), which is managed by the International Tropical Timber Organization (ITTO), aims to halt wanton logging in tropical forests. Although it has been adopted by countries representing 90 per cent of the world trade in tropical timber and is the first ever commodity agreement concerned with conservation, it has achieved very little.

Since 1987, WWF and other agencies have purchased part of the debts of such countries as Ecuador, Costa Rica, the Philippines, Zambia and Madagascar, at a discount. The debt has been redeemed in local currency and the money spent on conservation projects.

But neat, and valuable, though this is, it makes little impact on the destructiveness of the debt crisis, which continues to cause many countries to ravage their natural resources to try to keep up interest payments.

Saving the islands

Madagascar's unique wildlife is threatened by the wholesale destruction of its habitat. Belatedly, conservationists, the Government and local people are making massive efforts to stem the destruction. Under the National Conservation Strategy, 36 national parks and nature reserves have been created, and reforestation projects are underway. An environmental profile has been prepared which helps to educate people in how to use the country's resources sustainably. A large rural development program is being developed on the Masoala peninsula, which involves teaching sustainable agricultural methods, creating a national park, promoting tourism and providing training in biological diversity.

Last ditch efforts are also underway to preserve the unique genetic heritage of the Galapagos Islands, where

Major conservation initiatives

WILDLIFE & HABITAT — CITES, RAMSAR, BONN
MARINE — ITTA, LAW OF SEA, LONDON, PARIS, OSLO, MARPOL, REG. SEAS
LAND & ATMOSPHERIC POLLUTION — MONTREAL, BASEL, 30% CLUB
IUCN, WWF

N. & C. AMERICA
ANTIGUA & BARBUDA
BAHAMAS
BARBADOS
BELIZE
CANADA
COSTA RICA
CUBA
DOMINICA
DOMINICAN REP.
EL SALVADOR
GRENADA
GUATEMALA
HAITI
HONDURAS
JAMAICA
MEXICO
NICARAGUA
PANAMA
ST CHRIST. & NEVIS
ST LUCIA
ST VINCENT
TRINIDAD & TOBAGO
UNITED STATES

SOUTH AMERICA
ARGENTINA
BOLIVIA
BRAZIL
CHILE
COLOMBIA
ECUADOR
GUYANA
PARAGUAY
PERU
SURINAM
URUGUAY
VENEZUELA

EUROPE & USSR
ALBANIA
AUSTRIA
BELGIUM
BULGARIA
CZECHOSLOVAKIA
DENMARK
FINLAND
FRANCE
GERMANY
GREECE
HUNGARY
ICELAND
IRELAND
ITALY
LIECHTENSTEIN
LUXEMBOURG
MALTA
MONACO
NETHERLANDS
NORWAY
POLAND
PORTUGAL
ROMANIA
SAN MARINO
SPAIN
SWEDEN
SWITZERLAND
USSR††
UNITED KINGDOM
VATICAN CITY
YUGOSLAVIA††
EC

AFRICA
ALGERIA
ANGOLA
BENIN
BOTSWANA
BURKINA FASO
BURUNDI
CAMEROON
CAPE VERDE
CENTRAL AFR. REP.
CHAD
COMOROS
CONGO
COTE D'IVOIRE
DJIBOUTI
EGYPT
EQUATORIAL GUINEA
ETHIOPIA
GABON
THE GAMBIA
GHANA
GUINEA
GUINEA-BISSAU
KENYA
LESOTHO
LIBERIA
LIBYA
MADAGASCAR
MALAWI
MALI
MAURITANIA
MAURITIUS
MOROCCO
MOZAMBIQUE
NAMIBIA
NIGER
NIGERIA
RWANDA
SAO TOME & PRINCIPE
SENEGAL
SEYCHELLES
SIERRA LEONE
SOMALIA
SOUTH AFRICA
SUDAN
SWAZILAND
TANZANIA
TOGO
TUNISIA
UGANDA
ZAÏRE
ZAMBIA
ZIMBABWE

OCEANIA
AUSTRALIA
FIJI
KIRIBATI
NAURU
NEW ZEALAND
PAPUA NEW GUINEA
SOLOMON ISLANDS
TONGA
TUVALU
VANUATU
WESTERN SAMOA

* Other signatories: Cook Is. (NZ); Niue (NZ).
† Other party: Marshall Is.
‡ Other participating states: Belau (US); Cook Is. (NZ); Marshall Is.; Federated States of Micronesia; Niue (NZ); Palestine.
** UK signature is effective also for Hong Kong.

†† Note: It is uncertain at present which, if any, of the states of the former USSR and Yugoslavia recognize the treaty obligations entered into in the years before the disintegration of those nations.

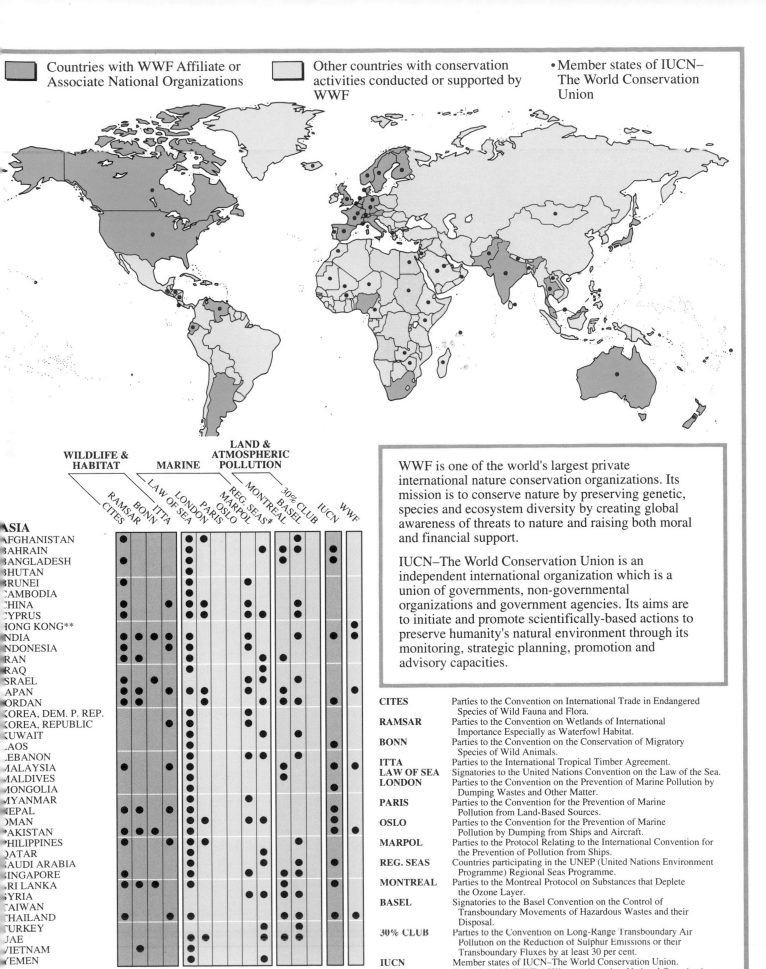

Countries with WWF Affiliate or Associate National Organizations

Other countries with conservation activities conducted or supported by WWF

• Member states of IUCN–The World Conservation Union

WILDLIFE & HABITAT — RAMSAR, CITES, BONN, ITTA
MARINE — LAW OF SEA, LONDON, PARIS, OSLO, MARPOL, REG. SEAS†
LAND & ATMOSPHERIC POLLUTION — MONTREAL, BASEL, 30% CLUB
IUCN
WWF

ASIA

AFGHANISTAN
BAHRAIN
BANGLADESH
BHUTAN
BRUNEI
CAMBODIA
CHINA
CYPRUS
HONG KONG**
INDIA
INDONESIA
IRAN
IRAQ
ISRAEL
JAPAN
JORDAN
KOREA, DEM. P. REP.
KOREA, REPUBLIC
KUWAIT
LAOS
LEBANON
MALAYSIA
MALDIVES
MONGOLIA
MYANMAR
NEPAL
OMAN
PAKISTAN
PHILIPPINES
QATAR
SAUDI ARABIA
SINGAPORE
SRI LANKA
SYRIA
TAIWAN
THAILAND
TURKEY
UAE
VIETNAM
YEMEN

WWF is one of the world's largest private international nature conservation organizations. Its mission is to conserve nature by preserving genetic, species and ecosystem diversity by creating global awareness of threats to nature and raising both moral and financial support.

IUCN–The World Conservation Union is an independent international organization which is a union of governments, non-governmental organizations and government agencies. Its aims are to initiate and promote scientifically-based actions to preserve humanity's natural environment through its monitoring, strategic planning, promotion and advisory capacities.

CITES	Parties to the Convention on International Trade in Endangered Species of Wild Fauna and Flora.
RAMSAR	Parties to the Convention on Wetlands of International Importance Especially as Waterfowl Habitat.
BONN	Parties to the Convention on the Conservation of Migratory Species of Wild Animals.
ITTA	Parties to the International Tropical Timber Agreement.
LAW OF SEA	Signatories to the United Nations Convention on the Law of the Sea.
LONDON	Parties to the Convention on the Prevention of Marine Pollution by Dumping Wastes and Other Matter.
PARIS	Parties to the Convention for the Prevention of Marine Pollution from Land-Based Sources.
OSLO	Parties to the Convention for the Prevention of Marine Pollution by Dumping from Ships and Aircraft.
MARPOL	Parties to the Protocol Relating to the International Convention for the Prevention of Pollution from Ships.
REG. SEAS	Countries participating in the UNEP (United Nations Environment Programme) Regional Seas Programme.
MONTREAL	Parties to the Montreal Protocol on Substances that Deplete the Ozone Layer.
BASEL	Signatories to the Basel Convention on the Control of Transboundary Movements of Hazardous Wastes and their Disposal.
30% CLUB	Parties to the Convention on Long-Range Transboundary Air Pollution on the Reduction of Sulphur Emissions or their Transboundary Fluxes by at least 30 per cent.
IUCN	Member states of IUCN–The World Conservation Union.
WWF	Countries with WWF Affiliate or Associate National Organizations.

UNEP, 1991; UNEP, 1991(a); IUCN membership list, 1991; WWF, pers. comm.

introduced species, mass tourism and brush fires have wreaked great damage. WWF and its Ecuador Affiliate Organization, Fundación Natura, are financing the protection projects through support for the Charles Darwin Foundation and Research Station. National parks are being created, conservation projects for endangered reptile species and management plans for marine areas are being developed, and programs have begun to control predators. The project is one of several being funded by a debt-for-nature swap deal.

Conservation strategies

In 1980, WWF, IUCN and the United Nations Environment Programme (UNEP) published the *World Conservation Strategy* – the first coherent international statement of how economic development and the conservation of natural resources depend upon each other. It has led more than 40 nations to integrate national conservation strategies, to varying extents, into their development plans. And in May 1992, the world agreed a convention on biological diversity which, despite opposition by the United States, was signed by over 150 countries at the Earth Summit in Rio de Janeiro in June.

The conservation and management of Antarctica – the world's last true wilderness – and its seas are covered by a variety of international treaties, and in 1991 the Antarctic Treaty nations agreed to ban mining there for at least 50 years.

Conservation education

Conservation projects will fail without a fundamental change in attitudes and so conservation organizations make education a major priority, with an unprecedented drive to build comprehensive education programs into primary school curricula.

In Germany, for example, schoolchildren are taught about the wildlife trade through the *Artenschutzkoffer* (the species protection suitcase), which contains examples of wildlife products, accompanied by audiovisual material about the trade. In Sarawak, WWF finances specialist conservationist education officers to work with teachers and other experts. In southern Africa, in Namibia, South Africa, Swaziland and Zimbabwe, students receive an education pack called "We Care!", which contains activity and discussion programs designed to focus attention on their natural surroundings and the human impact on them.

Fighting pollution

Pollution is an international problem, defying political boundaries. In 1979, the member states of the Economic Commission for Europe signed the Convention on Long-Range Transboundary Air Pollution, designed to reduce emissions of sulfur dioxide and, to a lesser extent, of nitrogen oxides. It came into force in 1983 and was subsequently amended in 1985 with the addition of the "30 Per Cent Club" – countries that agreed to reduce sulfur dioxide emissions from 1980 levels by 30 per cent by 1993.

Since the mid-1970s, UNEP has been developing a remarkable network of agreements, under which neighboring countries take joint action to conserve common seas. There are now 12 such regional seas programs – covering the Mediterranean, the Persian Gulf, the Caribbean, the Red Sea, the East Africa coast, the coasts of Central and West Africa, the Pacific coast of South America, the South Pacific islands, the East Asian and South Asian seas, the Northwest Pacific and the Black Sea. Countries participating in the programs include some old and present enemies, such as Israel and Syria, Greece and Turkey, Iran and Iraq, and Cuba and the United States, working together to protect their shared waters. The eight nations of the Zambesi River Basin also operate a joint conservation agreement.

Recent international treaties have offered hope that the world can combine to tackle global pollution problems. Thirty six countries agreed in March 1989 to control the trade in toxic waste by concluding the Basel Convention on the Control of Transboundary Movements of Hazardous Wastes and their Disposal. By the time it came into force in May 1992, 52 countries had signed the Convention and 21 countries had ratified it.

The world has taken rapid action against the chemicals that are destroying the ozone layer. In 1985, nations agreed an overall convention in Vienna to protect the earth's vital ozone shield. Two years later they agreed, in the Montreal Protocol, to cut use of the most damaging chemicals, chlorofluorocarbons (CFCs), by half by 1996. Then in June 1990, they decided to stop using them altogether by the year 2000, resolved to phase out other ozone-destroying chemicals, and agreed that a special fund should be set up to help Third World countries switch to less polluting substances. And in May 1992, the world agreed on a convention to combat global warming: widely seen as a first step, it was signed by 153 countries at the Earth Summit in Rio de Janeiro that June.

Wildlife Trade

The international trade in wild animals, plants and wildlife products is big business – worth up to $5 billion a year worldwide. Most of it is entirely legal, controlled by national laws and an international treaty. But about a third to a quarter of the trade – worth around $1.5 billion a year – is unlawful commerce in rare and endangered species, usually poached in the wild and smuggled across frontiers. The trade is one of the main ways in which species become threatened, endangered and driven to the verge of extinction.

Millions of live animals are shipped around the world every year to supply the pet trade. Furs, leather and ivory are also traded in vast quantities. Typically, a single year will see 50,000 live primates, tusk ivory from 70,000 African elephants, 4 million live birds, 10 million reptile skins, 15 million pelts, about 350 million tropical fish and about 1 million orchids bought and sold around the world.

Wildlife products are among the most expensive in the world. Daggers with handles fashioned from African rhino horn can fetch $12,000 each; white Asian rhino horn sells at $13,000 a pound for medicines in the Far East. A Clouded leopard coat can be worth over $100,000 in Japan, while Vicuña coats sell for around $20,000 in the US. Rare live specimens can command similar prices. Two smuggled Spix's macaw were offered for sale at $80,000 in Europe; a consignment of endangered parrots intercepted by Spanish customs officials was valued at $200,000. Rare species of cacti can fetch up to $15,000 for a single specimen.

Buyers, sellers and middlemen

The major markets for the wildlife trade are in the US, Japan and Europe. The US legal market, worth around $250 million per year, is outdistanced by an illegal trade worth a further $300 million. Primates, birds, tropical fish and reptile skins make up the bulk of the trade. Japan is regarded as the largest marketplace for illegal wildlife products, while Europe is also a major market for exotic birds, reptile skins, monkeys and small cats.

The main supplier regions are South America, Africa, East Asia and the US. In South America, Bolivia, Argentina, Brazil, Peru and Guyana are involved in the trade. In Africa, major exporters include Senegal, Tanzania, the Congo, the Sudan and South Africa. The Asian trade is centered on the Philippines, Taiwan, Thailand, Indonesia, Singapore and Japan.

Much of the world's illegal trade relies on a handful of countries which act as conduits. Until 1989, ivory from up to 30,000 poached elephants passed through Burundi on its way to the United Arab Emirates, where permits legitimizing the booty were issued. The Emirates represents one of the most important centers in the world for the illegal wildlife trade. In Latin America, French Guiana, Bolivia and Paraguay serve as middlemen for the illegal movement of up to 150,000 parrots per year to US pet shops. Taiwan offers a passage for poached and smuggled elephant tusks and rhino horn from Africa, as well as parrots and reptile skins from Latin America.

Drastic consequences

Worldwide, some 622 species of animals and plants face extinction as a result of the trade. In addition, about 2,300 animals and 24,000 plants are endangered. Asian demand for rhino horn probably presents the most serious single problem: the number of black rhinos has been cut by 95 per cent since 1970, mainly as a result of poaching, while the total population of the world's nine rhino species has fallen by 84 per cent.

Meanwhile the numbers of African elephants have been cut by over 50 per cent since 1979; about 70,000 are killed each year for their ivory. In the period from 1981 to 1989, Sudan lost 84 per cent of its elephants; Zambia lost 80 per cent; Kenya lost 75 per cent; Tanzania and Zaïre lost 70 per cent.

Other endangered species include leopards, giant otters, South American caiman and Nile crocodiles, fruit bats and medicinal leeches, hyacinth macaws and palm cockatoos, hummingbirds, gregarious peccaries, poison arrow frogs and the carnivorous pitcher plant.

Most animals in the wild are killed for their skins and products. Between 60 and 80 per cent of all live wild animals smuggled round the world die in transit. Customs officials at Madrid airport in Spain recently found 1,500 dead baby crocodiles in a shipment of crates containing 2,000.

As species become extinct, or their populations dwindle, the balance of nature is severely upset. The disappearance of a species can alter a food chain, upset the delicate balance between one species that preys on another, and cause insect pests to multiply. Malaria is on the upsurge in Asia, for example, because of over-harvesting of bullfrogs which eat the mosquitoes that carry the disease. Every year more than 200 million Asian bullfrogs are killed to provide frogs' legs for the trade.

Decimation of species also has devastating effects on a country's economy. In developing countries, local populations often depend on selling wild animal and plant products for their livelihood, and governments depend on the taxes and duties received from exporting these products.

Controls on wildlife trade

Most of the developed nations have some national legislation designed to prevent trade in endangered species, and provisions for its enforcement.

Wildlife trade

Members of CITES, October 1991:
(Convention on International Trade in Endangered Species of Wild Fauna and Flora)

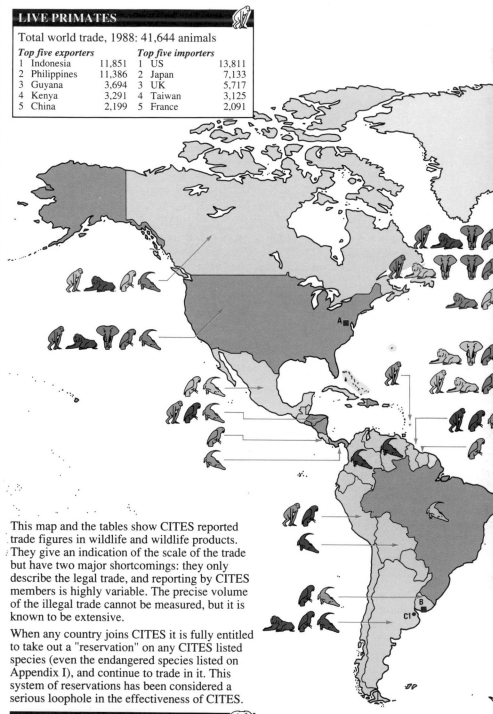

- Fulfilling 100% of their reporting requirement
- 75 – 99%
- 50 – 74%
- Less than 50%
- Figures not available

Parties to CITES are required to report annually all imports and exports of CITES listed species. Countries meeting less than 100% of their reporting requirement have not submitted trade records for all years since becoming party to CITES, or have submitted only partial records. The extent of the shortfall can be estimated by means of a comparative tabulation of all the reports which have been submitted, and this is expressed as a percentage figure.

- Top exporters (see boxes)
- Other major exporters (see below)
- Top importers (see boxes)
- Other major importers (see below)

Live primates, 1988
Major exporters and importers: over 500 animals

Cat skins, 1988
Major exporters and importers: over 1,000 skins

Ivory, 1988
Major exporters and importers: over 1,000 kilograms

Live parrots, 1988
Major exporters and importers: over 3,000 parrots

Reptile skins, 1988
Major exporters and importers: over 5,000 skins

CITES/TRAFFIC offices:
(Trade Records Analysis of Flora and Fauna in Commerce)

■ Regional headquarters

A. Washington; B. Montevideo; C. Brussels; D. Lilongwe; E. Petaling Jaya; F. Tokyo; G. Sydney.

● National representatives

C1. Buenos Aires; C2. Zeist; C3. Frankfurt; C4. Versailles; C5. Rome; C6. New Delhi; C7. Taipei.

LIVE PRIMATES

Total world trade, 1988: 41,644 animals

Top five exporters		Top five importers	
1 Indonesia	11,851	1 US	13,811
2 Philippines	11,386	2 Japan	7,133
3 Guyana	3,694	3 UK	5,717
4 Kenya	3,291	4 Taiwan	3,125
5 China	2,199	5 France	2,091

This map and the tables show CITES reported trade figures in wildlife and wildlife products. They give an indication of the scale of the trade but have two major shortcomings: they only describe the legal trade, and reporting by CITES members is highly variable. The precise volume of the illegal trade cannot be measured, but it is known to be extensive.

When any country joins CITES it is fully entitled to take out a "reservation" on any CITES listed species (even the endangered species listed on Appendix I), and continue to trade in it. This system of reservations has been considered a serious loophole in the effectiveness of CITES.

IVORY

Total world trade, 1988: 430,000 kilograms*

Top ten exporters		Top ten importers	
1 Hong Kong	135,938	1 Hong Kong	179,608
2 Singapore	100,568	2 Japan	100,985
3 Japan	33,149	3 China	50,879
4 Belgium	24,034	4 Belgium	38,730
5 Somalia	22,638	5 Singapore	17,722
6 Tanzania	22,581	6 US	4,803
7 Congo	18,806	7 France	4,486
8 Gabon	13,542	8 UK	4,222
9 Zaïre	11,009	9 India	3,879
10 Djibouti	10,901	10 Taiwan	2,955

* In 1989 a number of individual countries instituted bans on trade, and early in 1990 a general CITES ban came into force. Between them these measures have progressively reduced legal trade flows.

LIVE PARROTS

Total world trade, 1988: 625,595 parrots

Top ten exporters		Top ten importers	
1 Argentina	179,762	1 US	277,4
2 Indonesia	87,830	2 F.R. Germany	54,9
3 Tanzania	65,091	3 UK	40,7
4 Uruguay	40,537	4 Japan	35,0
5 Senegal	33,729	5 Spain	30,9
6 Guinea	29,878	6 France	30,1
7 Guyana	26,935	7 Italy	29,7
8 Honduras	24,515	8 South Africa	26,1
9 Taiwan	21,541	9 Singapore	12,1
10 Peru	16,760	10 Jordan	10,2

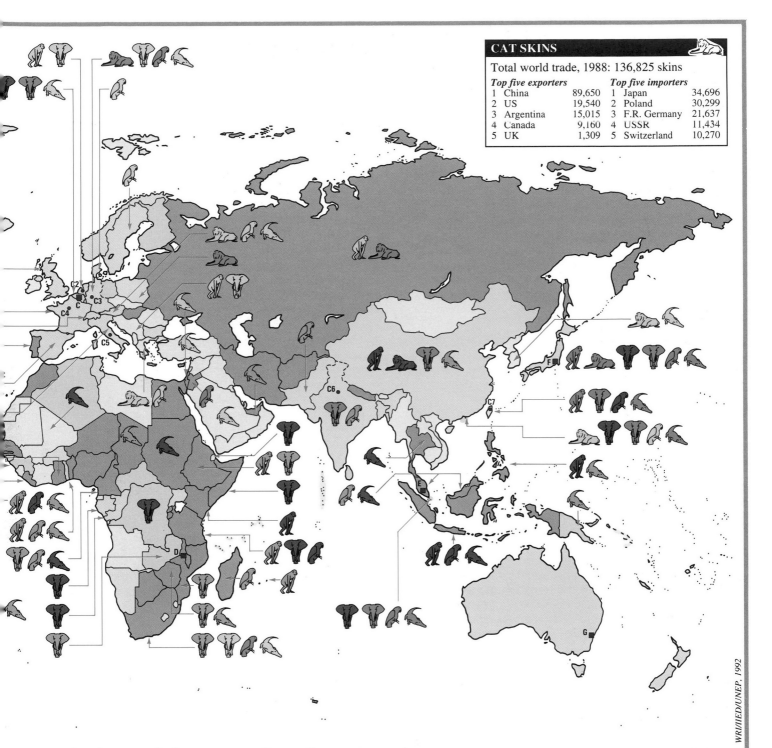

CAT SKINS

Total world trade, 1988: 136,825 skins

Top five exporters		Top five importers	
1 China	89,650	1 Japan	34,696
2 US	19,540	2 Poland	30,299
3 Argentina	15,015	3 F.R. Germany	21,637
4 Canada	9,160	4 USSR	11,434
5 UK	1,309	5 Switzerland	10,270

Populations of threatened and endangered species

The population figures given here are very crude, and are intended only to give some indication of what wildlife trade can do to populations. In the case of birds figures may refer to pairs or singing males only. MRE = most recent estimate.

MAMMALS

Black Rhino

1960s	Not available
1970s	Less than 15,000
1980s	Less than 10,000
MRE	3,400

Northern White Rhino

1960s	Less than 1,500
1970s	Less than 750
1980s	Less than 15
MRE	28

Scimitar Horned Oryx

1960s	Not available
1970s	Less than 4,500
1980s	Less than 2,500
MRE	Less than 1,000

African Elephant

1960s	Not available
1970s	1.3 million
1980s	764,000
MRE	610,000

BIRDS

Whooping crane

1960s	Less than 50
1970s	Less than 100
1980s	Less than 200
MRE	195

REPTILES

Kemp's Ridley Sea Turtle

1960s	Less than 5,500
1970s	Less than 2,000
1980s	Less than 600
MRE	Less than 1,000

WCMC, pers. comm.

REPTILE SKINS

Total world trade, 1988: 6,634,678 skins

Top ten exporters		Top ten importers	
1 Indonesia	3,032,189	1 US	1,641,308
2 Argentina	1,747,153	2 Japan	950,047
3 Mali	406,312	3 France	883,971
4 Thailand	260,080	4 Spain	625,457
5 Malaysia	238,205	5 UK	583,585
6 Cameroon	148,510	6 Singapore	445,873
7 Sudan	106,702	7 Italy	426,546
8 Bolivia	93,708	8 Hong Kong	267,393
9 Venezuela	92,294	9 Taiwan	162,888
10 Colombia	74,173	10 Canada	119,726

In 1975, an international treaty to govern trade in wildlife – the Convention on International Trade in Endangered Species of Wild Fauna and Flora (CITES) – came into force: by 1992, 107 countries adhered to its requirements. The treaty is designed to conserve endangered species while allowing trade in wildlife whose populations can stand it. CITES bans all commercial trade related to endangered species, which it lists in its Appendix I, and limits and monitors trade related to species at risk of becoming endangered, listed in Appendix II. In addition, any country may create another list of species to which it wants to give special protection and register it with CITES, or can ban wildlife trade altogether.

Enforcement of CITES is the responsibility of the member states and governments are required to submit reports and trade records to the CITES Secretariat. A CITES permit is the only legal permit recognized for international transit of a wild animal, plant or product.

CITES is assisted in its work by two other agencies: The Wildlife Trade Monitoring Unit, part of IUCN–The World Conservation Union, which collects and analyzes data on wildlife trade, and TRAFFIC (Trade Records Analysis of Flora and Fauna in Commerce), a worldwide network established by IUCN and WWF, which monitors trade in wild animals, plants and wildlife products and assists in implementing the treaty. TRAFFIC has 15 offices in key wildlife trade areas of the world. It carries out its own investigations and alerts both CITES and customs officials about suspected smugglers.

But species continue to be decimated by international trade. Many countries benefit economically from illegal trade and are unwilling to enforce the treaty. Any country, moreover, can legally continue to trade even in the most endangered species simply by telling CITES of its intention to do so, and entering a formal "reservation". By March 1987 there were 56 such reservations which avoided trade bans under Appendix I, and 45 avoiding controls under Appendix II.

In 1989, CITES faced a crisis over the ivory trade. Many countries, led by Kenya, wanted to ban it altogether by placing the elephant on Appendix I; but others with healthy elephant populations, led by Zimbabwe, insisted that they should be allowed to go on trading. After a stormy meeting, a majority of countries voted to stop the trade, but seven – Zimbabwe, South Africa, Botswana, Malawi, Zambia, China and the United Kingdom (on behalf of Hong Kong) – took out reservations. Nevertheless, the decision resulted in a rapid and dramatic reduction in the trade and, despite predictions by economists to the contrary, prices fell. The ban was reconfirmed at the next CITES meeting in Kyoto, Japan, in 1992. Overall, CITES has done a great deal both in calling attention to the wildlife trade, and in progressively tightening controls on it.

Captive breeding programs have been set up to try and increase the numbers of certain threatened species. Common marmosets are bred in the UK and Germany, while over 8,000 monkeys, most of them Rhesus monkeys, are born in the US each year. There are captive breeding centers for the Two-faced lovebird in Holland, Belgium and South Africa, and there are at least 10 crocodile ranching operations around the world. But these are not enough in themselves to save threatened species.

Stricter control measures are still needed both internationally and within countries to curb the trade further. It is particularly important to address the demand from consumer countries, as it is this that fuels the trade. If the demand exists for endangered species, it is almost impossible to stop them being exploited and smuggled across frontiers, especially in the Third World. When demand is halted – as with the ivory trade – poaching dies down. If people refused to buy ivory carving, rhino horn, rare parrots, spotted cat fur coats, reptile skin goods, tortoise shell jewelry, or rare cacti or orchids and the like, the trade in endangered species would rapidly dry up.

Migration

Many kinds of animals migrate – birds, fish, eels, mammals, reptiles, even butterflies and moths. Some, like birds and some fish, migrate to find food and a suitable year-round climate. Others, like salmon and eels, migrate to spawn.

Giant sea turtles, whales, caribou, wildebeests, Arctic foxes, and some frogs and salamanders have regular migration patterns. Cod, mackerel and pilchard forsake coastal waters when they get cold in winter, and move to warmer deeper parts of the sea. Dolphins, tunas, jacks and barracudas move southwards to warmer water as the northern winter approaches, and head back north again in the spring.

Migrating species achieve extraordinary feats of navigation and stamina. Salmon find their way back – through the open sea, rivers and tributaries – to spawn in the streams where they hatched years before. European and American eels spawn together in the Sargasso Sea and then go their separate ways; one to travel 5,000 kilometers over three years to Europe, the other to reach North America after a year and 1,600 kilometers. They grow at different speeds so that both are about the same size when they reach their destinations. Hundreds of millions of Monarch butterflies flock south from North America each year: Painted Ladies develop in North Africa and travel to Europe in the spring; some moths also migrate. Free-tailed bats summer in the United States and can travel 1,600 kilometers to winter in Mexico; others migrate hundreds of kilometers from northern to central Europe.

Birds, which build up stores of fat before their journeys, sometimes doubling their body weight, hold the long-distance records. The tiny Northern waterthrush journeys over 4,000 kilometers from New York State to Venezuela. Peregrine falcons have been known to travel over 16,000 kilometers from Alaska and Canada to Peru and Argentina. The Arctic tern breeds in Britain in the summer, roves as far north as the Arctic Circle, and then migrates the length of the earth to the Antarctic. No other living thing enjoys as much daylight every year, but it has to travel some 35,000 kilometers annually to get it.

Major migration routes

Migratory birds use many different North-South flyways, but only around 15 are considered major routes. Migration in the Americas follows four main paths. The 500-kilometer wide Atlantic flyway follows the east coast of North America south to Florida, and across the Caribbean to South America. Birds which breed in Canada and Alaska, but winter in Central and northern South America, use the Mississippi flyway with its surrounding wetlands as their main route. The Central flyway runs east of the Rocky Mountains south to the coastal wetlands of Texas and Louisiana; some species then go on to Central America. The Pacific flyway funnels birds south to Mexico and Central America from northwestern North America and along the Rocky Mountains.

Europe also has roughly four major migration routes south to the Mediterranean and Africa, with a number of crucial arteries and intersections. One brings migrants – mostly seabirds, ducks and geese – from Greenland and Iceland across to the British Isles, where they join throngs of local species going south along the Atlantic coasts of France and Spain, or through western and central europe to the Mediterranean and the Middle East. Birds from Siberia, northern Scandinavia and European Russia converge on the Wadden Sea before moving further south. Another flyway channels birds from Finland, the Baltic States and eastern Europe south along the Black Sea and the Bosphorus, where they join up with migrants from western Europe and follow the Levant coast to Egypt and the Red Sea; many head for East and southern Africa.

There are three major Asian migration routes. The most impressive funnels breeding birds from northern Scandinavia, Siberia, Korea and northern China south along four broad pathways: to Myanmar and the Andaman Islands, to the Malay Peninsula, through Indonesia to Australia and New Zealand, and island hopping along the Philippine and Indonesian archipelagoes to northern Australia. The second important flyway accommodates high fliers – mostly cranes, storks, ducks and geese – which cross the Himalayas in two broad fronts to winter in India and Sri Lanka. The third sends birds from northern and central Siberia west through the Aral and Caspian Seas to the Middle East and East Africa.

Environmental indicators

Birds are good environmental indicators, reflecting the health of natural ecosystems and giving early warnings of harmful changes.

Scientists estimate that 1,000 bird species, including many migratory birds, are threatened with extinction. More than two thirds of the earth's songbird species, including Redstarts, Sedge warblers, Sand martins and the Pied flycatcher, have declined drastically in numbers in recent years. The European populations of black storks and Dalmatian pelicans have shrunk to tiny threatened groups which grow more rare every year.

The human threat

Human activities pose the biggest threat. In the past 300 years, half of the bird species which have become

Migration

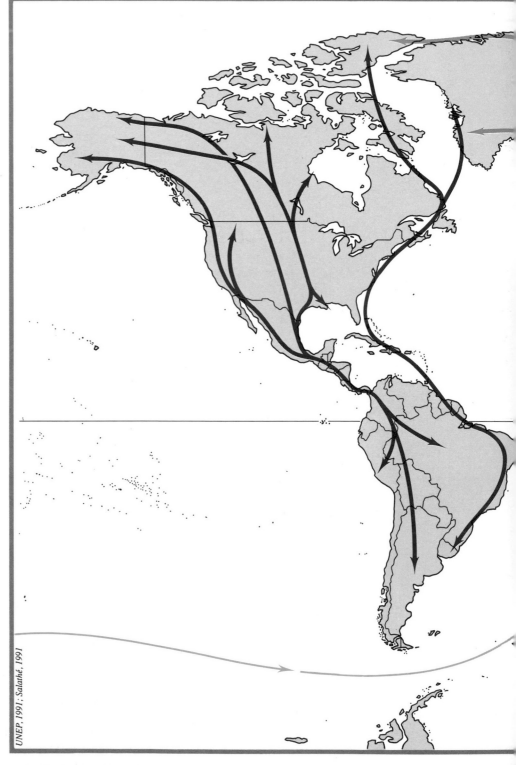

Parties to the Bonn Convention*
(Convention on the Conservation of Migratory Species of Wild Animals)

━━━ North American – South American flyways

━━━ European – African flyways

━━━ Asian flyways

The flyways shown are considered to be of major importance.

North America's flyways start in Alaska, Canada and Greenland and extend south in broad fronts to the continental US, the Caribbean, Mexico and Central and South America.

Europe's major flyways have a number of sub-routes. Depending on where they breed, populations of the same species – like the common crane and white stork – can reach their destination by different routes.

Asia's major flyways have many sub-routes and arteries linking them together.

――― Route of the Wandering albatross

* Includes European Community (not shown on map).

UNEP, 1991; Salathé, 1991

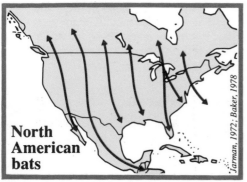

North American bats

Jarman, 1972; Baker, 1978

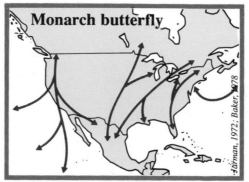

Monarch butterfly

Jarman, 1972; Baker, 1978

Bluefin tuna

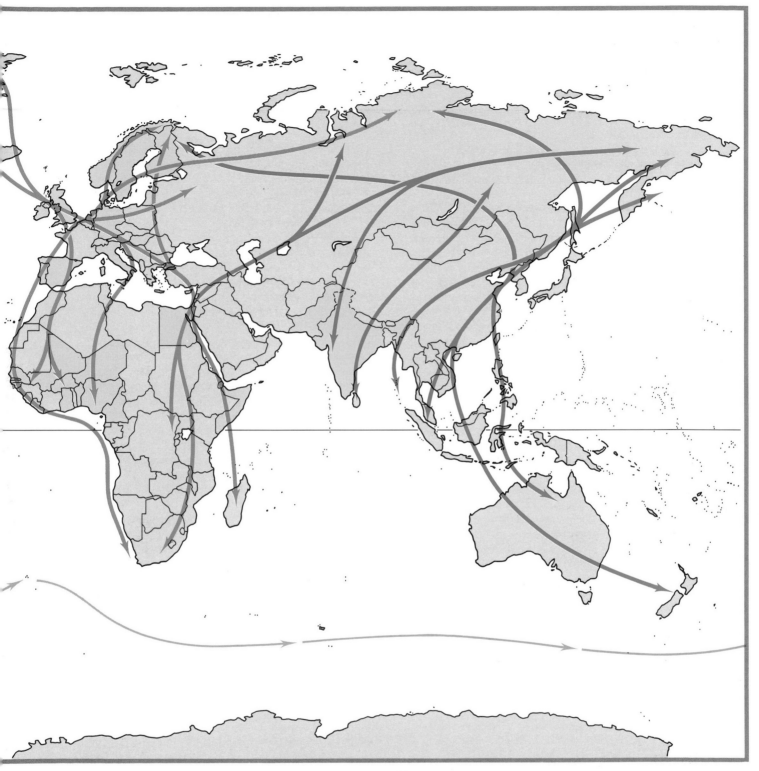

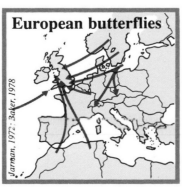

151

extinct have been killed by hunting or by the destruction of their habitat. The draining of marshes, wetlands and tidal estuaries for agricultural, recreational and industrial use has deprived migratory birds of the breeding, resting and feeding places vital for their survival.

Migrating shorebirds in the Americas funnel into a few key staging posts and feeding areas: at times just one area may contain four fifths of the entire population of the species. But such wetlands are increasingly under threat. In Europe and Africa many of the crucial stop-over sites on either side of geographical barriers like the Mediterranean and the Sahara desert are also threatened. In Asia, the mangrove swamps which provide sanctuary for birds are being converted into aquaculture ponds, while wetlands everywhere, from Hong Kong, to Pakistan to Indonesia are being turned into rice paddies or drained for urban expansion.

Modern agriculture has also upset the balance of nature. It suits small, fast-breeding migratory birds – such as finches, pigeons, starlings and parrots – so much so that they now reproduce too quickly for the slower-breeding species that prey on them and they have become serious pests. Meanwhile, pesticides such as DDT and dieldrin, which accumulate in the environment, build up in the bodies of animals and birds, taking their toll on populations.

Oil spills are another threat. More than 1.5 million metric tons of oil are discharged every year by tanker spills, and about the same amount comes from land-based sources. The spill from the Exxon Valdez in Alaska caused the deaths of at least 33,000 birds.

People have always hunted birds for food, eggs and feathers. An estimated 150 million birds are shot, "limed" or netted every year in the Mediterranean region, especially in Italy, Malta, Portugal, France and Cyprus. Liming – coating sticks with adhesive lime so that birds get stuck when they perch on them – kills up to 70 million birds annually in Italy. Netting is popular in Portugal and Cyprus, and kills around 10 million birds a year in France. North American casualties of overhunting include the Eskimo curlew and the Whooping crane. White storks are regularly hunted in the Sudan, and in the Philippines duck and wader species such as the Whimbrel are being decimated by overhunting.

International rescue

Threats to migratory species are now so serious that most developed and many developing nations have passed protective legislation. Many have also adopted laws to protect wetland habitats, and there are two international conventions designed to safeguard migratory species and their wetland habitats. The Convention on Wetlands of International Importance was signed at Ramsar in Iran in 1971 (see pages 153-156). The Convention on the Conservation of Migratory Species of Wild Animals (known as the Bonn Convention) came into force in 1983. Contracting parties agree to protect all migratory species deemed to be endangered through all or part of their range, by imposing strict conservation obligations. Species threatened with extinction are placed on Appendix I of the Convention. Hunting and trapping of the 51 species – 24 of them birds – is strictly forbidden. But its provisions are undermined by the limited number of members that adhere to it.

The Convention also seeks to persuade countries along migration routes to reach bilateral agreements in order to protect some 2,000 species listed in Appendix II. So far, agreements have been concluded between the former Soviet Union and India, the former Soviet Union and Japan, Japan and Australia, and Australia and China.

Many international conservation organizations promote the conservation of migratory birds. The oldest is the International Council for the Preservation of Birds, founded in 1922 – a federation of 330 organizations representing around 10 million people worldwide. The International Waterfowl and Wetlands Research Bureau, set up in 1954, has member organizations in 38 countries.

Both organizations joined IUCN–The World Conservation Union and WWF in the creation of the Conservation of Migratory Birds Programme in 1983. Under its auspices, attention has been focused especially on Eurasian-African migratory species, and international agreements have been signed with Morocco and Nigeria. A Migratory Birds Action Plan, to conserve migratory birds worldwide, is also being drawn up.

Despite these efforts, the future for migratory species looks uncertain. Numbers are in decline everywhere. A bird – the canary – was used by miners to give early warning of approaching danger. Today, hundreds of species are similarly sounding the alarm to humanity as a whole.

The World's Wetlands

Contrary to popular opinion, wetlands are not wastelands. Swamps, marshes, bogs, fens, estuaries and tidal flats are among the most fecund and productive ecosystems in the world. They provide critical habitats for thousands of species of plants and animals, yield up food, fiber and building materials, play important roles in regulating water cycles, filter pollution and guard shorelines from the depredations of the sea.

Wetlands cover about 6 per cent of the earth's land surface and are found in nearly every country and climate from tundra to tropics.

Productive wastelands

Scientists studying wetlands have been amazed at their biological productivity. A sub-tropical salt marsh can be twice as productive as tropical rainforest. A reed marsh may produce four to five times as much plant matter, per area, as the most fertile grasslands in the US Midwest. Some wetlands are capable of producing up to eight times as much as an average American wheat field.

Coastal wetlands – estuaries, salt marshes and tidal flats – are vital spawning and nursery areas for fish and shellfish. Two thirds of the fish caught worldwide are hatched in tidal zones. In the mid-1970s the yield of fish and shellfish dependent on wetlands was valued at over $700 million. The Wadden Sea, bordering parts of Germany, Denmark and the Netherlands, would produce far less marine life were it not for vast stretches of fertile salt marshes and tidal flats. In the Lower Mekong Basin, nearly half of the 500,000 metric tons of fish harvested every year comes from wetlands, providing roughly $90 million to the economy.

Inland wetlands: protein and cash

Inland wetlands connected to lakes and rivers also provide food and income for millions of people. The river swamps of Georgia, in the southern US, are farmed for fish. Over 25,000 poor villagers living around Lake Chilwa, on the Malawi-Mozambique border, derive most of their protein and cash income from a rich fishery concentrated around the edges of the lake.

The Inner Niger Delta in Mali has immense importance. Every year the river floods, spilling its nutrient-rich waters across 20,000 square kilometers of the driest land in the country. More than 2.5 million cattle, sheep and goats graze the delta's lush vegetation for six months of the year. And 90 per cent of Mali's fish catch comes from the delta, providing tribal fishermen with 100,000 metric tons of fish, worth around $5 million a year.

The delta's soils, renewed by flood waters, produce a wealth of food crops including millet and sorghum during the wet season. While the flood waters are still high, floating rice is grown.

At Lake Titicaca, 4,000 meters up in a bowl of the Andes Mountains, local Indians harvest submerged aquatic plants called "yacco", which are, in turn, fed to herds of cattle and vicuña pastured around the lake. Reeds are harvested for building materials, and around 6,000 metric tons of fish are caught and consumed, providing an important source of protein.

Wetland plants and sewage plants

Wetlands have a remarkable ability to filter a wide variety of pollutants. "Wetlands can transform, fix and render harmless viruses, coliform bacteria (from fecal matter) and suspended solids normally left after secondary sewage treatment," writes Professor Edward Maltby, in *Waterlogged Wealth*. In the southeastern part of the US, tidal marshes are used to filter and break down waste water from sewage treatment plants. This alone makes them worth around $123,000 per hectare (based on the cost of replacing them with artificial treatment facilities).

Hungary has long used peat bogs as natural filters for waste water from sewage plants and has plans to use more wetlands in this way.

Aquatic plants also have great potential as pollution filters. Studies carried out in Florida and elsewhere show that the noxious and persistent water hyacinth, a prolific weed, absorbs nutrients such as nitrogen, phosphorus, potassium and other substances directly from the water. It can also take out more than 75 per cent of lead in contaminated water over a 24-hour period, and has effectively filtered out cadmium, nickel, chromium, zinc, iron and other toxic wastes.

Not surprisingly, the greatest potential use for aquatic plants and wetlands for pollution control and waste water treatment is in the Third World, where the cost of building conventional treatment facilities is often prohibitive.

Wetlands are also vitally important breeding and feeding areas for thousands of species of birds, and serve as critical staging areas for migratory waterfowl. The great majority of endangered migratory bird species depend on them (for example, six out of nine species of migratory crane).

Wetlands under threat

Wetlands are in retreat nearly everywhere they are found. Reclamation schemes drain or fill them to make way for agricultural fields, or town and city expansion. They are drowned by dams and barrages built to provide

Wetlands

Parties to the
Ramsar Convention
(Convention on Wetlands of
International Importance especially
as a Waterfowl Habitat)

Ramsar sites:

● 500,000 hectares
and over

● 100,000 – 499,999 ha

· 20,000 – 99,999 ha

■ Non-Ramsar wetlands
under threat or of special
importance

* It is unclear at present whether all or any of
the states comprising the former USSR and
Yugoslavia will acknowledge their
commitments under the Ramsar Convention,
or whether wetlands in these nations can
still be considered as Ramsar sites.

Ramsar sites over 20,000 hectares

AUSTRALIA
1 Coongie Lakes
2 Kakadu National Park
3 Cobourg Peninsula Aboriginal
Land & Wildlife Sanctuary
4 Lakes Argyle & Kununurra
5 Coorong and Lakes Alexandrina
& Albert

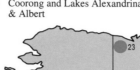

6 Ord River floodplain
7 Eighty-mile Beach
8 Western Port
9 Corner Inlet
10 Roebuck Bay
11 Gippsland Lakes
12 Western District Lakes
13 Riverland
14 Barmah Forest
15 Peel-Yalgorup system

AUSTRIA
16 Neusiedlersee (incl. Seewinkel)
& Donau/March/Auen

BURKINA FASO
17 Parc national du 'W'
18 La Mare d'Oursi

CANADA
19 Queen Maud Gulf Bird Sanctuary
20 Polar Bear Provincial Park
21 Whooping Crane Summer Range
22 Dewey Soper Bird Sanctuary
23 Old Crow Flats
24 Peace-Athabasca Delta
25 Rasmussen Lowlands
26 Polar Bear Pass
27 Quill Lakes
28 Hay-Zama Lakes
29 McConnell River Bird Sanctuary
30 Southern Bight, Minas Basin &
Malpeque Bay
31 Southern James Bay Migratory
Bird Sanctuaries

CHAD
32 Lac Fitri

DENMARK
33 Vadehavet (Wadden Sea)
34 Laeso
35 Vejlerne & Logstor Bredning
36 Horsens Fiord & Endelave
37 Sejero Bugt, Nekselo Bugt &
Saltbaek Vig
38 Randers & Mariager Fiords
39 South Funen Archipelago
40 Lillebaelt
41 Lolland & Falster
42 Waters SE of Fejo & Femo Is.
43 Ringkobing Fiord
44 Ulvedybet & Nibe Bredning

ECUADOR
45 Machalilla
46 Manglares-Churute

EGYPT
47 Lake Bardawil
48 Lake Burullus

FINLAND
49 Koitilaiskaira

FRANCE
50 La Brenne
51 Etangs de la Champagne humide
52 La Camargue
53 Marais du Cotentin et du Bessin,
Baie des Veys
54 Golfe du Morbihan

GABON
55 Parc national du Petit Loango &
Réserve de Sétté Cama
56 Réserve nationale de Wongha-
Wonghé

GERMANY
57 Schleswig-Holsteinisches
Wattenmeer
58 Ostfriesisches Wattenmeer &
Dollart
59 Wattenmeer, Jadebusen & Weser
Mundung
60 Wattenmeer, Elbe-Weser-Dreieck
61 Rügen/Hiddensee & eastern part
of Zingst Peninsula
62 Unterer Niederrhein

GREECE
63 Amvrakikos Gulf

GREENLAND
64 Eqalummiut Nunaat-
Nassuttuup Nunaa
65 Naternaq
66 Hochstetter Forland
67 Heden
68 Ikkatoq
69 Aqajarus-Sullorsuag
70 Kilen

GUATEMALA
71 Laguna del Tigre

GUINEA-BISSAU
72 Lagoa de Cufada

HUNGARY
73 Lake Balaton

ICELAND
74 Thjórsárver
75 Part of Myvatn-Laxá Region

INDIA
76 Chilka Lake
77 Loktak Lake
78 Sambhar Lake

IRAN
79 Lake Oroomiyeh
80 Shadegan Marshes & tidal mud flats
of Khor-al Amaya & Khor Musa
81 Neiriz Lakes & Kamjan Marshes
82 Khuran Straits
83 Hamoun-e-Saberi
84 Gavkhouni Lake & marshes of
lower Zaindeh Rud
85 Miankaleh Peninsula, Gorgan
Bay, Lapoo-Zaghmarz Ab-bandans

MALI
86 Walado Debo/Lac Debo
87 Séri

MAURITANIA
88 Banc d'Arguin

MEXICO
89 Rio Lagartos, Yucatan

NETHERLANDS
90 Dutch Wadden Sea
91 Oosterschelde

NIGER
92 Parc national du 'W'

PANAMA
93 Golfo de Montijo

ROMANIA
94 Danube Delta

SENEGAL
95 Le Delta du Saloum

SOUTH AFRICA
96 St Lucia System
97 Turtle beaches/coral reefs
of Tongaland

SPAIN
98 Doñana National Park

SWEDEN
99 Sjaunja-Kaitum
100 Tjålmejaure-Laisdalen

TURKS & CAICOS IS. (UK)
101 North, Middle & East Caicos Is.

USSR*
102 Volga Delta
103 Issyk-kul Lake
104 Lakes of Lower Turgay & Irgiz
105 Lake Khanka
106 Kourgal'dzhin & Tengiz Lakes
107 Kandalaksha Bay
108 Krasnovodsk & North Cheleken B
109 Kirov Bays
110 Intertidal areas of Dounai,
Yagorlits & Tendrov Bays
111 Matsalu Bay
112 Sivash Bay
113 Karkinitski Bay

UNITED KINGDOM
114 The Wash
115 Lough Neagh & Lough Beg

UNITED STATES
116 Everglades National Park
117 Izembek National Wildlife Refuge
118 Okefenokee National Wildlife
Refuge
119 Chesapeake Bay Wetlands System

URUGUAY
120 Banados del Este y Franja Costera

ZAMBIA
121 Bangweulu Swamps: Chikuni
122 Kafue Flats: Lochinvar & Blue
Lagoon

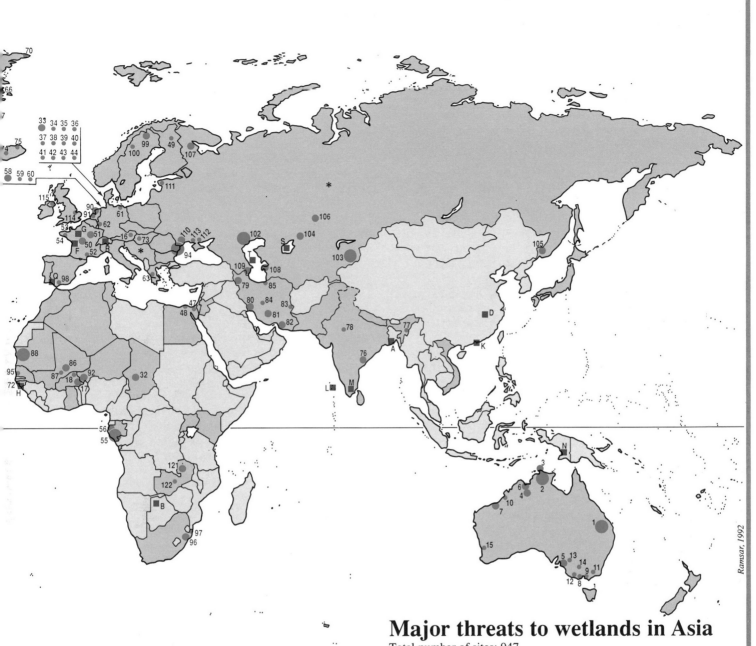

Ramsar, 1992

Non-Ramsar wetlands under threat or of special importance

BANGLADESH / INDIA
A Sundarbans

BOTSWANA
B Okavango Delta

BRAZIL
C Pantanal

CHINA
D Lake Poyang

COSTA RICA
E Tortuguero

FRANCE
F Poitevin Marsh
G Vernier Marsh

GUINEA-BISSAU
H Coastal zone

HONDURAS
J Quero y Salado

HONG KONG
K Mai Po Marsh

INDIA
L Laccadive Islands
M Vedanthangal

INDONESIA
N West Irian Jaya

MEXICO
P Sian Ka'an

SPAIN
Q Marismas de Guadalquivir

SWITZERLAND
R Rothenthurm

USSR*
S Aral Sea
T Caspian Sea

Major threats to wetlands in Asia

Total number of sites: 947
Number of sites for which information is available: 734
These activities represent typical major threats to wetland areas, although in other regions there may be variations. Many sites experience more than one of the following threats:

Threat	%
HUNTING	32%
GENERAL DISTURBANCE FROM SETTLEMENTS ETC.	27%
AGRICULTURAL DRAINAGE	23%
POLLUTION	20%
FISHING	19%
LOGGING/FORESTRY	17%
FUELWOOD	16%
DAMAGE FROM DEGRADED WATERSHEDS	15%
DIVERSION OF WATER	9%
CONVERSION FOR AQUACULTURE	9%
RECLAMATION FOR DEVELOPMENT	8%

Scott & Poole, 1989

155

power and irrigation, often at the cost of downstream fisheries and wildlife habitat. Or they are used as garbage dumps and junkyards. Others are poisoned by pollution. Between a quarter and a half of the world's swamps and marshes have already disappeared.

The US alone has lost half its wetlands. Between 1950 and 1970 it was losing 185,000 hectares a year, 87 per cent of them to agriculture. Iowa had ploughed over 99 per cent of its natural marshes by 1981, Nebraska lost 91 per cent by 1982. Florida's Everglades were progressively drained for agriculture for a century. Only a tenth of the 2.4 million wading birds that used them in the 1880s still remain. The state has now launched a massive campaign to try to bring the swamps back to the condition in which they began the 20th century by the year 2000 – the first ever attempt on this scale to restore, rather than merely preserve, a wetland.

Up to two thirds of Eastern Europe's wetlands have already disappeared, or are so saturated with industrial pollutants and toxic wastes that they are little more than biological graveyards. The greatest wetland in the region, the Danube Delta in Romania, has shrunk by a third, its rich marshlands converted to crops, fish ponds, managed forests and villages. The remaining 300,000 hectares are threatened by the tremendous loads of pollution – untreated sewage, agricultural poisons and industrial wastes – dumped in the river. Western Europe has lost nearly all its natural wetlands, and many tropical countries have lost over 80 per cent of them.

Africa's wetlands are its people's single most important source of animal protein; yet massive development projects threaten to destroy many of them. The Maga Dam on the river Logone in northern Cameroon has disrupted the lives of thousands of people living in the river's floodplain. Before the dam was built – to provide irrigation water for agriculture – the local communities depended on the floodplain in the same way as those living in the Inner Niger Delta. Now, the seasonal flooding has been checked and the irrigated fields have had only limited success in growing more food. Indeed, rice grown locally under the irrigation scheme is four times as expensive as imported rice.

Botswana's Okavango Swamp, the world's largest inland delta and one of its most remarkable wetlands, is threatened by irrigation schemes and cattle ranching projects. The Sudd Swamps of southern Sudan, home to the largest assembly of water birds in Africa, are threatened by proposals to build a giant canal. Meanwhile Pakistan, Myanmar, Bangladesh, Malaysia and northern India have all lost half their wetlands over the last century.

The Ramsar Convention

Wetlands are the only ecosystems to have their own international convention designed to protect them from further destruction. The Convention on Wetlands of International Importance especially as Waterfowl Habitat, known as the Ramsar Convention, was signed in 1971 and came into force in 1975. It calls on all signatories to conserve wetlands, promote their sound utilization, and set aside special areas as wildlife reserves. Every country is required to designate at least one wetland for inclusion on the List of Wetlands of International Importance, which is maintained by IUCN–The World Conservation Union in Gland, Switzerland. As of June 1991, it included 527 wetland sites, covering nearly 32 million hectares. More are added every year.

Unfortunately, many of the contracting parties to the Convention – there were 66 by mid-1992 – have done little to conserve wetlands other than declare one or two sites as conservation zones. Many sites are little more than descriptions on a piece of paper because countries have not developed programs for managing them. More encouragingly, developing countries are increasingly joining the Convention and now represent over half the total signatories.

The World's Fisheries

Homo sapiens remains a hunter-gatherer at sea, and for centuries it seemed as if exploitation could continue to expand indefinitely. The continents may have been largely hunted out, but the oceans, covering 70 per cent of the world's surface, seemed an inexhaustible cornucopia of riches. However, this is no longer so. Species after species is declining, as overfishing becomes chronic throughout most of the world's seas.

Sustainable yields exceeded

Fishermen rely on relatively few species. About 9,000 of the world's 20,000 known kinds of fish are caught. But only 22 of these are taken in large amounts, exceeding 100,000 metric tons a year. Just five groups of species – herrings, cods, jacks, redfishes and mackerels – make up half of the entire annual harvest from the world's seas.

The Food and Agriculture Organization of the United Nations (FAO) estimates that the world's fleets cannot catch more than 100 million metric tons a year of the species currently exploited without critically depleting stocks. Its figures suggest that commercial catches have increased fivefold since the Second World War, rising from 20 million metric tons in 1950 to 99.5 million metric tons in 1989 – suggesting that humanity is getting dangerously close to this maximum sustainable yield.

In fact, this level has already been exceeded because the FAO's figures do not include the fish taken by more than 8 million local fishermen whose catches are not recorded, because they are eaten by the fishermen's families or traded in their own communities. Such "artisanal" fishing is estimated to harvest about 24 million metric tons a year. So the world's seas are already being fished beyond their limits. In 1990, the world's catch fell by 4 million metric tons, the first major decline in nearly two decades.

Harvests in peril

Some seas and species are in particular trouble, because the world's marine harvest is taken unevenly from around the globe. The US National Fish and Wildlife Foundation says that 14 major fish species in US waters, yielding one fifth of the world's catch, are so depleted that even if all fishing stopped immediately stocks could take up to 20 years to recover. All fish stocks in the Northwest Atlantic have already been over-harvested. Total catches – dominated by the United States, Canada and Iceland – have slumped by 32 per cent since the early 1970s. Over the same period, the main stocks (and catches) of Northeast Atlantic fish – cods, hakes, haddocks, herrings, sardines and anchovies – have also declined significantly under the impact of fishing fleets from Norway, Denmark, the UK, Iceland, the former Soviet Union, Poland, Spain, France, the Netherlands and Germany.

Catches in the Northwest Pacific – dominated by Japan, the world's biggest fishing nation – are still increasing. But this will not continue for long, because the harvest is far beyond the maximum sustainable yield. The more fish that are taken beyond the stocks' availability to reproduce, the greater the crash will eventually be. The situation is the same in the Mediterranean and the Black Sea: increasing catches, far in excess of sustainability, courting inevitable disaster.

Fisheries in the Northeast, Southeast and Central Pacific, the Southeast Atlantic and the whole of the Indian Ocean are near or at their maximum sustainable yields: indeed more than 90 per cent of the world's commercial fisheries are heavily exploited. The FAO monitors about 280 fish stocks worldwide; only 25 of them are classed as under-exploited or moderately exploited.

A Third World crisis

Many Third World coasts are becoming fished out, partly by local fisheries, which are the largest single source of animal protein for several hundred million people. Artisanal fishermen in St Vincent and the Grenadines, in the West Indies, report that their catches have been cut in half since 1980 and that the size of fish they do manage to land is decreasing. Groupers, snappers and jacks are only about two thirds of the size they used to be. And the catch of lobsters and conches is down by more than a third. Similar reef fisheries have collapsed entirely in many parts of the world.

Most nearshore waters in South and Southeast Asia are over-exploited, as are many coastal areas around East and West Africa. Many poor, subsistence fishermen in Asia and Africa are resorting to dynamite and poisons to catch enough fish, compounding the problem by destroying habitat as well. In the Lingayen Gulf on the large Philippine island of Luzon, fish are so scarce that researchers from the University of the Philippines in Manila cannot even find enough specimens to do a proper study.

Pilchards used to be one of the mainstays of Namibia's artisanal fishermen, but the stock collapsed in 1974. At about the same time, there was a devastating decline in the anchovy fishery off Peru, which at 12 million metric tons a year used to account for a fifth of the world's entire fish catch. By 1973 it had fallen to 2 million metric tons and by 1980 to a mere 100,000 metric tons, as over-harvesting compounded the effect of changes brought on by the weather patterns known as

World fish production

Sea fish and shellfish

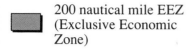
200 nautical mile EEZ (Exclusive Economic Zone)

FAO fishing area boundary

FAO AREA NUMBER
(Area name)
Total catch, 1989 (metric tons)
COUNTRIES CATCHING OVER 1 MILLION METRIC TONS IN 1989
Countries catching between 1 million and 200,000 metric tons in 1989

Major commercial catches

Selected species

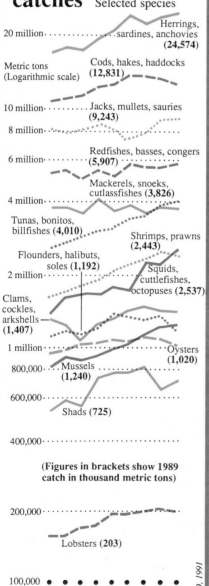

20 million · · · · · · · · · · · Herrings, sardines, anchovies **(24,574)**

Metric tons (Logarithmic scale)

Cods, hakes, haddocks **(12,831)**

10 million · · · · · · · · Jacks, mullets, sauries **(9,243)**

8 million · · · · · ·

Redfishes, basses, congers **(5,907)**

6 million · · · · · ·

Mackerels, snoeks, cutlassfishes **(3,826)**

4 million · · · ·

Tunas, bonitos, billfishes **(4,010)**

Shrimps, prawns **(2,443)**

Flounders, halibuts, soles **(1,192)**

Squids, cuttlefishes, octopuses **(2,537)**

2 million · · · ·

Clams, cockles, arkshells **(1,407)**

1 million · · · · · · · · Oysters **(1,020)**

Mussels **(1,240)**

800,000 · · · · ·

600,000 · · · · ·

Shads **(725)**

400,000 · · · · · · · · · ·

(Figures in brackets show 1989 catch in thousand metric tons)

200,000 · · · · ·

Lobsters **(203)**

100,000 · · · · · · · · · ·
1981   83   85   87   89

FAO, 1991

158

AREA 18
(Arctic Sea)
Comparable figures not available

57

AREA 67
(Northeast Pacific)
3,290,688
UNITED STATES
Canada

15

AREA 21
(Northwest Atlantic)
3,079,272
UNITED STATES, CANADA

6

AREA 31
(Western Central Atlantic)
1,791,885
UNITED STATES
Mexico, Venezuela

17

52

AREA 77
(Eastern Central Pacific)
1,705,154
Mexico, United States

53

39

25

4

22

AREA 87
(Southeast Pacific)
15,310,646
PERU, CHILE, USSR
Ecuador

5

AREA 41
(Southwest Atlant
2,254,213
Brazil, Argentina, USSR,

32

AREA 81
(Southwest Pacific)
990,609
Japan, New Zealand

AREA 88
(Antarctic Pacific)
1,110

FAO, 1991

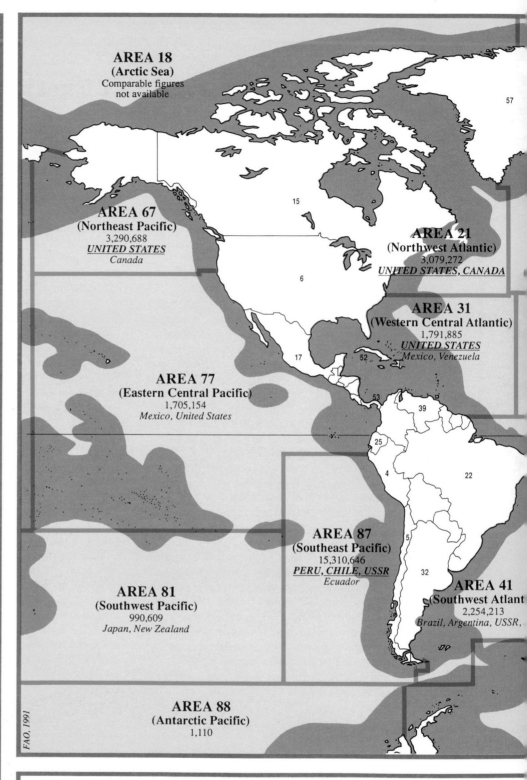

The world's top fishing nations

Countries with catches of at least 160,000 metric tons in 1989 (*estimate)

1	USSR	11,310,091	12	Denmark	1,927,493	22	Brazil	850,000
2	China	11,219,994	13	Norway	1,899,941	23	Bangladesh	832,791
3	Japan	11,174,464	14	Democratic People's		24	United Kingdom	797,259
4	Peru	6,832,465		Republic of Korea	1,700,100 *	25	Ecuador	723,624
5	Chile	6,454,142	15	Canada	1,554,233	26	Myanmar	702,700
6	United States	5,744,318	16	Iceland	1,504,771	27	Malaysia	608,967
7	India	3,618,919	17	Mexico	1,416,784	28	Poland	564,886
8	Republic of Korea	2,832,431	18	Spain	1,370,000	29	Italy	550,964
9	Thailand	2,822,530 *	19	South Africa	878,580	30	Morocco	520,354
10	Indonesia	2,700,000 *	20	France	875,839 *	31	New Zealand	514,000
11	Philippines	2,098,787 *	21	Vietnam	868,000 *	32	Argentina	486,631

FAO, 1991

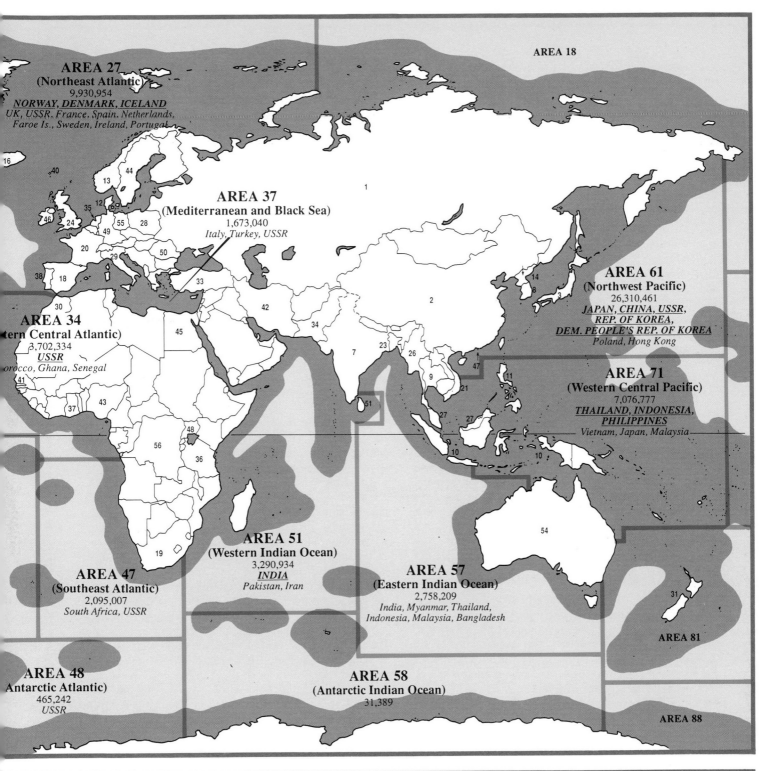

AREA 18

AREA 27
(Northeast Atlantic)
9,930,954
NORWAY, DENMARK, ICELAND
UK, USSR, France, Spain, Netherlands,
Faroe Is., Sweden, Ireland, Portugal

AREA 37
(Mediterranean and Black Sea)
1,673,040
Italy, Turkey, USSR

AREA 34
(...tern Central Atlantic)
...,702,334
USSR
...orocco, Ghana, Senegal

AREA 61
(Northwest Pacific)
26,310,461
JAPAN, CHINA, USSR,
REP. OF KOREA,
DEM. PEOPLE'S REP. OF KOREA
Poland, Hong Kong

AREA 71
(Western Central Pacific)
7,076,777
THAILAND, INDONESIA,
PHILIPPINES
Vietnam, Japan, Malaysia

AREA 51
(Western Indian Ocean)
3,290,934
INDIA
Pakistan, Iran

AREA 57
(Eastern Indian Ocean)
2,758,209
India, Myanmar, Thailand,
Indonesia, Malaysia, Bangladesh

AREA 47
(Southeast Atlantic)
2,095,007
South Africa, USSR

AREA 48
...Antarctic Atlantic)
465,242
USSR

AREA 58
(Antarctic Indian Ocean)
31,389

AREA 81

AREA 88

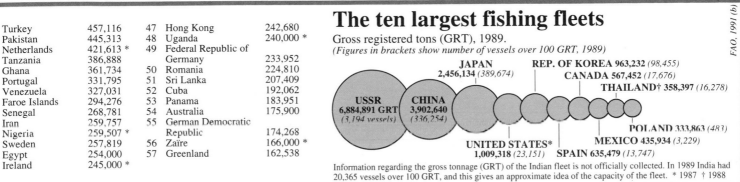

Turkey	457,116		47	Hong Kong	242,680
Pakistan	445,313		48	Uganda	240,000 *
Netherlands	421,613 *		49	Federal Republic of	
Tanzania	386,888			Germany	233,952
Ghana	361,734		50	Romania	224,810
Portugal	331,795		51	Sri Lanka	207,409
Venezuela	327,031		52	Cuba	192,062
Faroe Islands	294,276		53	Panama	183,951
Senegal	268,781		54	Australia	175,900
Iran	259,757		55	German Democratic	
Nigeria	259,507 *			Republic	174,268
Sweden	257,819		56	Zaïre	166,000 *
Egypt	254,000		57	Greenland	162,538
Ireland	245,000 *				

The ten largest fishing fleets

Gross registered tons (GRT), 1989.
(Figures in brackets show number of vessels over 100 GRT, 1989)

FAO, 1991 (b)

JAPAN
2,456,134 *(389,674)*

REP. OF KOREA 963,232 *(98,455)*

CANADA 567,452 *(17,676)*

THAILAND† 358,397 *(16,278)*

USSR
6,884,891 GRT
(3,194 vessels)

CHINA
3,902,640
(336,254)

POLAND 333,863 *(483)*

MEXICO 435,934 *(3,229)*

UNITED STATES*
1,009,318 *(23,151)*

SPAIN 635,479 *(13,747)*

Information regarding the gross tonnage (GRT) of the Indian fleet is not officially collected. In 1989 India had 20,365 vessels over 100 GRT, and this gives an approximate idea of the capacity of the fleet. * 1987 † 1988

El Niño, although there has since been a recovery. Other stocks particularly at risk include sardines off Northwest Africa and western India.

Increasing pressure

The pressure on the world's fisheries will continue to increase because consumption and trade continue to rise. World exports of fish and fish products now exceed $15 billion a year. Prices for fish and shellfish have risen faster than the prices of other foods, reflecting increasing demand. Much of the recent increase in consumption is attributed to rising demand in developing countries. Their catches and exports are both soaring. Their catches have been growing five times as fast as in developed countries and the Third World now boasts eight of the world's 11 biggest fishing countries: China, India, Indonesia, Thailand, the Philippines, South Korea, Peru and Chile. The sea is becoming an increasingly valuable source of foreign exchange for hard pressed economies, and governments want to make the most of it.

Yet the FAO reports that the time for spectacular and sustained increases in fisheries' catches is over, as fish stocks are depleted and overfished. Pollution and the destruction of mangroves, coral reefs and seagrasses are compounding the problem. Nearly all the world's catch is taken from within 200 miles of land, just the area most affected by pollution and degradation.

Expansion possibilities

The Southwest Atlantic, particularly off the Argentinian coast, is one of the very few parts of the world where there is room to expand conventional fisheries greatly; there are also more limited opportunities off New Zealand and in the Western Indian Ocean. Otherwise, the world will have to resort to more unusual ways of increasing its harvest from the seas.

One of these would be to catch different fish. Octopus and squid are underexploited and so are some deep ocean species like lantern fish. But the biggest potential of all, at least in theory, lies in the krill that throng the Antarctic Ocean. Trawlers can scoop up several tons of these shrimp-like creatures every hour, and they could increase the world's sustainable marine harvest by tens of million of metric tons every year. Unfortunately, they have to be processed quickly or they go bad; and they have to be caught far from the main consuming nations. So, at present, large-scale harvesting seems to be uneconomical.

Fish farming, in ponds and coastal cages, offers a more practical way of increasing production. It has been growing rapidly and now produces between 12 and 14 million metric tons of fish, crustaceans, mollusks and bivalves each year, and the FAO expects this to rise to about 18 million metric tons by the turn of the century. But creating fish farms is one of the major causes of the destruction of mangroves, which are vital nurseries for wild fish.

Better management

There have been attempts to manage the major international fisheries since the 1950s, with limited success. Literally dozens of agreements were supposed to safeguard the Northeast Atlantic, but they did not prevent severe overfishing. By the mid-1970s, 30 of the region's 33 fish stocks were found to be over-exploited or depleted. Eventually, strong controls were introduced, including bans on catching cod and herring in the North Sea, and some stocks did recover.

The Law of the Sea Convention has led to the adoption of Exclusive Economic Zones (EEZs), extending out 200 miles from national shores. The EEZs give coastal states sovereign rights over all fishing – including exploitation by foreign vessels – and this means that 99 per cent of the world's catch is under national jurisdictions.

This offers the chance of ending the fishing free-for-all, but it has not resulted in better management of stocks. It has given some developing countries, particularly the small island states and territories of the Pacific, more access to, and control over, marine resources within their EEZs. But conflicts remain. The scattered island nations of the South Pacific have such vast EEZs that the amount of open ocean has been reduced by 30 per cent. Deepwater fishing fleets from Japan, Europe and the US have come into conflict with South Pacific island governments, since they continue to trawl for tuna and other migratory fish within declared EEZs and have so far refused to pay fishing rights' fees.

The outlook for future fisheries is not good. Workable management plans need to be put into operation soon. The livelihoods of millions of people hang in the balance.

Whales and Whaling

Whales seem to have a unique hold on the human imagination. Perhaps it is their size – the Blue whale is much the biggest animal ever to have lived on earth. Perhaps it is their reputation for intelligence, or the haunting songs of the Humpbacks. Perhaps it is because whales are mammals which returned to the seas from which life originally emerged.

Liberated from natural enemies on land, whales have roamed the seas with extraordinary freedom for over 45 million years. But one fatal weakness has made them easy prey: their need to come to the surface to breathe.

The history of slaughter

The Basques were the first people to start whaling systematically. Possibly organizing mass hunts since 700 AD, they were the world's leading whalers for centuries. They first exploited the Right whales that lumbered through the Bay of Biscay. Always the easiest whales to kill because they swim slowly and tend to float when dead, they got their name because they were the "right whales" to catch. The Basques boiled down the blubber into oil, turned the bones into knives and other tools, and used the tough and flexible baleen, the filters that hang like curtains in the whale's mouth, for horsewhips and archers' bows.

Eventually the herds disappeared. The Basques had to go further afield to find their prey, pushing up past the Faroes to Spitzbergen and westward into the Atlantic.

In the 17th century, the British and Dutch expelled the Basques and set up in competition with each other around Spitzbergen and Jan Mayen, more than 1,200 kilometers north of the Arctic Circle. The Dutch soon had 300 ships and 18,000 men in the area, mainly killing Bowhead whales to use their baleen in bonnets, bustles and corsets. The hunters almost cleared the Arctic Ocean east of Greenland of whales, and pressed towards America.

There another group of seafarers were poised to dominate whaling. The people of Nantucket, south of Cape Cod, first hunted Right whales, and then pioneered the exploitation of Sperm whales. Sperm oil came to light cities and grease the wheels of industry. By 1740, 50 Nantucket ships were bringing back oil worth $25,000 a year; by 1774, 150 vessels were producing $500,000 a year. The whalers pushed progressively further south in the Atlantic to find, and plunder, new stocks. They then rounded Cape Horn and found the Pacific as rich in Sperm whales as the Atlantic had once been. Other New England towns joined in; by 1846, there were 736 ships at sea. But with the discovery of oil in Pennsylvania in 1859, whale oil became restricted to fewer and fewer uses.

The technology of death

New inventions saved the industry – and doomed a whole new class of whales, including Blue, Fin and Sei, which were too fast for men dependent on sail, rowing boats and hand-held harpoons, and which sank when killed. Fast propeller-driven ships, harpoon guns and inflation lances – which make dead whales float by pumping them full of air – ushered in the modern whaling industry. What had once been a risky and perilous hunt became a mass slaughter. Whale products would now only sell if enough were killed to keep prices low, and the new technology enabled the depletion of successive stocks to take on breakneck speed. Norway, the first country to adopt it, took over world whaling. It soon depleted northern seas and started exploiting the Antarctic, the last virgin ocean, where great assemblies of whales came to feed. It was joined by the UK and, in the 1930s, by Japan and Germany.

The powerful new whalers concentrated first on Blue whales, because they were the biggest. By 1925, 2,500 were being killed in the Antarctic each year. The next year, the first factory ship appeared and, by 1930, the catch had risen to 30,000. Between 1910 and 1966, a total of 330,000 were killed. As Blue whales became scarce, the whalers moved on to the second biggest species, Fin whales. They killed 30,000 a year in the 1950s and reduced them to a fraction of their former numbers. In the 1960s, they moved on to Sei whales and in the 1970s to Minke whales. In all, 1.5 million whales perished between 1925 and 1975. Most of this slaughter took place under controls supposed to protect whales from devastation. Whaling companies agreed to limit their catches after overkilling in 1931 swamped the market with whale oil, sending prices tumbling. This set a pattern for subsequent conservation measures: it followed, rather than prevented, a crisis.

Ineffective agreements

The first international agreement on protecting whales came into force in 1935, banning the killing of Right whales, which were by then down to 3 per cent of their original population. Two years later, the Gray whale was protected; its numbers had declined to around 250. In 1946, the International Whaling Commission (IWC) came into being, recognizing, in the words of its treaty: "The history of whaling has seen overfishing in one area after another to such a degree that it is essential to protect all species of whales from further overfishing." The fine words inaugurated another quarter of a century of virtually unregulated slaughter with the USSR, the US, the Netherlands, Australia and other nations joining Norway, Japan and the UK to devastate almost all the whale stocks remaining in the sea.

Whaling

Members of the
International Whaling
Commission, 1991:

(Alphabetical order)

1	Antigua & Barbuda	18	Korea, Rep. of
2	Argentina	19	Mexico
3	Australia	20	Monaco
4	Brazil	21	Netherlands
5	Chile	22	New Zealand
6	China	23	Norway
7	Costa Rica	24	Oman
8	Denmark	25	Peru
9	Ecuador	26	St Lucia
10	Finland	27	St Vincent & the Grenadines
11	France	28	Senegal
12	Germany	29	Seychelles
13	Iceland (withdrawing with effect from 30 June 1992)	30	South Africa
		31	Spain
		32	Sweden
		33	Switzerland
14	India	34	USSR
15	Ireland	35	United Kingdom
16	Japan	36	United States
17	Kenya	37	Venezuela

IWC, 1991

World catches of whales

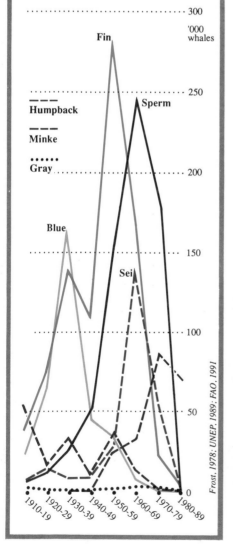

Fin

Humpback — — —

Minke — — —

Gray ••••••

300

'000 whales

250

Sperm

Blue

Sei

200

150

100

50

0

1910-19 1920-29 1930-39 1940-49 1950-59 1960-69 1970-79 1980-89

Frost, 1978; UNEP, 1989; FAO, 1991

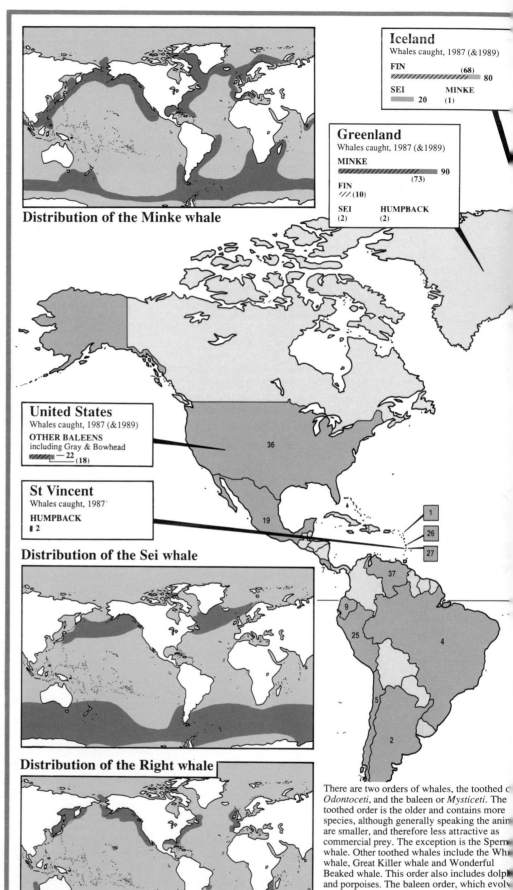

Distribution of the Minke whale

Iceland
Whales caught, 1987 (&1989)

FIN (68) 80

SEI 20 MINKE (1)

Greenland
Whales caught, 1987 (&1989)

MINKE 90 (73)

FIN (10)

SEI (2) HUMPBACK (2)

United States
Whales caught, 1987 (&1989)

OTHER BALEENS including Gray & Bowhead
22 (18)

St Vincent
Whales caught, 1987

HUMPBACK 2

Distribution of the Sei whale

Distribution of the Right whale

There are two orders of whales, the toothed or
Odontoceti, and the baleen or *Mysticeti*. The
toothed order is the older and contains more
species, although generally speaking the animals
are smaller, and therefore less attractive as
commercial prey. The exception is the Sperm
whale. Other toothed whales include the White
whale, Great Killer whale and Wonderful
Beaked whale. This order also includes dolphins
and porpoises. The baleen order, which evolved
from the toothed whales, contains fewer species
but individual animals tend to be large. They
include the Minke, Fin, Sei and Bryde's whale,
the Great Right whale, the Blue, the Gray, the
Bowhead and the Humpback.

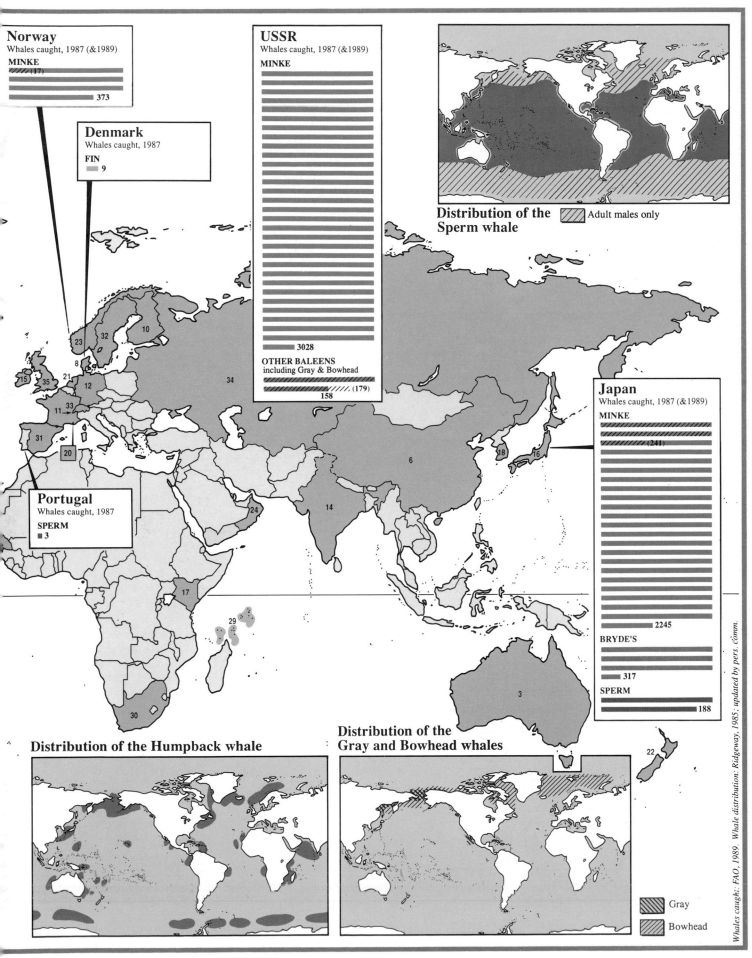

Norway
Whales caught, 1987 (&1989)
MINKE
(17)
373

Denmark
Whales caught, 1987
FIN
9

USSR
Whales caught, 1987 (&1989)
MINKE
3028
OTHER BALEENS
including Gray & Bowhead
(179)
158

Portugal
Whales caught, 1987
SPERM
3

Distribution of the Sperm whale
Adult males only

Japan
Whales caught, 1987 (&1989)
MINKE
(241)
2245
BRYDE'S
317
SPERM
188

Distribution of the Humpback whale

Distribution of the Gray and Bowhead whales
Gray
Bowhead

23 32 10 8 21 12 15 35 11 33 31 20 34 24 14 18 16 6 17 29 30 3 22

Whales caught: FAO, 1989. Whale distribution: Ridgeway, 1985; updated by pers. comm.

The IWC rapidly jettisoned two of the conservation measures enshrined in its treaty – a ban on killing Humpbacks and the creation of a whale sanctuary in the Antarctic. Humpbacks were not protected for another 13 years, when only a few thousand remained. The IWC then was effectively a club for whaling nations. It repeatedly ignored strong warnings about over-exploitation from its Scientific Committee. It set over-generous quotas, which were regularly exceeded in the first 10 years. It only agreed to stop killing Blue whales when almost none could be found to catch. Any nation could legally ignore any controls the IWC adopted.

Conservation and the whale

The rising conservation movement of the 1960s seized on whaling as a symbol of wanton environmental destruction. The US stopped commercial whaling in 1971. The next year, the United Nations Conference on the Human Environment in Stockholm voted for a 10-year moratorium on commercial whaling. The IWC rejected this, but did adopt a new, more scientific procedure for deciding quotas. The balance of power on the IWC was changing as many member countries gave up whaling either on principle or because it no longer paid. Japan, which had emerged as the world's leading whaling nation in the 1960s, bought up the fleets and quotas of other nations. By the 1970s, a majority of members of the IWC were non-whalers and increasingly conservationist. But they could not force major changes because they did not have the three-quarters majority needed, under the IWC rules, to make decisions.

At this point Lyall Watson, a bestselling author, persuaded the Seychelles to join the IWC, as any nation is entitled to do, and campaign for a moratorium. A close-fought battle between the Seychelles, led by Watson, and Japan followed. Each tried to recruit new members to the Commission to support its stand and to persuade those already there; Watson had to find three votes for every one secured by Japan. He had his first major victory in 1981 when, after intricate maneuvering, the IWC decided that no more Sperm whales should be taken. And in 1982 it voted for a moratorium on commercial whaling, to take effect in 1985.

But whaling continued: Japan, Norway, Peru and the USSR exempted themselves from the moratorium. Over the next years Japan, Norway and Iceland continued to kill whales under the controversial guise of "scientific whaling", exploiting a loophole in the treaty that allows whales to be killed for research. In 1991, Iceland announced that it would leave the IWC, spectacularly walking out of the next annual meeting in 1992.

When that meeting ended, in July 1992, the IWC was under unprecedented threat. It formally agreed to extend the whaling ban for another year, but Norway announced that it would resume commercial whaling nevertheless. Iceland and Norway set up an alternative whaling organization for the North Atlantic, while Japan announced that it would set up a similar one for the North Pacific.

The US fails to act

Only the US could force an end to whaling. Two provisions in its fishery laws – the Pelly and Packwood-Magnuson amendments – would stop countries that "diminish the effectiveness" of the IWC from fishing in US waters or exporting fish to its markets. This is a powerful sanction: Japan takes fish worth nearly 20 times its whaling industry from Alaskan waters each year. But the Reagan administration entered an agreement with Japan, upheld by the Supreme Court, which allowed it to go on whaling.

Provisional (and highly contentious) IWC figures have hinted at the result of centuries of devastation. Of the million Sperm whales that once roamed the oceans, only 10,000 are thought to be left. Humpbacks seem to be down from 200,000 to 4,000, Fin whales from more than 100,000 to 2,000, Blue whales from 250,000 to around 500. No one knows when, if ever, their numbers will recover.

Mangroves, Coral Reefs and Seagrasses

Mangroves, coral reefs and seagrasses are among the world's most important – and most endangered – ecosystems. They both nurture the seas and protect the land. They provide vital breeding grounds and habitats for fish and shellfish, stabilize shorelines, protect coasts from the effects of storms, and inhibit erosion. But they have been increasingly under attack for the past three decades, a crisis that has been virtually ignored.

Nearly two thirds of all fish caught throughout the world are hatched in mangrove and tidal areas. Roughly 90 per cent of all commercial species of fish and shellfish taken in the Gulf of Mexico and the Caribbean, for example, depend on mangroves, seagrasses and coral reefs at critical stages in their life cycles. Some 80 per cent of the Indian fish catch from the lower delta region of the Ganges and Brahmaputra rivers comes from the mangrove swamps of the Sundarbans, which cover 6,000 square kilometers.

Mangroves in retreat

Mangrove forests supply important habitats for over 2,000 species of fish, invertebrates and plants. They are at least as productive as good farmland. Some 55 species of salt-tolerant mangrove trees and shrubs cover roughly 240,000 square kilometers of coastal land throughout the world.

Everywhere they are in retreat. Millions of hectares worldwide have been cut down for timber, fuelwood and wood chips, destroyed to create fish and shellfish ponds, or to expand urban areas and agricultural land. They are poisoned by pesticides washed off agricultural fields and smothered by sediment from coastal development, the deforestation of upland watersheds and mining operations.

The worst destruction is taking place in Asia. The mangrove area of the Philippine archipelago was reduced from 5,000 square kilometers to only 380 between 1920 and 1988. Most of it was clear-cut for its valuable timber, exploited for tannin, or converted into fish ponds and rice paddies. More than 2,000 square kilometers of Indonesia's mangrove forests are being exploited for the woodchip industry, producing 250,000 cubic meters a year for export to Japan; another 10,000 square kilometers of Indonesian mangroves have been converted into brackish water ponds in order to cultivate prawns, shrimp and milkfish.

Nearly the entire north coast of Java – once lined with mangroves – is now lined with these ponds, called "tambaks". Here, as in many areas, mangrove destruction has begun a cycle of coastal degradation.

If properly managed, mangrove swamps could produce huge annual yields of fish, shrimp, crabmeat and mollusks. But as they are destroyed, fish catches decline so fishing communities create more tambaks, destroying more mangroves. Often shrimp farms depend on larvae from wild stocks at sea, which disappear as the mangroves are cut down. Within a few years, the ponds have become uneconomic and the whole process is repeated elsewhere.

Mangroves stabilize shorelines; when they disappear their protection is gone and the land is vulnerable to the sea. Both Thailand and the United States specially cultivate mangroves to protect coasts from erosion – the US reforesting thousands of hectares of sea-ravaged land along the Gulf coasts of Florida, Mississippi and Louisiana. They also trap silt washing off the land and filter out pollution.

Coral reefs under attack

Coral reefs are among the oldest and richest living communities of plants and animals on earth. Most are between 5,000 and 10,000 years old: many form thin living veneers on older, much thicker reef structures several million years old.

The world's 600,000 square kilometers of coral reefs are the marine equivalents of tropical rainforests. A single barrier reef surrounding the small island of Belau in the South Pacific has 300 species of corals and 2,000 varieties of fish. Coral communities can yield valuable medicines; Didemnin, which comes from sea squirts, is active against a broad range of viruses – from colds to herpes and meningitis. Potential anti-cancer drugs are also being found. The Australian Institute of Marine Science has isolated a compound which protects coral from sunburn. It has great potential in the new, stronger sunscreen products that will be needed as the ozone layer thins out, increasing skin cancers.

Nearly one third of all fish species live on coral reefs, while others are dependent on reefs and seagrass beds at various stages in their life cycles. Although estimates as to the amount of fish that can be harvested from reefs vary, it should be possible to take some 15 metric tons per square kilometer per year without destroying stocks. Almost 90 per cent of all fish caught by artisanal fishermen in Indonesia depend on coral reefs, as do some 55 per cent of the fish consumed by Filipinos.

The reefs calm the energy of the waves, and protect the shores against storms. When one reef in Sri Lanka was destroyed, the shoreline was pushed back some 300 meters by the unrestrained sea. They also bring tourists: more than half of the GNP of the Bahamas comes from people holidaying on its coasts, and the Great Barrier Reef earns Australia $90 million a year.

Yet coral reefs are being degraded and destroyed at unprecedented rates, throughout the tropics. Of the 109 countries with significant coral communities, 93 are damaging them. In over 50 countries coral is being

Areas under threat

Mangroves

A *Niger River delta, Nigeria*
Exploited for timber, fuel, fodder and urban expansion.

B *Kenya and Tanzania*
Cleared for fuelwood, building materials and tourist resorts.

C *Indus River mouth, Pakistan*
Over-exploitation for fuel, fodder and building material.

D *India, Sri Lanka*
Cleared for fuelwood, agriculture and fish ponds

E *Sundarbans, India and Bangladesh*
Over-exploitation for fuel, fodder, timber and fishponds.

F *Malaysia and Gulf of Thailand*
Destruction for fish and shellfish ponds and agricultural land.

G *Philippines*
Destruction for timber, tannin, fuelwood and fish and shellfish ponds.

H *Indonesia*
Massive destruction for logging and woodchip industries, fish and shellfish ponds and for building materials and fuelwood.

I *Queensland, Australia*
Town and tourist development.

J *US south coast, Texas to Florida*
Over-development of coastline for urban expansion, resorts, housing estates. Also used as garbage dumps.

K *Panama*
Cleared for fish and shrimp ponds.

L *Ecuador*
Cleared for fish and shrimp ponds.

M *Caribbean*
All mangrove stands disturbed.
Main threats: tourism and coastal land development.

Seagrasses

A *East Africa*
Under threat from heavy sedimentation of shallow coastal waters caused by erosion of agricultural lands.

B *Southeast Asia*
Under threat from loss of mangroves, coastal development, urban expansion and bucket dredging for tin.

C *Caribbean and Gulf of Mexico*
Under threat from dredge and fill operations, loss of mangroves, coastal development for tourism, oil production.

Coral reefs

A *East Africa*
Coral mining for building materials, blast fishing, tourist trade and sedimentation.

B *The Gulf*
Oil and industrial pollution, sedimentation.

C *Thailand and Malaysia*
Tourist resorts, bucket dredging for tin, over-fishing.

D *Philippines*
Blast fishing, coral mining, collection for tourist trade and use of poisons.

E *Southern Japan – Ryukyu Archipelago*
Destroyed by coastal development and sedimentation.

F *Indonesia*
Destroyed by blast fishing, coral mining, tourist trade and coastal development.

G *South Pacific*
Tourism, sedimentation from coastal development.

H *Wider Caribbean*
Collection for tourist trade, coastal development, mangrove destruction/sedimentation and damage by boat anchors.

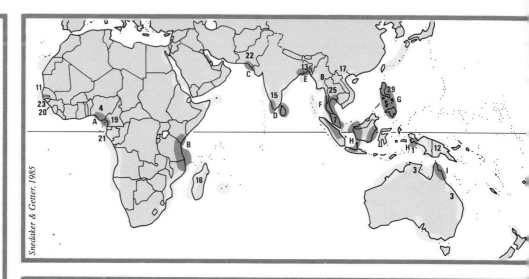

Snedaker & Getter, 1985

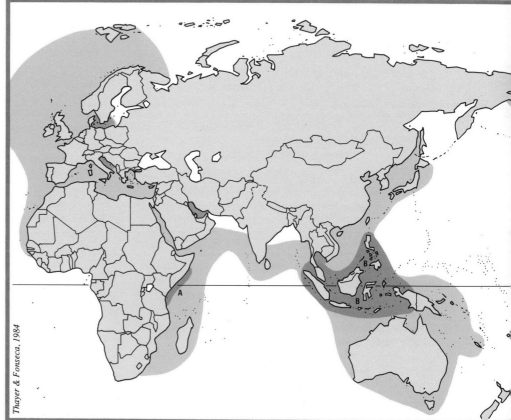

Thayer & Fonseca, 1984

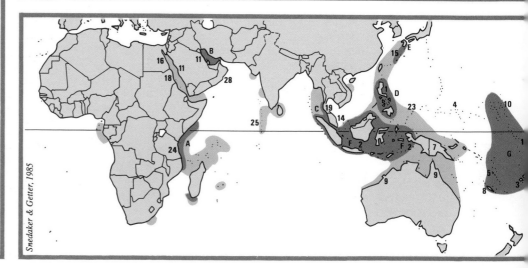

Snedaker & Getter, 1985

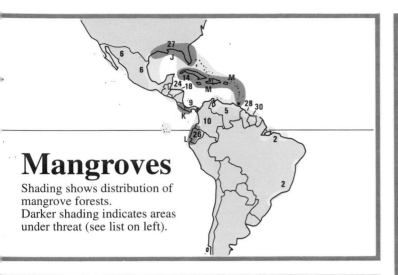

Mangroves

Shading shows distribution of
mangrove forests.
Darker shading indicates areas
under threat (see list on left).

Mangrove forests
Countries with the largest areas of mangrove forest

Square kilometers, early 1980s

1	Indonesia	36,000	16	Madagascar	3,207
2	Brazil	25,000	17	Vietnam	3,200
3	Australia	11,617	18	Honduras	3,000
4	Nigeria	9,730	19	Cameroon	2,720
5	Venezuela	6,736	20	Guinea	2,600
6	Mexico	6,600	21	Gabon	2,500
7	Malaysia	5,700	22	Pakistan	2,495
8	Myanmar	5,171	23	Guinea-Bissau	2,430
9	Panama	4,860	24	Belize	2,400
10	Colombia	4,400	25	Thailand	2,240
11	Senegal	4,400	26	Ecuador	2,158
12	Papua New Guinea	4,116	27	US: Gulf of Mexico states	2,050
13	Bangladesh	4,050	28	Guyana	1,500
14	Cuba	4,000	29	Philippines	1,461
15	India (peninsular)	3,565	30	Surinam	1,150

Since they represent the extent of mangroves in the early
1980s, most of these figures are now considered
over-estimations. In some cases, such as the Philippines, the
loss of mangrove forests over the past decade has been
dramatic. Whereas in 1980 the Philippines had 1,461 square
kilometers, or 146,000 hectares of mangrove forests, today it
has a mere 38,000 hectares. Indonesia, Thailand, India and
Pakistan have also lost much of their mangrove cover over
the past 10 years. So too have countries in Africa and Latin
America.

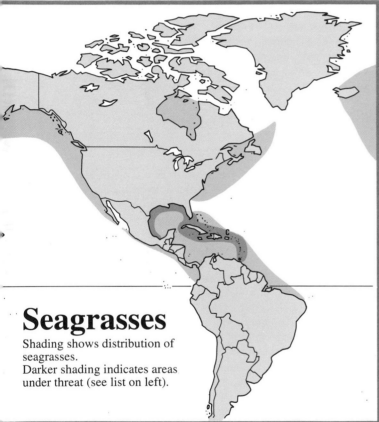

Seagrasses

Shading shows distribution of
seagrasses.
Darker shading indicates areas
under threat (see list on left).

Coral reefs
Countries with the greatest length of coral reef

Kilometers, early 1980s

1	Philippines	22,450	16	Egypt	1,100
2	Indonesia	17,500	17	Cuba	1,046
3	Fiji	5,360	18	Sudan	963
4	Fed. States of Micronesia	3,915	19	Thailand	900
5	Vanuatu	3,792	20	US: Florida	866
6	French Polynesia	3,788	21	Mexico	820
7	Papua New Guinea	3,684	22	Lesser Antilles	796
8	New Caledonia	3,381	23	Belau	750
9	Australia	3,000	24	Tanzania	750
10	Marshall Islands	2,531	25	Maldives	644
11	Saudi Arabia	1,955	26	Tonga	628
12	Bahamas	1,727	27	American Samoa	628
13	Kiribati	1,715	28	Oman	625
14	Malaysia	1,670	29	Colombia	500
15	Japan: Nansei Shoto	1,200	30	Belize	474

No accurate scientific survey of the extent of coral reefs has
been carried out. These estimates date from the early 1980s
or before and are acknowledged by WWF to be inadequate.
In many cases the inventories on which they are based failed
to distinguish between live and dead corals, so most of the
figures are probably over-estimations of live coral cover.
Since the early 1980s the situation has deteriorated. A survey
of reefs in the Philippines carried out in 1982 found that of
632 reefs sampled, nearly 70% had less than 50% live coral
cover. A WWF global survey of coral reefs has been
launched, and will provide accurate data in due course.

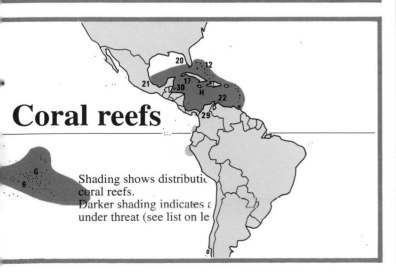

Coral reefs

Shading shows distributi[on of]
coral reefs.
Darker shading indicates a[reas]
under threat (see list on le[ft).]

smothered by silt. As forests are cut down inland, soil washes down the rivers and into the sea; the mangroves that once trapped the silt have often disappeared, so it is carried out to the reef.

In nearly 70 countries, reefs have been affected by dredging and land reclamation, or the building of harbors, airports and hotels. The tourists who fly in to stay in the hotels do further damage by grounding boats on the reefs and collecting coral. Imports of ornamental corals to the United States have risen more than sevenfold since 1960. In at least 10 countries, fishing with dynamite has laid waste the reefs; it has helped cause one entire Tanzanian island to disappear, and caused beach erosion that imperils tourist hotels. Mining corals for building materials, another widespread practice, is equally devastating. Corals recover very slowly; it takes 20 years for a brain coral colony to grow as big as a man's head.

A survey of 632 reefs in the Philippines in 1982 found that two thirds had less than 50 per cent live coral cover. Since then, over-exploitation and the increasing use of poisons and dynamite by subsistence fishermen helped ensure that less than 10 per cent of the country's reefs have healthy, undamaged coral communities.

Seagrass vulnerability

Seagrasses are the only land plants that have returned to the sea. Their underwater "meadows" are found in a wide swath around the world in both temperate and tropical seas. They grow close to shore in shallow water and, like mangroves, trap sediment; so they slow down erosion and they make the waters clearer, benefiting sea life. But this also makes them particularly vulnerable to pollution, and to overloads of sediment that overwhelm them.

They, too, are particularly rich in fish, providing both nurseries and life-long habitats for different species. Five times as many fish live over seagrass beds as over seafloors made up of mud, shells and sand, according to one United States study, while a single hectare of seagrass in the Arabian Gulf can yield nearly a metric ton of shrimp every year.

There is little information on what is happening to seagrasses around the world, but dredging and industrial wastes are known to do great damage. One dredge-and-fill operation in Florida, which destroyed only a fifth of the seagrass bed, reduced the number of fish species by four fifths and cost nearly $1.5 million in lost catches.

Effects of the Gulf War

The 6 million barrels of oil spilled into the Persian Gulf as a result of the 1991 Gulf War, coated 560 kilometers of Kuwaiti and Saudi coastal areas. All the mangroves, and more than 90 per cent of the salt marshes and nearshore seagrass beds were killed or severely damaged, though most seagrass communities and coral reefs have so far escaped serious damage. Recovery is likely to take a decade.

Management and protection

Less than 1 per cent of the world's entire mangrove forests receives any form of official protection, and there are no reserves anywhere in the world specially set aside for seagrasses. Coral reefs are slightly better protected, because of their importance for tourism, but only about half the countries with reefs have any reserves to safeguard them.

Although some 50 countries now have coastal management plans, only a handful have left the "paper stage". Only one international organization has a program which deals with these problems on a regional level – the United Nations Environment Programme's Oceans and Coastal Areas Programme Activity Centre in Nairobi. But UNEP's "regional seas" programs are in danger of failing as more urgent economic concerns head government agendas. If humankind is to preserve these remarkable ecosystems, and the wealth of species they harbor, well-coordinated national and regional programs will be needed.

Mineral Resources of the Sea

Countless multitudes of minute floating plants and animals sank to the bottom of prehistoric tropical oceans and turned, in time, into gas and oil. As the shape of the continents changed, the once tropical deposits moved as far as Antarctica and Alaska and many of the ancient seabeds were heaved onto dry land. Terrestrial oil fields, naturally, were the first to be exploited, but 2 trillion barrels of oil – about half the presently known onshore resources – may be recovered from beneath the sea.

Offshore oil and gas

The world's first offshore wells were sunk off the coast of California in the 1890s, using wooden jetties built out from the land. Development began in the Gulf of Mexico in 1938: production first flourished in the shallow, relatively sheltered seas because the oil was the cheapest to extract. But as terrestrial fields became more expensive to exploit – and as the oil shocks of the 1970s multiplied prices and made nations look for secure supplies – the search for offshore oil spread. The wild North Sea became the test bed of new technology. Giant rigs, standing in over 300 meters of water, drilled up to 8,000 meters beneath the seabed.

About 15 million barrels of oil a day – more than 20 per cent of the world's oil production – now comes from beneath the sea. Natural gas is often exploited at the same time, as it tends to occur in the same areas. Offshore oil and gas are being tapped by more than 40 of the world's countries, including Brazil, Nigeria, India, Egypt, Australia, and Trinidad and Tobago.

Oil has now been found underneath 3,000-4,000 meters of sea in the Gulf of Mexico and vast reserves lie in deeper waters around the world. Special submarine production units have been developed to exploit them. The industry will also attempt to push deep into the Arctic. About 40 per cent of US oil reserves lie beneath the icy Beaufort Sea and the former Soviet Union has enormous reserves off the Arctic Siberian coast.

Galleries from coal mines sunk on land have long been extended to pursue seams that run underneath the sea. Drilling for North Sea gas has found large coal resources more than 7,000 meters beneath the seabed, and enormous amounts similarly lie beneath continental shelves in many parts of the world. At present, there is no way of extracting them, but in future the coal may be converted into gas, to be piped to the surface.

Salt, sand and shells

Another mineral – salt – has been extracted from seawater for 4,000 years. Six million metric tons are produced every year. In all, more than 80 elements have been found in seawater, but only two are extracted on a large scale. Over half the world's output of magnesium metal and two thirds of its production of bromide come from seawater.

Much the most valuable marine mining is for sand and gravel, used in construction. In the UK a tenth of the entire production is dredged from the sea, and there are estimated to be 500 billion metric tons off the US Atlantic coast alone. Shells are dredged off Iceland for use in making cement, and special sands, rich in calcium carbonate, are taken from the Bahamas Bank.

Placers and oozes

Other minerals are carried down the world's rivers and dumped as "placer" deposits on the seafloor. Prospectors rushed to Nome, Alaska, at the end of the 19th century when deposits of gold were found on its beaches. Offshore placers of tin are mined in Indonesia, which is thought to have about 650,000 metric tons of the metal off its coasts. Copper, iron, tungsten, silver, titanium, platinum and even diamonds have been found off the world's shores.

Sulfur is mined from salt domes in the Gulf of Mexico. Massive sulfide deposits have been found in the East Pacific. Phosphate materials are found on many continental shelves; one deposit off the west coast of Mexico is thought to contain 20 billion metric tons of recoverable phosphate rock. But these deposits are economically unattractive as there are large reserves of phosphate on land.

The most extensive – and most inaccessible – mineral reserves lie on the deep ocean floor. Calcareous oozes, made of the shells and skeletons of marine creatures, cover 130 million square kilometers. Different oozes cover another 40 million square kilometers and yet another 100 million square kilometers are blanketed with red clay. Theoretically, these materials could be put to use in cement and concrete, filters, soil conditioners and clay products. In practice, it will almost certainly never be worthwhile.

One of the greatest mineral discoveries of all time may lie at the bottom of the Red Sea. Sediments rich in metals lie beneath pools of hot brine about 2,000 meters beneath the surface. The largest pool, the Atlantis II Deep halfway between the Sudan and Saudi Arabia near Jiddah, is thought to contain about $2 billion worth of silver, gold, copper and zinc. The sediments, between 2 and 23 meters thick, could probably be pumped to the surface. The two countries have set up a joint commission to administer exploitation and there are hopes that commercial extraction may begin.

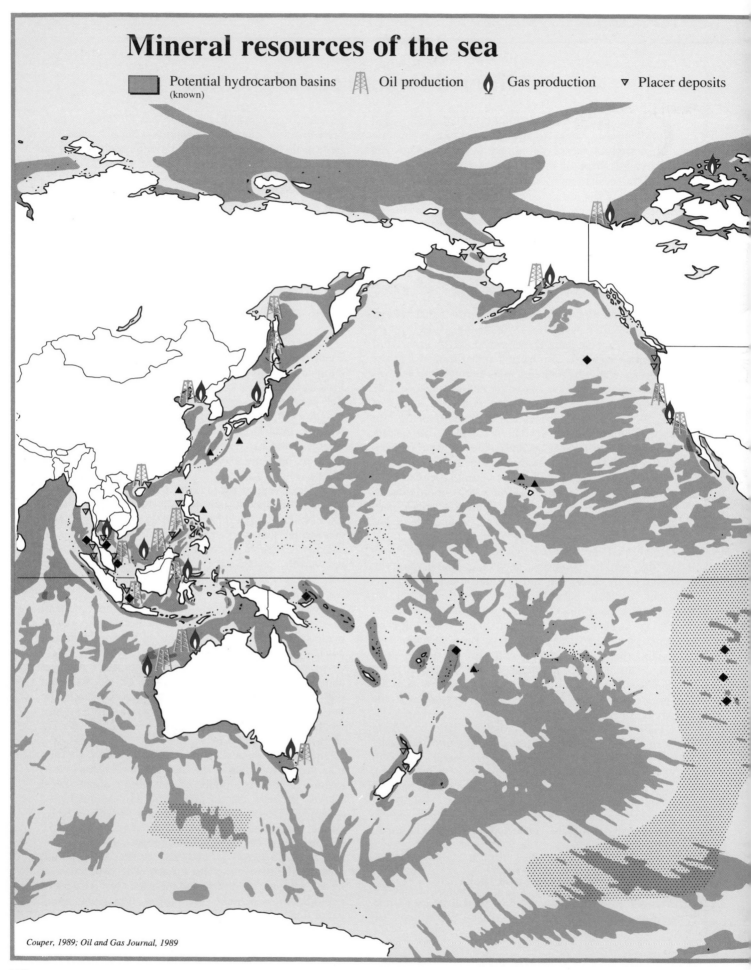

Mineral resources of the sea

Potential hydrocarbon basins (known)   Oil production   Gas production   ▽ Placer deposits

Couper, 1989; Oil and Gas Journal, 1989

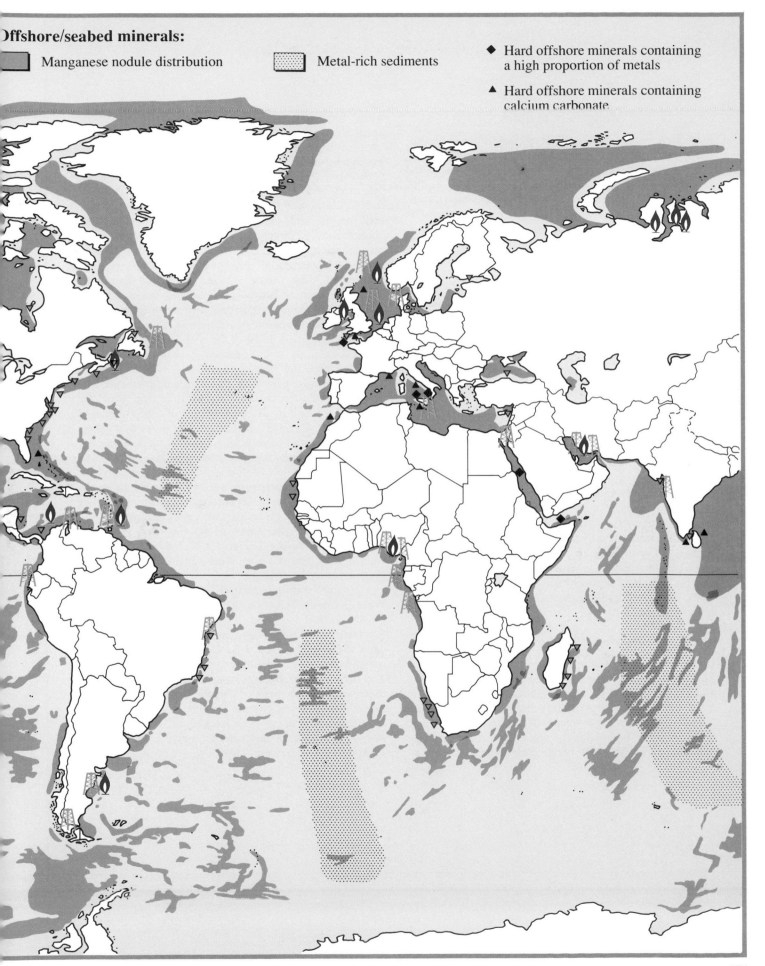

Offshore/seabed minerals:

- ▭ Manganese nodule distribution
- ░ Metal-rich sediments
- ◆ Hard offshore minerals containing a high proportion of metals
- ▲ Hard offshore minerals containing calcium carbonate

Manganese nodules

Even these riches, however, seem insignificant beside the wealth of the strange potato-shaped manganese nodules that litter the depths of the oceans. There are an estimated 1,500 trillion metric tons of these at the bottom of the Pacific alone, and they are thought to cover about 20 per cent of the floor of the world's oceans. Extraordinarily rich in metals, they can contain as much as 2.5 per cent copper, 2 per cent nickel, 0.2 per cent cobalt and 35 per cent manganese – high grade ores if found on land – as well as significant concentrations of titanium, aluminum, potassium, molybdenum, lead, strontium and other substances. Fascinatingly, they are forming all the time out of seawater, building up around nuclei such as sharks teeth, fish bones or pieces of pumice. They grow by only about a millimeter every 100,000 years. But the resource is so great that this produces an increase of some 10 million metric tons a year, far higher than world consumption of the metals.

Several components of this apparently inexhaustible treasury are becoming scarce on land. Onshore reserves of nickel, in particular, are falling while consumption has been rising rapidly. In theory at least, it should not be too difficult to recover them: the nodules could be sucked off the ocean floor by giant vacuums or scooped up by long chains of buckets. The operation would take place at great depths – the best nodules seem to lie more than 4,000 meters below the surface – but consortia, formed to exploit them, have tested mining techniques, and claim that nickel, copper, cobalt and other metals can be recovered economically. None has yet started work, partly because of uncertainty about the problems of international law.

Disputed exploitation rights

In the past, seabed resources have belonged to nobody. This led to the suggestion that they belong to everybody, and should be used to benefit humanity as a whole. In 1970, the UN General Assembly resolved that they were "the common heritage of mankind" and should be mined according to international agreement and control. Negotiations on the Law of the Sea opened and, for a while, there was hope that the world's nations might agree to distribute the ocean's wealth internationally, giving most of the revenues to poorer countries. But this vision has progressively faded.

Both the US and the former Soviet Union originally proposed quite restricted limits to national jurisdiction that would have left a large proportion of the shelves, and the oil and gas beneath them, to international control. But, led largely by Third World countries, many nations soon insisted that coastal states should have exclusive rights to all resources within 200 nautical miles from their shores. These "Exclusive Economic Zones" (EEZs) swallowed up nearly all the continental shelf, all the then proved oil and gas reserves, plus 87 per cent of probable reserves and, indeed, almost one third of the entire oceans. Most of the seabed thus given away was assigned to rich countries – with the US and Australia much the greatest gainers.

Coastal states were also given the right to mine the continental shelves beyond the 200-mile line, provided they shared a small proportion of the revenues internationally. Rich countries insisted that their companies should have a guaranteed right to mine the deep seabed for nodules, and a complicated regime was negotiated which would divide promising areas in half, giving one to the companies and retaining the other for exploitation by an international authority. Further concessions were given to rich-country interests and finally, in 1982, the Law of the Sea was ready for completion. But the US and UK led a last minute campaign against it, citing ideological objections to the common heritage concept.

A group of mainly industrialized countries – including West Germany, Italy, Japan and France, as well as the US and the UK – refused to sign or ratify the Law of the Sea treaty and reached agreement among themselves on deep sea mining. The US, UK and West Germany issued licenses to companies to explore the deep seabed. The nations that adhere to the treaty have denounced these activities as illegal. With the world split into two camps, the political climate for the mining of the vast resources of the ocean floor is likely still to be unstable when the operation becomes technically feasible and economically attractive.

Pollution of the Seas

The seas have always been the ultimate depository of humanity's wastes. The Mediterranean off the Nile Delta, the waters of the Bosphorus and the canals of Venice have been health hazards for centuries: but until recently pollution was relatively localized. Now it has spread around the world's coasts and is beginning to affect the open ocean.

Enclosed seas are most endangered; Russia's Aral Sea is already effectively dead. Semi-enclosed seas, like the Mediterranean, the Black Sea and the Baltic are highly vulnerable and increasingly polluted. So are areas where currents are weak or tend to concentrate contaminants – such as the German Bight in the North Sea. Pollution is generally worst in river estuaries and harbors, while coastal waters are dirtier than the open sea. The deep oceans are still relatively clean, though oil and litter foul sea lanes and chemical contaminants can be found in even the remotest waters.

Pollution: trouble at sea

Of the total amount of pollution at sea, about 44 per cent is discharged either directly to the sea or via rivers. Another third is fallout from air pollution. About 12 per cent comes from shipping, and another 10 per cent is deliberately dumped; offshore oil and gas production contributes perhaps 1 per cent.

Pollution from dumping and offshore installations is usually confined to relatively small areas of the open seas. Discharges mainly affect coastal waters, and pollution from shipping is concentrated along transportation routes. Fallout from air pollution is the most widely spread, though this too varies with the proximity of industrialized countries: roughly speaking, airborne pollution in the North, Baltic, and Mediterranean Seas is 10 times as high as in the open North Atlantic, 100 times as high as the tropical North Pacific, and 1,000 times as high as the South Pacific.

The menace of algae

Vast blooms of algae have emerged as the most serious effect of marine pollution. In the summer of 1988, a slick of toxic algae more than 10 meters deep and 10 kilometers wide, spread through the Kattegat and Skagerrak, which separate the coasts of Sweden, Norway and Denmark. More than 200 kilometers of coastline was blighted, beaches were closed, millions of fish are thought to have died; it became known as the "marine Chernobyl". In that same year, and in 1989, massive blooms clogged the northern Adriatic, spreading slime over the Italian and Yugoslavian coasts and cutting tourist revenues by a third. By 1992, 200 blooms of toxic algae were occurring in Japan's Seto Inland Sea every year.

Nitrogen and phosphorus compounds from sewage discharges, fertilizers and some industrial wastes nourish the algae, causing their populations to explode and release poisons. Worldwide, the amount of these nutrients in seawater is doubling; off the Dutch coast nitrogen concentrations increased fourfold between 1930 and 1980, while in the northern Adriatic phosphate levels are 10 times those in other areas of the Mediterranean, as a result of pollution from the river Po. Eventually, the bottom of the sea can be starved of oxygen; 100,000 square kilometers of the Baltic, comprising nearly half its deep waters, is virtually dead. So too is much of the Black Sea since 1970; its fish catch has dropped from 26 commercial species to five. The United Nations Environment Programme (UNEP) fears that in the coming decades, such problems could spread worldwide.

Poisons from the algae contaminate seafood. Paralytic shellfish poisoning, which suffocates victims, is increasing. Another toxin strikes down around 50,000 people a year with neurological, cardiovascular and gastro-intestinal symptoms. A third poison caused an outbreak of memory loss and death in Canada in 1987.

Seafood contaminated with mercury from a local factory killed over 400 people at Minimata Bay in Japan in the 1950s: by 1990, 2,900 people had been formally registered as suffering brain damage from mercury pollution. International limits for mercury in food were set after the damage was discovered; as a result the consumption of herrings caught in the North Sea has had to be banned periodically. Other heavy metals found in seas around the world include lead from vehicle exhausts, and cadmium from river pollution.

Sewage, chemicals, disease and death

Three quarters of the sewage from Britain's coastal towns and cities is discharged raw, without even minimal treatment; so is two thirds of the sewage pumped into the Mediterranean. More than a quarter of Britain's beaches regularly fail European Commission safety standards. Bathing in water contaminated by sewage causes gastro-intestinal infections, particularly in children under five, and may lead to ear, eye, respiratory and skin diseases. In 1988 medical waste, including syringes, was washed up on beaches in Northeast United States, raising fears that holidaymakers might catch AIDS. The horror was repeated the following year on the French and Italian Rivieras.

Two miracle chemicals of the 1950s and 1960s, long discontinued as unsafe in most industrialized countries, still cause severe sea pollution. DDT from pesticides and PCBs, used in everything from glues to electrical appliances, are both toxic and persistent. They have

Pollution of the sea

Areas of severe coastal pollution
(very concentrated pollution for all or part of the year, seriously affecting marine life and ecosystems)

Areas of persistent coastal pollution
(long-term, though not necessarily severe, pollution)

 Deep-sea dump sites

Millions of tons of industrial and municipal wastes, and dredge spoils are dumped at sea. The real totals may never be known, for only a handful of countries actually report the kinds of wastes dumped and where. It is thought that thousands of tons of hazardous wastes – for example PCBs, DDT and other dangerous pesticides, mine tailings, radioactive wastes, etc. – are also dumped illegally in clandestine offshore sites.

However, the wastes dumped for which data exist, consist mostly of dredge spoils from harbors, rivers, canals and coastal areas; selected industrial wastes, mostly liquid wastes and sludges; and sewage sludge (which can be highly infectious and toxic). Attempts are being made to control the dumping at sea of hazardous industrial wastes and other toxic garbage (such as infectious medical wastes). But the two major Conventions that do set limits on the kinds and quantities of wastes that can be dumped do not go far enough.

It should be remembered, however, that the amounts of wastes and debris flushed into coastal waters from land-based sources dwarf the amounts dumped at sea directly.

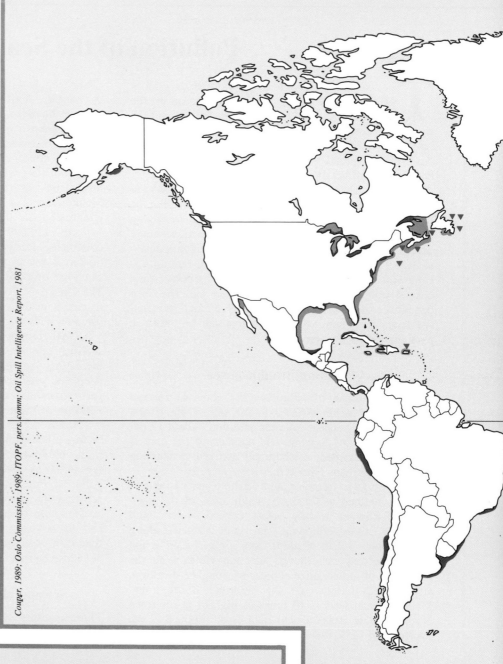

Couper, 1989; Oslo Commission, 1989; ITOPF, pers. comm; Oil Spill Intelligence Report, 1981

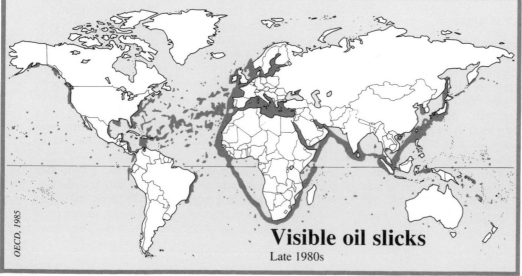

OECD, 1985

Visible oil slicks
Late 1980s

Oil pollution is shown separately on the left. It is often persistent but not usually severe, except in cases of a major spill.

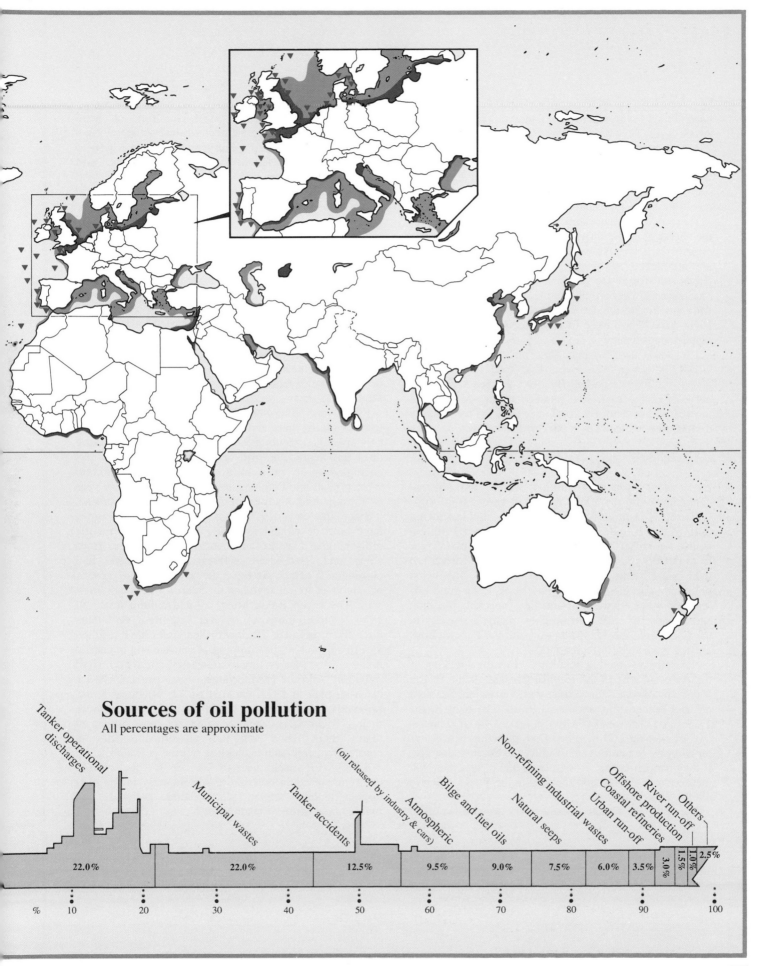

Sources of oil pollution

All percentages are approximate

Tanker operational discharges

Municipal wastes

Tanker accidents

Atmospheric (oil released by industry & cars)

Bilge and fuel oils

Natural seeps

Non-refining industrial wastes

Coastal refineries

Urban run-off

Offshore production

River run-off

Others

| 22.0% | 22.0% | 12.5% | 9.5% | 9.0% | 7.5% | 6.0% | 3.5% | 3.0% | 1.5% | 1.0% | 2.5% |

| % | 10 | 20 | 30 | 40 | 50 | 60 | 70 | 80 | 90 | 100 |

been found everywhere, in Antarctic penguins and Arctic seals, in whales in the Pacific and rat tail fish 3,000 meters down in the abyssal depths. Concentrations in most northern seas declined slowly for a while, following discontinuance, but have now stabilized. They are increasing around Third World countries where the chemicals are still widely used.

These two substances account for about 5 per cent of the chemical contamination of fish; the cause of the rest remains largely unknown. Hundreds of new chemicals are introduced every year; TBT, used for defouling paints, illustrated the dangers, having devastating and unforeseen effects on marine life. Diseases that affect a fifth to a third of the dab in polluted parts of the German Bight have been linked to effluents; similar effects are suspected elsewhere.

The Aral Sea, in what was formerly the southern USSR, is dying. The rivers that feed it have been diverted to irrigate cotton. By 1990, the Sea had lost 60 per cent of its water; the remainder was so salty that most fish have died. By 2010, it is likely to have totally disappeared, replaced by a poisoned, salt desert.

Oil pollution

Every year around 4 million metric tons of oil contaminate the sea. Only about half comes from ships – the rest is from land-based pollution – and less than a third of that is spilt by accident. More than 1.1 million metric tons of oil is deliberately discharged every year from tankers: they pump out oily waters and wash out their tanks before taking on new cargoes. The worst pollution is on the main shipping lanes. Even quite low levels of oil can damage marine life, especially plankton and larvae. Tar balls ruin beaches; many, especially in the Caribbean, the Mediterranean and the Persian Gulf, are now unusable for recreation.

Naturally it is the big accidents – like the wrecking of the Amoco Cadiz in the English Channel or the Exxon Valdez in Alaska – that attract world attention. Seabirds die and beaches are smothered in oil. The biggest, the 1979 Ixtoc blow-out off Mexico, leaked 400,000 metric tons of oil over a 10-month period, polluting beaches as far away as Texas. But small spills can do great damage, if they happen in the wrong place: one of less than 200 metric tons killed 40,000 seabirds in the Wadden Sea in 1969.

The Gulf War resulted in one of the worst oil spills in history. Around 6 million barrels of crude oil were released into the Gulf; most of it washed ashore along a 560-kilometer stretch of Saudi coast from Al Khafji to Abu'Ali Island, annihilating salt marshes and mangroves in the area and killing up to 30,000 seabirds. Long-term damage is difficult to predict, but it could take a decade before these ecosystems recover.

Dumping at sea

Dumping at sea has done great damage to limited areas, but is now decreasing. Radioactive and industrial wastes, sewage sludge and dredge spoils have all been dumped at sea. Radioactive waste dumping was stopped in 1983, and the dumping of industrial wastes and sewage sludge is being phased out – as is the incineration of toxic waste at sea. About 90 per cent of all dumping results from dredging; most causes few problems, but a tenth of the spoil is contaminated with PCBs, DDT, heavy metals or other pollutants.

Every year fishermen discard about 150,000 metric tons of plastic nets and lines, while countless plastic containers and bands are tossed overboard. The plastic does not degrade easily; sea creatures eat it or get entangled – up to a million seabirds and 100,000 whales, seals and dolphins are thought to die every year as a result. And this form of dumping is on the increase.

Since the early 1970s, many international treaties have been developed to regulate dumping, forbid the jettisoning of plastics, and control oil discharges from ships and land-based effluents. They have had considerable effect. So have the agreements of groups of countries to get together to clean up their common seas. The seven Baltic States agreed a joint treaty in 1974; concentrations of several important pollutants have since declined. The North Sea states have reached agreements to halt the dumping of sewage and industrial material, stop incinerating toxic wastes, and reduce river pollution. And the Mediterranean countries launched a clean-up plan in 1975, the first of 12 "Regional Seas" programs coordinated by UNEP, now involving more than 120 countries. Though some are more advanced than others, such regional agreements offer a constructive approach to tackling marine pollution.

Yet UNEP itself warned in 1990: "the marine environment could deteriorate significantly in the next decade unless strong, coordinated national and international action is taken now."

The Arctic

In the dark months of winter, a sulfurous haze settles over the top of the world. The North Pole lies thousands of kilometers from the industrial heartlands of Europe, North America and Russia and it draws pollution to it. In the early 1980s, DDT and PCBs were found in cod in the Barents Sea. In 1990, Caesium 137 from Chernobyl lingered on in the lichen-covered rangelands of northern Scandinavia, where half a million reindeer used to graze.

A fragile environment

Unlike Antarctica, a continent surrounded by icy seas, the Arctic is an ocean, hemmed in by the northern tips of three continents: North America, Asia and Europe. Its southern limits are hard to define. The Arctic Circle, drawn round the lands of the midnight sun, encompasses areas of Scandinavia warm enough to support reptiles. Some scientists set the boundary of the Arctic at the point where the taiga – the earth's largest forest – gives way to the sparse scrub of the tundra. The taiga covers more than half of Canada, much of Alaska and Scandinavia and 8 million square kilometers of Russia. The coldest temperatures in the northern hemisphere have been recorded in the depths of the taiga, over 1,500 kilometers south of the North Pole.

Trees yield to tundra somewhere near the point where the long winters leave them less than three months' growing time a year. The tundra, which covers 15 per cent of the earth's land surface, is a wilderness of frozen earth, rock-hard to a depth of over a kilometer. Only the top 50 centimeters or so thaw each summer. With global warming, scientists fear the permafrost could melt altogether, destroying tundra ecosystems and releasing vast quantities of methane gas to accelerate the greenhouse effect worldwide.

The tundra soil is poor, acidic and low in nutrients. Vegetation grows slow and short. In the south, the tundra is covered with a scrub of tiny willows and birches. "You realize suddenly that you are wandering around on top of a forest," writes Barry Lopez in *Arctic Dreams*. In the north, the growing season contracts to three weeks, and some lichens grow for only two or three days a year.

The pressures on Arctic wildlife

Arctic animals too are up against time. Only some dozen species of birds winter in the tundra, but nearly a sixth of the birds of the northern hemisphere go there to breed. While birds further south can lay again if a first brood fails, Arctic ones have only a few weeks to lay and hatch their eggs, molt and build up the reserves of fat they will need to fly south.

Survival through the Arctic winter depends on energy conservation. Animals cut their activity to a minimum – muskoxen, for instance, will stand immobile for long periods of time. If an animal is disturbed and has to flee, the exertion can be fatal. A small rodent may lose up to two fifths of its summer weight during the winter while the Svalbard reindeer works its way through a 10-centimeter layer of blubber as it digs through the snow for its food. The species which overwinter in the Arctic are well adapted to withstand the cold. Under their long hair, muskoxen are protected by a dense underfur, eight times warmer than sheep's wool, which makes the animals look much larger than they are. The polar bear is insulated by a thick layer of blubber and its transparent white hairs draw the sun's rays through to its black skin, where their heat is trapped by the fur.

The vital seas

While the Arctic's land creatures have to put up with temperatures of –40°C, its waters rarely fall far below 0°C. The third of the Arctic Ocean which is underlaid by continental shelves is one of the world's richest fishing grounds, yielding about a tenth of the annual global catch. In the spring three quarters of a million marine mammals – whales, seals and walruses – swim through the Bering Strait to summer in the rich waters to the north. Meanwhile 24 million birds fly in to nest in the delta of Alaska's Yukon and Kuskokwim rivers.

Exploration and exploitation

Since the earliest days, Europeans' interest in the Arctic has been commercial. In the Middle Ages sailors hunted its waters for the narwhal, whose spiral tusks sold for 20 times their weight in gold as "unicorn horn". In the 16th and 17th centuries Dutch and English explorers, sailing north to search for the Northwest and Northeast Passages which, they believed, would give them access to the riches of the spice islands, found other sources of wealth – whales in the sea and furs on ice and land. During the same period, the Russian Tsars were spreading their empire eastwards. By 1650 almost a third of Russia's income came from furs from its Siberian north.

The traders and explorers brought death with them. In 27 years, between 1741 and 1768, Steller's sea cow was "discovered" and exterminated. By 1875, the spectacled cormorant, the great auk and the Labrador duck were all extinct. By 1914, only 40,000 of some 300,000 walruses were left and the Bowhead whale was almost extinct.

Indigenous populations: the human cost

The Arctic's human population fared little better than its wildlife. Enslaved by the traders and infected by their

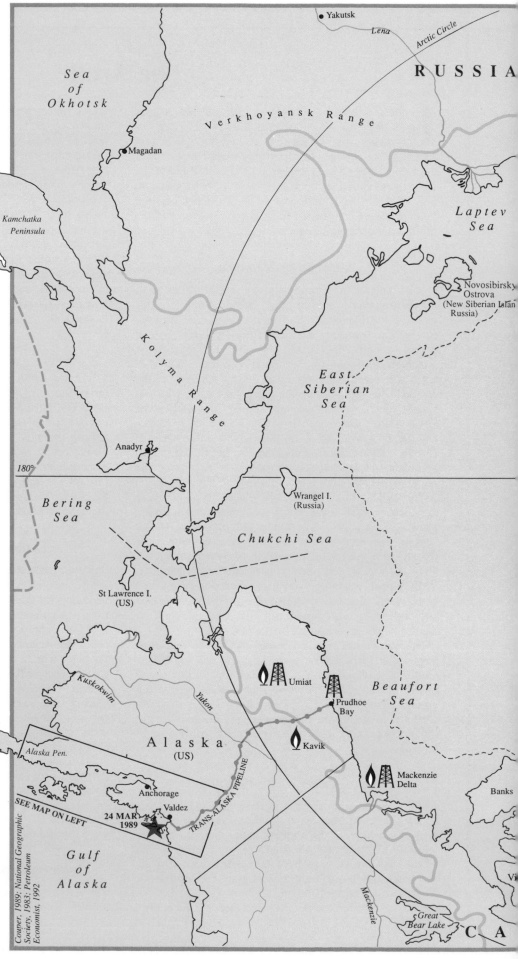

The Arctic

Fish catches are shown for selected Arctic nations. As the figures for the USSR, US and Canada include a large proportion of fish caught away from Arctic waters they have not been shown (but see the map on pages 158-159 for 1989 total catches by these countries).

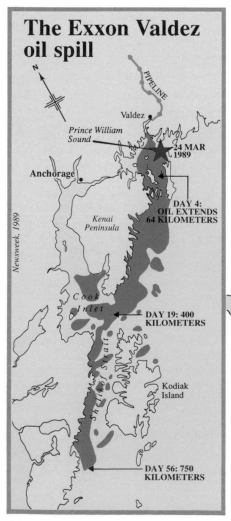

The Exxon Valdez oil spill

Valdez

PIPELINE

Prince William Sound

Anchorage

24 MAR 1989

DAY 4: OIL EXTENDS 64 KILOMETERS

Kenai Peninsula

Cook Inlet

DAY 19: 400 KILOMETERS

Shelikof Strait

Kodiak Island

DAY 56: 750 KILOMETERS

Newsweek, 1989

Sea of Okhotsk

Magadan

Verkhoyansk Range

Yakutsk

Lena

Arctic Circle

RUSSIA

Laptev Sea

Kamchatka Peninsula

Kolyma Range

East Siberian Sea

Novosibirsky Ostrova (New Siberian Islan Russia)

Anadyr

180°

Bering Sea

Wrangel I. (Russia)

Chukchi Sea

St Lawrence I. (US)

Kuskokwim

Yukon

Alaska (US)

Umiat

Prudhoe Bay

Beaufort Sea

Kavik

Mackenzie Delta

Banks

Alaska Pen.

Anchorage

Valdez

24 MAR 1989

TRANS-ALASKA PIPELINE

SEE MAP ON LEFT

Gulf of Alaska

Couper, 1989; National Geographic Society, 1983; Petroleum Economist, 1992

Mackenzie

Great Bear Lake

C A

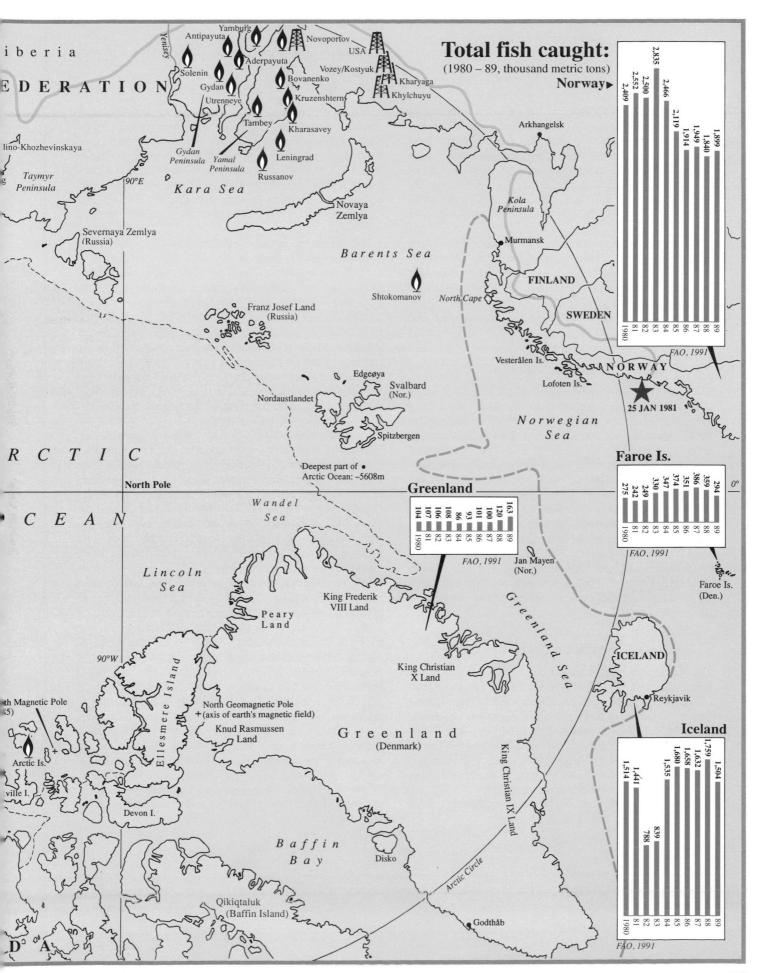

iberia

 E D E R A T I O N

lino-Khozhevinskaya

Taymyr Peninsula

90°E

Kara Sea

Antipayuta
Yamburg
Aderpayuta
Solenin
Gydan
Utrenneye
Gydan Peninsula
Yamal Peninsula
Tambey
Russanov
Leningrad
Bovanenko
Kruzenshtern
Kharasavey

Novoportov
USA
Vozey/Kostyuk
Kharyaga
Khylchuyu

Severnaya Zemlya
(Russia)

Novaya Zemlya

Barents Sea

Shtokomanov

Arkhangelsk

Kola Peninsula

Murmansk

FINLAND

SWEDEN

North Cape

Vesterålen Is.

NORWAY

Lofoten Is.

★

25 JAN 1981

Norwegian Sea

Franz Josef Land
(Russia)

Edgeøya

Svalbard
(Nor.)

Nordaustlandet

Spitzbergen

• Deepest part of •
Arctic Ocean: −5608m

R C T I C

North Pole

O C E A N

Wandel Sea

Lincoln Sea

90°W

Ellesmere Island

Peary Land

King Frederik
VIII Land

King Christian
X Land

th Magnetic Pole
5)

D

Arctic Is.

ville I.

Devon I.

North Geomagnetic Pole
+ (axis of earth's magnetic field)

Knud Rasmussen
Land

G r e e n l a n d
(Denmark)

Disko

*B a f f i n
B a y*

Qikiqtaluk
(Baffin Island)

Godthåb

Jan Mayen
(Nor.)

Greenland Sea

ICELAND

Reykjavik

King Christian IX Land

Arctic Circle

A

Total fish caught:
(1980 – 89, thousand metric tons)
Norway ▶

2,409 | 2,552 | 2,500 | 2,835 | 2,466 | 2,119 | 1,914 | 1,949 | 1,840 | 1,899
1980 | 81 | 82 | 83 | 84 | 85 | 86 | 87 | 88 | 89

FAO, 1991

0°

Faroe Is.

275 | 242 | 249 | 330 | 347 | 374 | 351 | 386 | 359 | 294
1980 | 81 | 82 | 83 | 84 | 85 | 86 | 87 | 88 | 89

FAO, 1991

Faroe Is.
(Den.)

Greenland

104 | 107 | 106 | 108 | 86 | 93 | 101 | 100 | 120 | 163
1980 | 81 | 82 | 83 | 84 | 85 | 86 | 87 | 88 | 89

FAO, 1991

Iceland

1,514 | 1,441 | 788 | 839 | 1,535 | 1,680 | 1,658 | 1,632 | 1,759 | 1,504
1980 | 81 | 82 | 83 | 84 | 85 | 86 | 87 | 88 | 89

FAO, 1991

179

diseases, communities were decimated. A few decades after Bering discovered the Aleutian islands, their population had crashed from more than 16,000 to 2,000. When the Polar Inuit first set eyes on white people in 1818, they thought they had come from the moon: by 1860, more than half had died. Of 1,000 Mackenzie Inuit on the coasts of the Beaufort Sea in 1888 only 100 were left in 1908.

As the white man has rushed to "civilize" the north and to exploit the oil and mineral wealth of the region, the indigenous people have suffered again from the disruption of their traditional lifestyles. One in 10 of all deaths in Greenland is alcohol-related, while suicide rates among Canada's native peoples are double and violent deaths triple the national level.

A balance upset

Early Alaskan artists used a hand with a large hole in its palm to show that a hunter must let some of his prey escape if he is to survive. For centuries indigenous people have hunted Arctic species, mostly sustainably. Extensive commercial hunting from the south and modern technology in the hands of the local people have changed this balance and brought indigenous communities and conservationists into conflict. "We see the animal rights laws systematically destroy our way of life and violate our right as aboriginal peoples to our traditions and values," Rhoda Inuksu, President of the Inuit Tapirisat of Canada, told the World Commission on Environment and Development. "Yet our people need development. The challenge is to find strategies for development that meet the needs of the people and the environment."

Mineral wealth: the Arctic in danger

The Arctic's mineral potential is vast. Two thirds of the former Soviet Union's gas reserves lie north of the Arctic Circle and Alaska supplies 25 per cent of the United States' oil. There have been croylite mines on Greenland since 1856 and coal mines in Spitsbergen since 1906. Most of Alaska's and Canada's oil and gas reserves are as yet untapped – and the Arctic holds some of the world's largest deposits of coal, iron, copper, lead, zinc, uranium, gold, diamonds, phosphates and rare metals.

Russia in particular is already exploiting these reserves, with a heavy cost to the environment. On the Kola Peninsula, about 100 kilometers south of Murmansk near the Finnish border, sulfur pollution from the nickel works has destroyed nearly 2,000 square kilometers of forest. In 1989, Finland and the USSR signed an agreement to clean up the peninsula's two nickel smelters – and to exploit its untapped mineral resources.

The discovery of oil near Prudhoe Bay, Alaska, in 1968, marked the birth of the US's environmental and indigenous rights movements. For seven years they fought the oil companies' plans to build a 1,200-kilometer pipeline from the North Slope to Valdez, America's most northerly icefree port. In 1971, the state's native people were given roughly 12 per cent of Alaska and nearly $1 billion in cash. Construction of the pipeline began in 1975 – and cost eight times as much as expected, partly due to the strict environment rules won by conservationists. In the years since then, Alaska's population has doubled and 85 per cent of the state's revenue now comes from oil.

In 1980, the US Congress declared 65 million hectares of Alaska as a permanent wilderness, including the Arctic National Wilderness Refuge, west of Prudhoe Bay near the Canadian border. Since 1987, government, developers and environmentalists have been locked in debate over whether 60,000 hectares of it should be opened to oil exploration. A similar debate, won by the environmentalists, raged in Canada in the 1980s over proposals to ship liquid natural gas south from Melville Island in vast icebreaking tankers.

On Good Friday, 1989, the Exxon Valdez ran aground in the Prince William Sound, spilling 11 million gallons of crude oil and polluting over 1,500 kilometers of pristine coastline. After six months of clean-up, costing $1.3 billion and providing 10,000 jobs, the State of Alaska reckoned that less than 200 kilometers of beaches were fit for plant and animal life. Scientists believe that the 980 dead sea otters, 146 bald eagles and 33,126 other seabirds found represent only a fraction of the numbers killed by the oil.

The disaster carries a warning for one of the world's great wilderness and wildlife areas. "Never in the millennium of our tradition have we thought it possible for the water to die," commented an Alaskan chief, Walter Meganeck. "But it is true. We walk our beaches and instead of gathering life, we gather death."

Antarctica

In 1774, the British explorer James Cook, searching for a continent south of the Antarctic Circle, came up against an ice wall, beyond which he thought there was land. It was nearly 50 years before Antarctica itself was first sighted. It took another 75 years before the first party set foot on its shores.

Since then, humankind has derived material wealth from Antarctica's rich seas and scientific knowledge from its pristine ice and air. There are probably also large mineral and oil reserves, as yet unexplored.

Antarctica is half as big again as the United States and makes up a tenth of the earth's land surface. It is covered with ice up to 3 kilometers thick, containing 70 per cent of the planet's freshwater supplies. It holds the world record for cold (-89.2°C, recorded in 1983) and for wind speed (322 kilometers per hour).

Only 2 per cent of the continent's surface is free of ice. This includes about a twentieth of the coast, the peaks of mountains and the Dry Valleys, where no rain has fallen for at least 2 million years. Over nine tenths of the coastline is made up of ice cliffs or shelves which stretch out from the land over the sea, breaking off into icebergs. One, sighted in 1956, was the size of Belgium.

In winter, the seas freeze and the size of the continent doubles. The ice sheet helps to power the circulation of the earth's atmosphere, while Antarctica's cold waters drive the circulation of the oceans. Together, these two mechanisms control the world's climate.

Antarctica's vast, unspoiled expanses hold vital interest for scientists studying the earth's evolution and atmosphere. Layers of ice, compacted over the millennia, provide a history of the earth's climate, stretching back half a million years. By studying cores drilled out of the ice, scientists can detect temperature changes over the centuries. Trapped air bubbles record changes in the concentration of carbon dioxide in the atmosphere, essential to our understanding of the greenhouse effect. Ice cores have also shown up how radioactive and lead pollution have increased in the atmosphere since 1945.

Antarctica's living resources

Most of Antarctica's land species are microscopic: the largest indigenous land animal is a wingless midge, no larger than a centimeter long. The continent's fragile vegetation includes only two flowering plants: the Antarctic hair grass (*Deschampsia antarctica*) and the Antarctic pearlwort (*Colobenthos subulatus*). Otherwise its flora is dominated by mosses and 350 species of lichen.

The land may be barren, but its waters are prolific. Vast blooms of phytoplankton, which burgeon in the spring and summer sunlight, provide the primary food source for krill, a small shrimp-like creature, whose swarms dye the oceans red by day and shine with phosphorescence at night.

Krill provides the main food source for five species of whale (whose migration is linked to its lifecycle), three species of seal, 20 species of fish, three species of squid and many birds, including penguins. These different predators feed at different stages in the krill's lifecycle, at different times of the year, in different places and at different depths, thus achieving a delicate balance of supply and demand.

Antarctica's marine species are finely tuned to their environment. Most of the region's fish belong to the sub-order *Notothenioidea*, whose bodies contain anti-freeze molecules. During the winter months, Weddell seals live permanently under the ice, using sonar to locate their food and find their way back to their breathing holes. The Emperor penguin breeds at the end of the summer, broods its eggs through winter and hatches them in the spring, so that the chicks become independent just when food is most abundant in the summer seas.

Since Cook first reported vast seal colonies, human predators have flocked across the Antarctic Circle. Sealers almost destroyed the fur seals of the Southern Ocean during the 19th century. It has taken 80 years for the species to recover. Whalers have killed an estimated 99 per cent of the region's Blue whales, 97 per cent of Humpbacks and 80 per cent of Fin whales since early this century – and continue to kill Minke whales in defiance of an internationally-agreed moratorium.

Territorial claims

By 1945, seven countries had registered claims to wedges of Antarctica. Britain, Australia, New Zealand, Norway and France all acknowledged each other's claims but Argentina, Chile and Britain, whose claims overlapped, did not. A series of incidents between Britain and Argentina highlighted the need for agreement.

Meanwhile, scientists were allowed to ignore territorial claims and establish research stations wherever they wanted. Twelve countries – including the seven states which claimed territory – organized research programs: scientists from 67 countries took part, and over 40 stations were set up.

The Antarctic Treaty

On December 1, 1959, Argentina, Australia, Belgium, Chile, France, Japan, New Zealand, Norway, South Africa, the UK, US and USSR signed the Antarctic Treaty, which came into force in 1961. It set Antarctica aside for peaceful uses only, established "freedom of scientific investigation", banned nuclear testing and

Antarctica

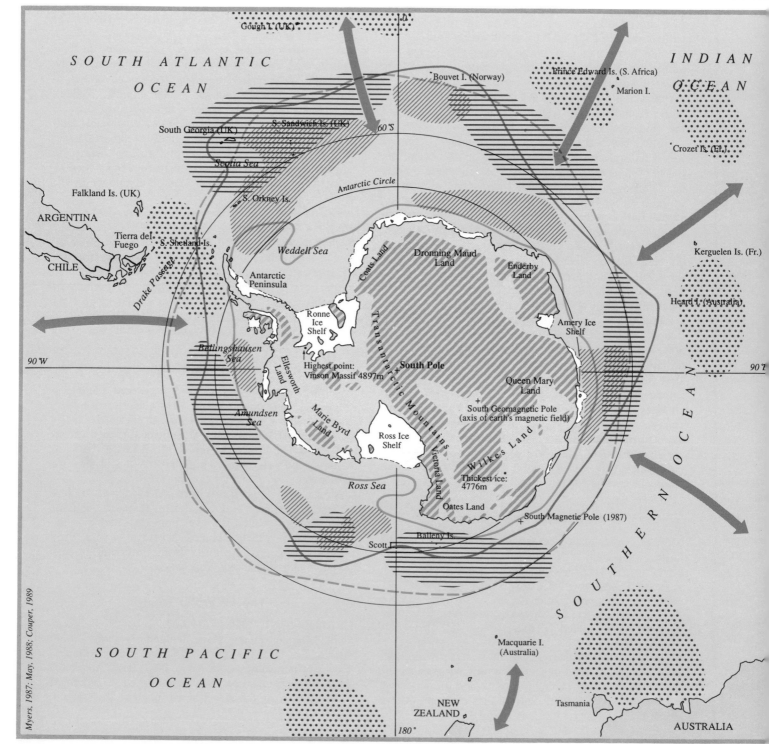

Legend:
- Land above present sea level
- Antarctic ice cap
- Ice shelf
- March pack ice boundary
- September pack ice boundary
- Major concentration of krill
- Northern limit of krill distribution
- Migratory routes of humpback whales
- Distribution of sei whales
- Distribution of blue and fin whales

Relative sizes of Antarctica and the United States

SOUTH ATLANTIC OCEAN

INDIAN OCEAN

Gough I. (UK)

Bouvet I. (Norway)

Prince Edward Is. (S. Africa)

Marion I.

South Georgia (UK)

S. Sandwich Is. (UK)

60°S

Crozet Is. (Fr.)

Scotia Sea

Antarctic Circle

Falkland Is. (UK)

S. Orkney Is.

ARGENTINA

Tierra del Fuego

S. Shetland Is.

Weddell Sea

Coats Land

Dronning Maud Land

Enderby Land

Kerguelen Is. (Fr.)

CHILE

Antarctic Peninsula

Ronne Ice Shelf

Amery Ice Shelf

Heard I. (Australia)

Drake Passage

Bellingshausen Sea

Ellesworth Land

Highest point: Vinson Massif 4897m

South Pole

Queen Mary Land

90°W

90°E

Amundsen Sea

Marie Byrd Land

Ross Ice Shelf

Transantarctic Mountains

Victoria Land

South Geomagnetic Pole (axis of earth's magnetic field)

Wilkes Land

Thickest ice: 4776m

Ross Sea

Oates Land

South Magnetic Pole (1987)

Scott I.

Balleny Is.

SOUTHERN OCEAN

SOUTH PACIFIC OCEAN

Macquarie I. (Australia)

NEW ZEALAND

180°

Tasmania

AUSTRALIA

Myers, 1987; May, 1988; Couper, 1989

Minerals

- ☐ Chromium
- ⬚ Coal
- ◇ Cobalt
- ▽ Copper
- △ Gold
- ◆ Iron
- L Lead
- ⬮ Manganese
- ☆ Molybdenum
- N Nickel
- O Platinum
- △ Silver
- ■ Tin
- ✳ Titanium
- U Uranium
- Z Zinc

• Antarctic bases open during 1989 austral winter (* Sub-Antarctic bases)

1 Gough I. (South Africa)*
2 Marion I. (South Africa)*
3 Bird I., South Georgia (UK)*
4 King Edward Point, South Georgia (UK)*
5 Signy, South Orkney Is. (UK)
6 Orcadas, Laurie I. (Argentina)
   7-14. King George Island
7 Artigas (Uruguay)
8 Teniente Rodolfo Marsh Martín (Chile)
9 Bellingshausen (USSR)
10 Great Wall (China)
11 King Sejong (Korea, Republic of)
12 Commandante Ferraz (Brazil)
13 Henryk Arctowski (Poland)
14 Teniente Jubany (Argentina)

15 Capitán Arturo Prat, Greenwich I. (Chile)
16 General Bernardo O'Higgins, (Chile)
17 Esperanza (Argentina)
18 Vicecomodoro Marambio, Seymour I. (Argentina)
19 Palmer (US)
20 Faraday, Argentine Is. (UK)
21 Rothera, Adelaide I. (UK)
22 General San Martín, Barry I. (Argentina)
23 Russkaya (USSR)
24 General Belgrano II (Argentina)
25 Halley (UK)

26 Georg von Neumayer (Germany)
27 SANAE III & IV (South Africa)
28 Dakshin Gangotri (India)
29 Novolazarevskaya (USSR)
30 Georg Forster (Germany)
31 Asuka (Japan)
32 Syowa (Japan)
33 Molodezhnaya (USSR)
34 Mawson (Australia)
35 Progress (USSR)
36 Zhongshan (China)
37 Davis (Australia)
38 Alfred-Faure, Crozet Is. (France)*

39 Port-aux-Français, Kerguelen Is. (France)*
40 Martin-de-Viviès, Amsterdam I. (France)*
41 Amundsen-Scott (US)
42 Vostok (USSR)
43 Mirnyy (USSR)
44 Casey (Australia)
   45-47: Ross Island
45 Scott Base (New Zealand)
46 McMurdo (US)
47 Cape Evans (Greenpeace)
48 Leningradskaya (USSR)
49 Dumont d'Urville (France)
50 Campbell I. (New Zealand)*
51 Macquarie I. (Australia)*

In addition to these 51 winter scientific stations, there are upwards of 20 summer-only bases. In many cases, these consist of little more than a supply hut or a camp suitable for stop-overs of a few days.

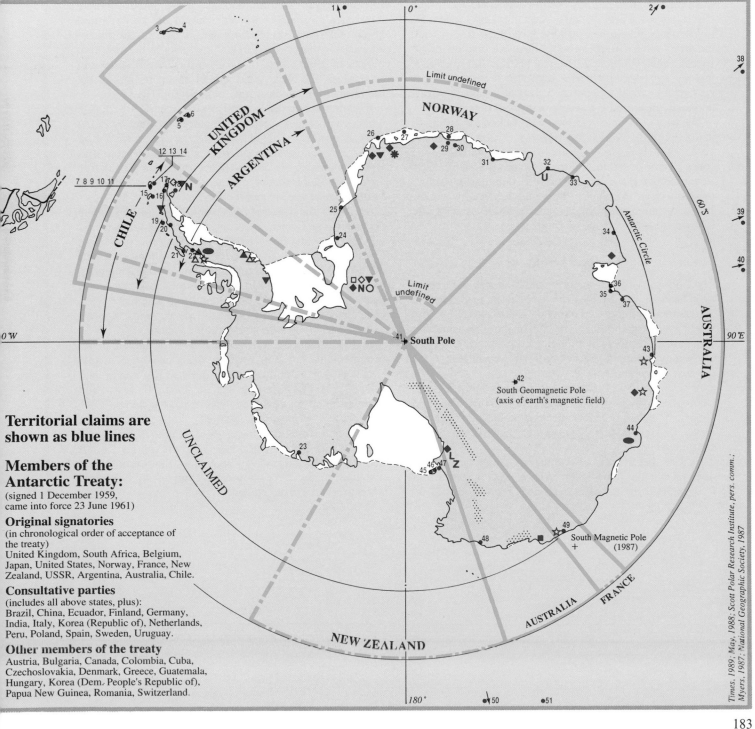

Territorial claims are shown as blue lines

Members of the Antarctic Treaty:
(signed 1 December 1959, came into force 23 June 1961)

Original signatories
(in chronological order of acceptance of the treaty)
United Kingdom, South Africa, Belgium, Japan, United States, Norway, France, New Zealand, USSR, Argentina, Australia, Chile.

Consultative parties
(includes all above states, plus):
Brazil, China, Ecuador, Finland, Germany, India, Italy, Korea (Republic of), Netherlands, Peru, Poland, Spain, Sweden, Uruguay.

Other members of the treaty
Austria, Bulgaria, Canada, Colombia, Cuba, Czechoslovakia, Denmark, Greece, Guatemala, Hungary, Korea (Dem. People's Republic of), Papua New Guinea, Romania, Switzerland.

Times, 1989; May, 1988; Scott Polar Research Institute, pers. comm.; Myers, 1987; National Geographic Society, 1987

waste disposal, and suspended the issue of territorial rights. New countries could join and become Consultative Parties if they were involved in "substantial scientific research" on the continent. By 1991, the treaty had 26 Consultative Parties and 14 other non-voting adherents, who abide by the treaty but are not sufficiently involved in research to become full members.

The treaty has provided the framework for environmental protection, but has been widely criticized. Since Consultative Party status depends on scientific investment, developing nations have seen it as a rich country's club. In the 1980s, nine nations including Brazil, India, China and Uruguay became full members, and by 1991 its signatories represented nearly 80 per cent of the world's population. But all of black Africa and many developing countries were still outside it.

The treaty states have also been attacked for failing to abide by conservation agreements. Some abandoned bases have been left as rubbish dumps while wastes from some active bases have polluted their surroundings. Divers say that a seabed dumpsite used by the American McMurdo base is "essentially dead". Work on an airstrip near France's Dumont d'Urville base has destroyed some nesting sites.

Mineral resources

The political and the environmental debates both came to a head in the 1980s over an attempt to produce a convention on mineral exploration and exploitation. The world's largest coalfield may lie in the Transantarctic Mountains and there is thought to be enough iron ore elsewhere to supply global needs, at present levels, for the next 200 years.

Antarctica may be rich in lead, copper and uranium. Geologists believe there may be over 900 major mineral deposits – but only a few of these would be in icefree areas. US researchers believe that there is oil in the continental shelf.

No one yet knows what minerals Antarctica has to offer or whether it will ever be economic or practicable to exploit them, but it is clear that regulation will be needed. The treaty states spent the 1980s thrashing out a convention to regulate mineral prospecting and eventual exploitation. Meanwhile, developing nations campaigned at the United Nations for the continent to be regarded as the common heritage of humankind whose resources would be exploited for the benefit of all.

The threat to the environment

Environmental groups, on the other hand, believe Antarctica's mineral resources should not be exploited at all. Offshore oil drilling could bring supertanker accidents – there were three shipwrecks in the first two months of 1989 alone. If an oil-spill occurred at the end of summer, the slick could spread under the ice for six months before any clean-up could begin.

Exploitation risks serious atmospheric pollution which could wreck scientific research. In some areas the snow is now so polluted by air traffic that it cannot be used for research. Both offshore and onshore activities would increase the pressure on Antarctica's icefree areas – already fiercely contended between breeding birds, seals, scientists and tourists.

In 1989, the minerals convention collapsed when Australia and France, convinced by the environmental arguments, refused to ratify it. The Antarctic and South Ocean Coalition, an environmental alliance which includes WWF and Greenpeace, worked for Antarctica to be declared a World Park, where mineral exploration and exploitation would be banned and scientific work and tourism carefully controlled – and, against expectations, they won a remarkable victory.

In 1991, the Antarctic Treaty nations agreed to an Environment Protocol, which designates the whole continent a "natural reserve, devoted to peace and science". It bans mining for at least 50 years and imposes much-improved environmental measures.

Once this Protocol is ratified, Antarctica will, in practice, be close to being a World Park, fulfilling the vision of Sir Peter Scott – son of the explorer Robert Scott who died on his way back from the South Pole – who used to say: "We should have the wisdom to know when to leave a place alone."

Units and Conversion Tables

This example shows how to use the conversion tables:

centimeter		inch
2.54	**1**	0.39

Read the number in the center column. Look left to convert Imperial to Metric, and look right to convert Metric to Imperial. In this case:

1 inch = 2.54 centimeters   1 centimeter = 0.39 inches

Numbers

Million	= 1,000,000
Billion	= thousand million = 1,000,000,000
Trillion	= million million = 1,000,000,000,000
Quadrillion	= million billion = 1,000,000,000,000,000

Distance

10 millimeters (mm) = 1 centimeter (cm)
100 centimeters   = 1 meter (m)
1,000 meters   = 1 kilometer (km)

meter		yard	kilometer		mile
0.9144	**1**	1.0936	1.6093	**1**	0.6214
1.8288	**2**	2.1872	3.2187	**2**	1.2427
2.7432	**3**	3.2808	4.8280	**3**	1.8641
3.6576	**4**	4.3744	6.4374	**4**	2.4855
4.5720	**5**	5.4680	8.0467	**5**	3.1069
9.1440	**10**	10.936	16.093	**10**	6.2140
45.720	**50**	54.680	80.467	**50**	31.069
91.440	**100**	109.36	160.93	**100**	62.140
914.40	**1,000**	1,093.6	1,609.3	**1,000**	621.40

1 nautical mile = 1.15 miles = 1.852 kilometers

Speed

1 mile per hour (mph) = 1.6093 kilometers per hour (kph)

Area

100 square millimeters (mm²) = 1 square centimeter (cm²)
10,000 square centimeters   = 1 square meter (m²)
10,000 square meters   = 1 hectare (ha)
100 hectares   = 1 square kilometer (km²)

hectare		acre	square kilometer		square mile
0.4047	**1**	2.4710	2.590	**1**	0.3861
0.8094	**2**	4.9421	5.180	**2**	0.7722
1.2141	**3**	7.4131	7.7699	**3**	1.1583
1.6187	**4**	9.8842	10.36	**4**	1.5444
2.0234	**5**	12.355	12.95	**5**	1.9305
4.0470	**10**	24.710	25.90	**10**	3.8610
20.234	**50**	123.55	129.50	**50**	19.305
40.470	**100**	247.10	259	**100**	38.610
404.70	**1,000**	2,471	2,590	**1,000**	386.10

Volume

1,000 cubic millimeters (mm³)	= 1 cubic centimeter (cm³)
1,000,000 cubic centimeters	= 1 cubic meter (m³)
1,000,000,000 cubic meters	= 1 cubic kilometer (km³)

cubic meter	cubic foot	liter		gallon (US)	
0.0283	**1**	35.315	3.7853	**1**	0.2642
0.0566	**2**	70.629	7.5706	**2**	0.5284
0.0849	**3**	105.94	11.356	**3**	0.7925
0.1133	**4**	141.26	15.141	**4**	1.0567
0.1416	**5**	176.57	18.926	**5**	1.3209
0.2830	**10**	353.15	37.853	**10**	2.6420
1.4160	**50**	1,765.7	189.26	**50**	13.209
2.8300	**100**	3,531.5	378.53	**100**	26.420
28.300	**1,000**	35,315	3,785.3	**1,000**	264.20

1 gallon (US) = 0.8327 gallon (UK)
1 gallon (UK) = 1.2009 gallons (US)

1 barrel of oil = 159 liters
= 42 gallons (US)
= 35 gallons (UK)

Weight

1,000 milligrams (mg) = 1 gram (g)
1,000 grams   = 1 kilogram (kg)
1,000 kilograms   = 1 metric ton (tonne) (t)

kilogram		pound (lb)
0.4536	**1**	2.2046
0.9072	**2**	4.4092
1.3608	**3**	6.6139
1.8144	**4**	8.8185
2.2680	**5**	11.023
4.5360	**10**	22.046
22.680	**50**	110.23
45.360	**100**	220.46
453.60	**1,000**	2,204.6

Power

Electrical power is measured in watts

1,000 watts (W) = 1 kilowatt (kW)
1,000 kilowatts = 1 megawatt (MW)
1,000 megawatts= 1 gigawatt (GW)
1,000 gigawatts = 1 terawatt (TW)

The kilowatt hour (kWh) measures the amount of electrical energy supplied or consumed.

1,000 kWh = 1 megawatt hour (MWh)
1,000 MWh = 1 gigawatt hour (GWh)
1,000 GWh = 1 terawatt hour (TWh)

Energy Equivalents

1 million metric tons (t) of oil is equivalent to:
   1.5 million tons of coal
   1.2 billion cubic meters of natural gas
   2.5 million tons of fuelwood
   4 terawatt hours of electricity
   2 metric tons of uranium (fast reactors)

This is expressed as million tons of oil equivalent (mtoe)

Temperature

Degrees Centigrade (Celsius) °C
Degrees Fahrenheit °F

°C	=	°F		°F	=	°C
−40		−40		−40		−40
−20		−4		0		−17.78
0		32		32		0
20		68		75		23.89
40		104		100		37.78
75		167		150		65.56
100		212		212		100

To convert from Fahrenheit to Celsius $(°F − 32) ÷ 1.8$.
To convert from Celsius to Fahrenheit $(°C \times 1.8) + 32$.
(The normal human body temperature is 37°C or 98.6°F.)

Radiation

The effect ionizing radiation (from X-rays and radioactive sources) has on humans is measured in sieverts (Sv).

1,000 microsieverts = 1 millisievert
1,000 millisieverts = 1 sievert

1 millisievert = total background radiation every human is exposed to each year from outer space and natural sources.

Isotopes of an element are different forms of that element, eg uranium-235 and uranium-238. Some isotopes, such as uranium-238, are radioactive.

Radioactive isotopes are made up of atoms with cores or nuclei that are unstable. These nuclei change or decay spontaneously, giving out radiation as they do so. Radioactivity, the rate at which nuclei change, is measured in becquerels (Bq).

1 becquerel = 1 nucleus changing and emitting radiation each second.

1 curie = 37,000,000,000 becquerels.

The pH Scale

A logarithmic scale measuring acidity and alkalinity (see page 86).

Richter Scale

A scale measuring energy released at the focus of an earthquake.

Richter scale	Effect	Relative magnitude of energy release
1	Very weak	
2	Detectable	250 x Richter 1
3	Feeble	250 x Richter 2
4	Moderate	250 x Richter 3
5	Strong	250 x Richter 4
6	Destructive	250 x Richter 5
7	Disastrous	250 x Richter 6
8	Catastrophic	250 x Richter 7

Ozone Layer

Dobson unit = unit of measurement of ozone concentration using a Dobson spectrophotometer.

Economics

Gross domestic product (GDP) = total value of goods and services produced by a country (residents and non-residents) per annum.

Gross national product (GNP) = GDP + income residents receive from abroad for labor and investments, less similar payments made to non-residents who contributed to the domestic economy.

Per capita GNP = GNP divided by the country's population total.

Bibliography

Sources

Acid News (1991), 4, Göteborg & Hvalstad: the Swedish & Norwegian NGO Secretariats on Acid Rain.

Allaby, M. (1989), *Green Facts: The Greenhouse Effect and Other Key Issues*, London: Hamlyn.

ASIWPCA (1985), *America's Clean Water*, Washington D.C.: Association of State and Interstate Water Pollution Control Administrators.

Baker, R.R. (1978), *Evolutionary Ecology of Animal Migration*, London: Hodder & Stoughton.

Batisse, M. (1986), "Developing and forming the biosphere reserve concept", *Nature and Resources, 22(3)*.

Bolin, B. *et al* (1986), *The Greenhouse Effect, Climate Change and Ecosystems*, SCOPE Report 29, Chichester: J. Wiley.

BP (1992), *BP Statistical Review of World Energy*, London: The British Petroleum Company.

Burger, J. (1987), *Report from the Frontier*, London: Zed Books.

CCC/IPCC (1990), Canadian Climate Centre model, in *Climate Change: The Intergovernmental Panel on Climate Change Scientific Assessment*, Cambridge: Cambridge University Press.

Collins, M. (1990), *The Last Rain Forests*, London: Mitchell Beazley in association with IUCN–The World Conservation Union.

Costa, L.G. *et al* (1987), *The Toxicology of Pesticides*, Berlin: Springer Verlag.

Couper, A. (1989), *The Times Atlas and Encyclopedia of the Seas*, London: Times Books.

ECE (1991), *Action on Transboundary Air Pollution*, Geneva: Economic Commission for Europe.

Ekholm, E. *et al* (1984), *Fuelwood, the Energy Crisis That Won't Go Away*, London: Earthscan.

Falkenmark, M. (1977), "Water and mankind – a complex system of mutual interaction", *Ambio, 6(1)*.

FAO (1989), *Yearbook: Fishery Statistics, Catches and Landings 1987*, Rome: Food and Agriculture Organization of the United Nations.

FAO (1991), *Yearbook: Fishery Statistics, Catches and Landings 1989*, Rome: Food and Agriculture Organization of the United Nations.

FAO (1991a), *Yearbook: Forest Products 1989*, Rome: Food and Agriculture Organization of the United Nations.

FAO (1991b), *Bulletin of Fishery Statistics, Fishery Fleet Statistics*, Rome: Food and Agriculture Organization of the United Nations.

Frost, Sir S. (1978), *Whales and Whaling*, Vol.1, Canberra: Australian Government Publishing Service.

GEMS (1987), *Forest Damage and Air Pollution*, Geneva & Nairobi: World Health Organization/United Nations Environment Programme.

Georghiou, G.P. & Saito, T. (1983), *Pest Resistance to Pesticides*, New York: Plenum Press.

Hardoy, J. & Satterthwaite, D. (1989), *Squatter Citizen: Life in the Urban Third World*, London: Earthscan.

Hinrichsen, D. (1990), *Our Common Seas*, London: Earthscan.

Hoyt, E. (1988), *Conserving the Wild Relatives of Crops*, Rome & Gland: International Board for Plant Genetic Resources/IUCN–The World Conservation Union/WWF International.

IAEA (1991), *Nuclear Power Reactors in the World*, Vienna: International Atomic Energy Agency.

IAEA (1992), *Nuclear Power Reactors in the World*, Vienna: International Atomic Energy Agency.

IBPGR (1989), *Annual Report 1989*, Rome: International Board for Plant Genetic Resources.

Jarman, C. (1972), *Atlas of Animal Migration*, London: Aldus Books.

Mackinnon, J. & K. (1986), *Review of the Protected Areas System in the Afro-Tropical Realm*, Gland: IUCN–The World Conservation Union.

Mackinnon, J. & K. (1986a), *Review of the Protected Areas System in the Indo-Malayan Realm*, Gland: IUCN–The World Conservation Union.

May, J. (1988), *The Greenpeace Book of Antarctica*, London: Dorling Kindersley.

McNeely, J. A. *et al* (1990), *Conserving the World's Biological Diversity*, Gland & Washington D.C.: IUCN–The World Conservation Union/World Resources Institute/Conservation International/World Wildlife Fund US/World Bank.

Milliman, J.D. & Meade, R.H. (1983), "Worldwide delivery of river sediment to the oceans", *Journal of Geology, 91*: 1-21.

Montalembert, M.R. & de Clément, J. (1983), *Fuelwood Supplies in the Developing Countries*, FAO Forestry Paper 42, Rome: Food and Agriculture Organization of the United Nations.

Myers, N. (1987), *The Gaia Atlas of Planet Management*, London: Pan Books.

National Geographic Society (1983), *The Arctic* (map), Washington D.C.: National Geographic Society.

National Geographic Society (1987), Supplement to the *National Geographic, 171(b)*: 556A.

Newsweek (September 18, 1989).

OECD (1985), *The State of the Environment 1985*, Paris: Organization for Economic Co-operation and Development.

OECD (1990), *Uranium Resources, Production and Demand 1990*, Paris: Organization for Economic Co-operation and Development.

OECD (1991), *Environmental Data Compendium 1991*, Paris: Organization for Economic Co-operation and Development.

Oil and Gas Journal (March 20, 1989), "Build-up predicted for China's offshore oil flow".

Oil Spill Intelligence Report (1981), *4(13)*.

Oslo Commission (1989), *Thirteenth Annual Report*, London: The Oslo Commission.

Petroleum Economist (1992), *Energy Map of the CIS and other Former Soviet Republics*, London: Petroleum Economist.

Prescott-Allen, C. & R. (1983), *The First Resource*, London & New York: Yale University Press.

Reid, W. & Miller, K.R. (1989), *Keeping Options Alive: The Scientific Basis for Conserving Biodiversity*, Washington D.C.: World Resources Institute.

Ridgeway, S.H. & Harrison, Sir R. (1985), *Handbook of Marine Mammals*, London: Academic Press.

Rodhe, H. & Herrera, R. (1988), *Acidification in Tropical Countries*, SCOPE Report 36, Chichester: J. Wiley.

Salathé, T. (1991), *Conserving Migratory Birds*, ICBP Technical Publication No. 12, Cambridge: International Council for Bird Preservation.

Scott, D. & Poole, C.M. (eds) (1989), *A Status Overview of Asian Wetlands*, Slimbridge: Asian Wetlands Bureau.

Smithsonian Institution (1988), *Tropical Rainforests: A Disappearing Treasure*, Washington D.C.: Smithsonian Institution Traveling Exhibition Service.

Snedaker, S.C. & Getter, C.D. (1985), *Coasts: Coastal Resources Management Guidelines*, Columbia: Research Planning Institute Inc.

Stolarski, R.S. (1992), Chapter 2 in Watson, R.J. *et al, Scientific Assessment of Ozone Depletion: 1991*, Washington D.C.: United Nations Environment Programme.

Survival International (1988), *Annual Report 1988*, London: Survival International Secretariat.

Survival International (n.d.), *Forest Indians of South America*, London: Survival International Secretariat.

Tarling, D.H. & M.P. (1972), *Continental Drift*, London: Penguin Books.

Thayer, F.W. & Fonseca, M.S. (1984), *The Ecology of Eel-Grass Meadows of the Atlantic Coast*, Washington D.C.: US Department of the Interior.

Times (1989), *Times Atlas of the World*, London: Times Books.

UN (1977), *United Nations Map of World Desertification*, Nairobi: Food and Agriculture Organization of the UN, UN Educational, Scientific and Cultural Organization and the World Meteorological Organization.

UNEP (1987), *UNEP Profile*, Nairobi: United Nations Environment Programme.

UNEP (1989), *Environmental Data Report*, Oxford: Basil Blackwell.

UNEP (1989a), *Action on Ozone*, Nairobi: United Nations Environment Programme.

UNEP (1991), *Register of Treaties and Other Agreements in the Field of the Environment*, Nairobi: United Nations Environment Programme.

UNEP (1991a), *Status of Regional Agreements Negotiated in the Framework of the Regional Seas Programme*, Nairobi: United Nations Environment Programme.

UNEP (1991b), *Status of Desertification and Implementation of the UN Plan of Action to Combat Desertification*, Nairobi: United Nations Environment Programme.

UNESCO MAB (n.d.), *Biosphere Reserves* (map), Paris: Man and the Biosphere Programme Secretariat, United Nations Educational, Scientific and Cultural Organization.

UNFPA (1991), *Population, Resources and the Environment: The Critical Challenges*, New York: United Nations Population Fund.

UNFPA (1992), *State of World Population 1992*, New York: United Nations Population Fund.

UNHCR (1990), *World Map of Refugees*, London: United Nations High Commissioner for Refugees.

UNICEF (1989), *The State of the World's Children 1989*, New York: United Nations Children's Fund.

UNICEF (1990), *The State of the World's Children 1990*, New York: United Nations Children's Fund.

UNICEF (1991), *The State of the World's Children 1991*, New York: United Nations Children's Fund.

UN Population Division (1987), *The Prospects for World Urbanization*, Study 21, New York: United Nations Population Division.

UNSCEAR (1988), *Reports to the General Assembly, with Appendices*, New York: United Nations Scientific Committee on the Effects of Atomic Radiation.

US Congressional Budget Office (1985), *Hazardous Waste Management: Recent Changes and Policy Alternatives*, Washington D.C.: US Government Printing Office.

US National Academy of Sciences (1984), *Toxicity Testing: Strategies to Determine Needs and Priorities*, Washington D.C.: National Academy Press.

USOFDA (1991), *Disaster History: Significant Data on Major Disasters Worldwide, 1900-Present*, Washington D.C.: Office of US Foreign Disaster Assistance.

Vallette, J. & Spalding, H. (eds) (1990), *The International Trade in Waste: a Greenpeace Inventory*, Washington D.C.: Greenpeace International.

Water Power and Dam Construction (June 1991), "The world's major dams and hydro plants."

WEC (1989), *Survey of Energy Resources 1989*, London: World Energy Conference.

Wilson, E.O. (1988), *Biodiversity*, Washington D.C.: National Academy Press.

World Bank (1991), *The World Bank Atlas 1991*, Washington D.C.: World Bank.

World Bank (1991a), *World Debt Tables 1991-92*, Washington D.C.: World Bank.

World Bank (1991b), *Social Indicators of Development, 1990*, Baltimore and London: Johns Hopkins University Press for the World Bank.

World Bank (1992), *World Development Indicators 1992*, pre-publication draft.

World Bank (1992a), *World Development Report*, Oxford: Oxford University Press.

WHO (1983), *Management of Hazardous Waste*, Copenhagen: World Health Organization Regional Office for Europe.

WHO (1992), *Weekly Epidemiological Record*, (January 17, 1992), Geneva: World Health Organization.

WHO/UNEP (1988), *Urbanization and its Implications for Child Health*, Geneva: World Health Organization.

WRI/IIED/UNEP (1988), *World Resources 1988-89*, New York: Basic Books.

WRI/IIED/UNEP (1990), *World Resources 1990-91*, New York: Oxford University Press.

WRI/IIED/UNEP (1992), *World Resources 1992-93*, New York: Oxford Univeristy Press.

WWF (n.d.), *Climate Change*, WWF Special Report 2, Gland: WWF International.

Further Reading

Audubon Society (1989), *Audubon Wildlife Report 1989/90*, San Diego: Academic Press.

Ayensu, E.S. *et al* (1984), *Our Green World: The Wisdom To Save It*, Cambridge: Cambridge University Press/Smithsonian Institution.

Boyle, S. & Ardill, J. (1989) *The Greenhouse Effect*, London: Hodder & Stoughton.

Brown, L. *et al* (1991 and other years), *State of the World 1991*, New York: W.W. Norton for Worldwatch Institute.

Brown, L. & Jacobson, J. (1987), *The Future of Urbanization*, Washington D.C.: Worldwatch Institute.

Bull, D. (1982), *A Growing Problem: Pesticides and the Third World Poor*, Oxford: Oxfam.

Caufield, C. (1985), *In the Rainforest*, London: Heinemann.

Caufield, C. (May 14, 1990), "The ancient forest", *New Yorker*.

CEC (1987), *The State of the Environment in the European Community*, Luxembourg: Commission of the European Community.

Cherfas, J. (1988), *The Hunting of the Whale*, London: Bodley Head.

Commonwealth Secretariat (1989), *Climate Change: Meeting the Challenge*, London: Commonwealth Secretariat.

CSE (1985), *The State of India's Environment, 1984-85: The Second Citizens' Report*, New Delhi: Centre for Science and Environment.

Dankleman, I. & Davidson, J. (1988), *Women and the Environment in the Third World*, London: Earthscan.

Davies, S.H. (1986), *Plants in Danger: What Do We Know?*, Gland: IUCN–The World Conservation Union.

Deudney, D. & Flavin, C. (1983), *Renewable Energy: The Power to Choose*, Washington D.C.: W.W. Norton for Worldwatch Institute.

Donaldson, D. *et al*, (1990), *Nuclear Power and the Greenhouse Effect*, London: United Kingdom Atomic Energy Agency.

Dudley, N. (1990), *Nitrates: The Threat to Food and Water*, London: Green Print.

ECE (1987), *Environment Statistics in Europe and North America*, Geneva: United Nations Economic Commission for Europe.

ECE/FAO (1985), *The Forest Resources of the ECE Region*, Geneva: United Nations Economic Commission for Europe/Food and Agriculture Organization of the United Nations.

EPA (1988), *National Air Quality and Emission Trends Report 1986*, North Carolina: US Environmental Protection Agency.

FAO (1981), map of the fuelwood situation in developing countries, Rome: Food and Agriculture Organization of the United Nations.

FAO (1982), *Atlas of the Living Resources of the Sea*, Rome: Food and Agriculture Organization of the United Nations.

FAO (1983), *Wood for Energy*, Forestry Topics Paper 1, Rome: Food and Agriculture Organization of the United Nations.

FAO (1990 and other years), *Production Yearbook 1990*, Rome: Food and Agriculture Organization of the United Nations.

FAO (1991 and other years), *Bulletin of Fishery Statistics, Fishery Fleet Statistics*, Rome: Food and Agriculture Organization of the United Nations.

FAO (1991 and other years) *Yearbook: Fishery Statistics, Catches and Landings 1989*, Rome: Food and Agriculture Organization of the United Nations.

FAO (1991 and other years), *Yearbook: Forest Products, 1989*, Rome: Food and Agriculture Organization of the United Nations.

Farman, J. *et al* (1988), "Large losses of total ozone in Antarctica reveal seasonal C10x/N0x interaction", *Nature 315*.

Flemming, N.C. (ed) (1977), *The Undersea*, New York: Macmillan.

FOE (1978), *Whale Manual*, London: Friends of the Earth.

Forkasiewicz, J. & Margat, J. (1980), *Tableaux Mondiales de Données Nationales d'Economie de l'Eau*, Orléans: Département Hydrologie.

George, S. (1990), *Ill Fares the Land*, London: Penguin Books.

Gibbons, J.H. *et al* (1990), "Strategies for energy use", in Scientific American: *Managing Planet Earth*, New York: Freeman.

Glantz, M. *et al* (eds) (1987), *Climate Crisis*, Nairobi & Boulder: United Nations Environment Programme/National Center for Atmospheric Research.

Global Environment Monitoring System (1987), *Global Pollution and Health*, n.p.: World Health Organization/United Nations Environment Programme.

Goldemburg, J. *et al* (1987), *Energy for a Sustainable World*, Washington D.C.: World Resources Institute.

Goldsmith, E. & Hildyard, N. (1990), *Earth Report 2*, London: Mitchell Beazley.

Gradwohl, J. & Greenberg, R. (1988), *Saving the Tropical Forests*, London: Earthscan.

Gribbin, J. (1988), *The Hole in the Sky: Man's Threat to the Ozone Layer*, London: Corgi.

Gribbin, J. & Kelly, M. (1989), *Winds of Change: Living in the Global Greenhouse*, London: Hodder & Stoughton.

HABITAT (1987), *Global Report on Human Settlements*, Oxford & New York: Oxford University Press for United Nations Centre for Human Settlements (HABITAT).

Harrison, P. (1987), *The Greening of Africa*, London: Paladin.

Hawkes, N. *et al* (1987), *Chernobyl: The End of the Nuclear Dream*, New York: Vintage Books.

Hazarika, S. (1987), *Bhopal: The Lessons of a Tragedy*, New Delhi: Penguin Books.

Hewitt, A. & Wells, B. (1987), *Growing Out of Debt*, London: Overseas Development Institute.

Holdgate, M. *et al* (1982), *The World Environment: A Report by United Nations Environment Programme*, Dublin: Tycooly.

Huxley, A. (1984), *Green Inheritance: The WWF Book of Plants*, London: Collins.

IAEA (1990 and other years), *Annual Report*, Vienna: International Atomic Energy Agency.

IBPGR (1989 and other years), *Annual Report*, Rome: International Board of Plant Genetic Resources.

ICIHI (1986), *The Encroaching Desert: The Consequences of Human Failure*, London: Zed Books for the Independent Commission on International Humanitarian Issues.

ICIHI (1987), *Indigenous Peoples: A Global Quest for Justice*, London: Zed Books for the Independent Commission on International Humanitarian Issues.

IPCC (1989), *Reports of IPCC Working Groups I and II*, Geneva: Intergovernmental Panel on Climate Change.

IUCN (1990), *Directory of Wetlands of International Importance: Sites Designated under the Ramsar Convention*, Gland: IUCN–The World Conservation Union.

IWGIA (1989), *Yearbook 1988*, Copenhagen: International Working Group for Indigenous Affairs.

Kallio, M. *et al* (eds) (1987), *The Global Forest Sector: An Analytical Perspective*, New York: J. Wiley.

Khalil, M.A.K. & Rasmussen, R.A. (1987), "Atmospheric methane: trends over the last 10,000 years", *Atmospheric Environment, 21(11)*.

Khalil, M.A.K. & Rasmussen, R.A. (1988), "Nitrous oxide: trends and global balance over the last 3,000 years", *Annals of Glaciology, 10*.

Kidron, M. & Segal, R. (1987), *The New State of the World Atlas*, London: Pan Books.

Lanly, J.P. (1982), *Tropical Forest Resources*, Forestry Paper 30, Rome: Food and Agriculture Organization of the United Nations.

Lappé, F.M. & Collins, J. (1988), *World Hunger: 12 Myths*, London: Earthscan.

Lashof, D.A. & Tirpak, D.A. (eds) (1989), *Policy Options for Stabilizing Global Climate Change*, Washington D.C.: US Environmental Protection Agency.

Layfield, F. (1987), *Sizewell B Public Enquiry*, Vols 1-8, London: HMSO.

Leach, G. & Mearns, R. (1989), *Beyond the Woodfuel Crisis: People, Land and Trees*, London: Earthscan.

Lean, G. (1978), *Rich World, Poor World*, London: George Allen & Unwin.

Lean, G. (1986), *Paradise Lost*, London: Earthlife/Observer.

Lipton, M. (1977), *Why Poor People Stay Poor: Urban Bias in World Development*, London: Maurice Temple Smith.

Lopez, B. (1986), *Arctic Dreams*, London: Macmillan.

Maltby, E. (1986), *Waterlogged Wealth*, London: Earthscan.

Manabe, S. & Wetherald, R. (1986), "Reduction in summer soil wetness induced by an increase in atmospheric carbon dioxide", *Science, 232*.

"Map of fuelwood supplies in the Third World" (1981), *Ambio, 10(5)*.

Maskrey, A. (1989), *Disaster Mitigation: A Community Based Approach*, Oxford: Oxfam.

Maurits la Riviera, J. (1990), "Threats to the world's water", in Scientific American: *Managing Planet Earth*, New York: Freeman.

McCormick, J. (1989), *Acid Earth: The Global Threat of Acid Pollution*, London: Earthscan.

Meadows, D. *et al* (1991), *Beyond the Limits*, London: Earthscan.

Merrick, T.W. (1986), "World population in transition", *Population Bulletin, 42(2)*.

Meybeck, M. *et al* (1989), *Global Environmental Monitoring System: Global Freshwater Quality, a First Assessment*, Oxford & Cambridge, Mass: B. Blackwell for World Health Organization/United Nations Environment Programme.

Mistry, P. (1990), *The Problem of "Official" Debt Owed by Developing Countries*, The Hague: European Campaign on Debt and Development.

Moss, B. (1988), *Ecology of Fresh Waters: Man and Medium*, 2nd ed., Oxford: Blackwell Scientific.

Myers, N. (1983), *A Wealth of Wild Species*, Colorado: Westwood Press.

Myers, N. (1984), *The Primary Source: Tropical Rainforests and Our Future*, New York: W.W. Norton.

Nectoux, F. & Kuroda, Y. (1989), *Timber from the South Seas*, Gland: WWF International.

Nilsson, S. & Duinker, P. (1987), "The extent of forest decline in Europe", *Environment, 29(9)*.

OECD (1983), *World Economic Interdependence and the Evolving North-South Relationship*, Paris: Organization for Economic Co-operation and Development.

OECD (1988), *Energy Balances of OECD Countries*, Paris:

Organization for Economic Co-operation and Development.

OECD (1989), *Agricultural Policies, Markets and Trade*, Paris: Organization for Economic Co-operation and Development.

OECD (1991 and other years), *Environmental Data Compendium 1991*, Paris: Organization for Economic Co-operation and Development.

OECD (1991 and other years), *The State of the Environment 1991*, Paris: Organization for Economic Co-operation and Development.

Pains, J.E. *et al* (1986), *Impact of Sea Level Rise on Society*, Proceedings of a workshop 27-29 August 1986, Delft: Delft Hydraulics Laboratory.

Patterson, W.C. (1983), *Nuclear Power*, London: Penguin Books.

Pearce, F. (1992) "Soils spoilt by farming and industry", *New Scientist, 134*.

Pitt, D. & Stone, P. (eds) (1989), *Towards Alpine Action: The Bellerive Initiative*, Geneva: Bellerive Foundation.

Plucknett, D.L. *et al* (eds) (1987), *Gene Banks and the World's Food@*, Princeton: Princeton University Press.

Rennie, J.K. (1988), *Population, Resources and Development: A Guidebook*, Gland: IUCN–The World Conservation Union.

Repetto, R. (1987), "Population, resources, environment: an uncertain future", *Population Bulletin, 42(2)*.

Shapley, D. (1985), *The Seventh Continent@*, Washington D.C.: Resources for the Future Inc.

Soysa, C.H. *et al* (eds) (1982), *Man, Land and Sea*, Bangkok: Agricultural Development Council.

Speidel, D.H. *et al* (1988), *Perspectives on Water Uses and Abuses*, New York: Oxford University Press.

Stumm, W. (1986), "Water, an endangered ecosystem", *Ambio, 15(4)*.

Sundborg, A. & Rapp, A. (1986), "Erosion and sedimentation by water: problems and perspectives", *Ambio, 15(4)*.

Timberlake, L. (1988), *Africa in Crisis: The Causes, the Cures of Environmental Bankruptcy*, London: Earthscan.

Timberlake, L. & Thomas, L. (1990), *When the Bough Breaks: Our Children, Our Environment*, London: Earthscan.

Toke, D. (1990), *Green Energy: A Non-Nuclear Response to the Greenhouse Effect*, London: Green Print.

Tolba, M.K. (1992), *Saving Our Planet: challenges and hopes*, London: Chapman & Hall.

Trzyna, T.C. (1989), *World Directory of Environment Organizations*, 3rd Edition, Claremont: California Institute of Public Affairs.

Udvardy, M.D.F. (1975), *A Classification of the Biogeographical Provinces of the World*, Gland: IUCN–The World Conservation Union.

UN (1986), *Urban and Rural Population Projections 1950-2025, the 1984 Assessment*, New York: United Nations.

UN (1989 and other years), *Demographic Yearbook*, New York: United Nations Statistical Office.

UN (1989), *World Population Prospects 1988*, New York: United Nations.

UNCTD (1985), *The Least Developed Countries*, New York: United Nations Conference on Trade and Development.

UNCTD (1989), *Trade and Development Report 1989*, New York: United Nations Conference on Trade and Development.

UNDP (1991 and other years), *Human Development Report*, New York: Oxford University Press.

UNDRO (1990), *Preliminary Study on the Identification of Disaster-Prone Countries Based on Economic Impact*, Geneva: Office of the United Nations Disaster Relief Co-ordinator.

UNEP (1982), *Conservation of Coastal and Marine Ecosystems and Living Resources of the East African Region*, Nairobi: United Nations Environment Programme.

UNEP (1987), *The Greenhouse Gases*, UNEP/GEMS Environment Library 1, Nairobi: United Nations Environment Programme.

UNEP (1987), *The Ozone Layer*, UNEP/GEMS Environment Library 2, Nairobi: United Nations Environment Programme.

UNEP (1990), *The State of the Marine Environment*, Regional Seas Reports and Studies 115, Nairobi: United Nations Environment Programme.

UNEP (1991 and other years), *Environmental Data Report*, Oxford: Basil Blackwell.

UNEP (1991 and other years), *The State of the Environment*, Nairobi: United Nations Environment Programme.

UNESCO (1978), *The World Heritage* (map), Paris: United Nations Educational, Scientific and Cultural Organization.

UNESCO (1985), *Action Plan for Biosphere Reserves*, Paris: United Nations Educational, Scientific and Cultural Organization.

UNESCO (1988), *Compendium of Statistics on Illiteracy*, Paris: United Nations Educational, Scientific and Cultural Organization.

UNFPA (1989), *For Future Generations: The Amsterdam Population Plan*, New York: United Nations Population Fund.

UNFPA (1989), *Safeguarding the Future*, New York: United Nations Population Fund.

UNFPA (1990), *Meeting the Population Challenge*, New York: United Nations Population Fund.

UNFPA (1991), *Population and the Environment: the Challenges Ahead*, New York: United Nations Population Fund.

UNFPA (1992 and other years), *State of World Population 1992*, New York: United Nations Population Fund.

UNSCEAR (1989 and other years), *Reports to the General Assembly, with Appendices*, New York: United Nations Scientific Committee on the Effects of Atomic Radiation.

UN Statistical Office (1988 and other years), *1986 Energy Statistics Yearbook*, New York: United Nations Statistical Office.

US National Academy of Sciences (1984), *Toxicity Testing: Strategies to Determine Needs and Priorities*, Washington D.C.: National Academy Press.

USOFDA (1991 and other years), *Disaster History: Significant Data on Major Disasters Worldwide, 1900-Present*, Washington D.C.: Office of US Foreign Disaster Assistance.

"The vanishing jungle: ecologists make friends with economists" (March 15, 1988), *The Economist*.

Vavilov, N.I. (1951), *The Origin, Variation, Immunity and Breeding of Cultivated Plants*, New York: Ronald Press.

Viedna, C. (May 1988), "A health and nutrition atlas – food for thought", *World Health*.

Walton, D.W.H. (1987), *Antarctic Science*, Cambridge: Cambridge University Press.

Ward, B. (1979), *Progress for a Small Planet*, London: Maurice Temple Smith.

Watson, L. (1981), *Sea Guide to Whales of the World*, London: Hutchinson.

WCED (1987), *Our Common Future*, Oxford: Oxford University Press for World Commission on Environment and Development.

WCMC (1989), *List of Biosphere Reserves*, Cambridge: World Conservation Monitoring Centre.

Whales of the World (1976), Washington D.C.: National Geographic Society.

WHO (1984), *The International Drinking Water Supply and Sanitation Decade: Review of National Baseline Data, December 1980*, Geneva: World Health Organization.

WHO (1984), *Urban Air Pollution 1973-84*, Geneva: World Health Organization.

WHO (1987), *Air Quality Guidelines for Europe,* European Series 23, Copenhagen: World Health Organization Regional Office for Europe.

WHO (1987), *Evaluation of the Strategy for Health for All by the Year 2000*, Geneva: World Health Organization.

WHO (1989), *World Health Statistics Annual*, Geneva: World Health Organization.

WHO/UNEP (1987), *Global Pollution and Health*, London: Yale University Press for World Health Organization and United Nations Environment Programme.

Wijkman, A. & Timberlake, L. (1984), *Natural Disasters, Acts of God or Acts of Man?*, London: Earthscan.

Wolf, E.C. (1987), *On the Brink of Extinction: Conserving the Diversity of Life*, Washington D.C.: Worldwatch Institute.

World Bank (1989), *Sub-Saharan Africa: From Crisis to Sustainable Growth,* Washington D.C.: World Bank.

World Bank (1990 and other years), *Social Indicators of Development*, Baltimore & London: Johns Hopkins University Press.

World Bank (1991 and other years), *The World Bank Atlas 1991*, Washington D.C.: World Bank.

World Bank (1991 and other years), *World Debt Tables 1991-92*, 2 vols, Washington D.C.: World Bank.

World Bank (1992 and other years), *World Development Report 1992*, Washington D.C.: World Bank.

WRI (1985), *Tropical Forests: A Call for Action*, Washington D.C.: World Resources Institute.

WRI/IUCN/UNEP (1992), *Global Biodiversity Strategy*, n.p.

WRI/IIED/UNEP (1992 and other years), *World Resources 1992-93*, A Report by the World Resources Institute, the International Institute for Environment and Development in collaboration with the United Nations Environment Programme, New York: Oxford University Press.

WWF (1988), *Acid Rain and Air Pollution*, WWF Special Report 2, Gland; WWF International.

WWF (1989), *Tropical Forest Conservation*, Position Paper 3, Gland: WWF International.

WWF (n.d.), *Acid Rain*, WWF Special Report 1, Gland: WWF International.

WWF (n.d.), *The Importance of Biological Diversity*, Gland: WWF International.

WWF/UN/ECE: (1988), *Conservation, Acidification and the Convention on Long-Range Transboundary Air Pollution*, WWF Position Paper 2, Gland: WWF International.

WWF (1989), *Wetlands*, WWF Special Report 1, Gland: WWF International.

WWF UK/The Observer (May 1992), Forest Supplement in *The Observer*.